Bayesian Inference in the Social Sciences

Bayesian Inference in the Social Sciences

Edited by

Ivan Jeliazkov
Department of Economics
University of California, Irvine
California, USA

Xin-She Yang
School of Science and Technology
Middlesex University
London, UK

WILEY

Published by John Wiley & Sons, Inc., Hoboken, New Jersey.
Published simultaneously in Canada.

Library of Congress Cataloging-in-Publication Data:

Bayesian inference in the social sciences / edited by Ivan Jeliazkov, Department of Economics, University of California, Irvine, California, USA, Xin-She Yang, School of Science and Technology, Middlesex University, London, United Kingdom.
pages cm
Includes bibliographical references and index.
ISBN 978-1-118-77121-1 (hardback)
1. Social sciences—Statistical methods. 2. Bayesian statistical decision theory. I. Jeliazkov, Ivan, 1973- II. Yang, Xin-She.
HA29.B38345 2014
519.5'42—dc23 2014011437

Printed in the United States of America.

10 9 8 7 6 5 4 3 2 1

CONTENTS

PREFACE

No researcher is an island. Indeed, in scientific inquiry, we are generally motivated by, and build upon, the work of a long line of other researchers to produce new techniques and findings that in turn form a basis for further study and scientific discourse. Interdisciplinary research can be particularly fruitful in enhancing this interconnectedness despite the intrinsic difficulty of having to carefully sail the unfamiliar waters of multiple fields.

Bayesian Inference in the Social Sciences was conceived as a manifestation, if any were needed, of the major advances in model building, estimation, and evaluation that have been achieved in the Bayesian paradigm in the past few decades. These advances have been uneven across the various fields, but have nonetheless been widespread and far-reaching. Today, all branches in the social sciences make use of the tools of Bayesian statistics. In part, this is due to the conceptual simplicity and intellectual appeal of the Bayesian approach, but it has also much to do with the ability of Bayesian methods to handle previously intractable problems due to the computational revolution that started in the 1990s. The book provides chapters from leading scholars in political science, sociology, economics, marketing and finance, and offers clear, self-contained, and in-depth coverage of many central topics in these fields. Examples of novel theoretical developments and important applications are found throughout the book, aiming to appeal to a wide audience, including readers with a taste for conceptual detail, as well as those looking for genuine practical applications.

Although the specific topics and terminology differ, much common ground can be found in the use of novel state-of-the-art computational algorithms, elaborate hierarchical modeling, and careful examination of model uncertainty. We hope that this book will enhance the spread of new ideas and will inspire a new generation of applied social scientists to employ Bayesian methodology, build more realistic and flexible models, and study important social phenomena with rigor and clarity.

We wish to thank and acknowledge the hard work of the contributing authors and referees, and the production team at Wiley for their patience and professionalism.

IVAN JELIAZKOV AND XIN-SHE YANG

July, 2014

CHAPTER 1

BAYESIAN ANALYSIS OF DYNAMIC NETWORK REGRESSION WITH JOINT EDGE/VERTEX DYNAMICS

ZACK W. ALMQUIST[1] AND CARTER T. BUTTS[2]

[1]University of Minnesota, USA.
[2]University of California, Irvine, USA.

1.1 Introduction

Change in network structure and composition has been a topic of extensive theoretical and methodological interest over the last two decades; however, the effects of endogenous group change on interaction dynamics within the context of social networks is a surprisingly understudied area. Network dynamics may be viewed as a process of change in the edge structure of a network, in the vertex set on which edges are defined, or in both simultaneously. Recently, Almquist and Butts (2014) introduced a simple family of models for network panel data with vertex dynamics—referred to here as *dynamic network logistic regression* (DNR)—expanding on a subfamily of temporal exponential-family random graph models (TERGM) (see Robins and Pattison, 2001; Hanneke et al., 2010). Here, we further elaborate this existing approach by exploring Bayesian methods for parameter estimation and model assessment. We propose and implement techniques for Bayesian inference via both maximum *a posteriori* probability (MAP) and Markov chain Monte Carlo (MCMC) under several different priors, with an emphasis on minimally informative priors that can be employed in a range of empirical settings. These different approaches are compared in terms of model fit and predictive model assessment using several reference data sets.

This chapter is laid out as follows: (1) We introduce the standard (exponential family) framework for modeling static social network data, including both MLE and Bayesian estimation methodology; (2) we introduce network panel data models, discussing both MLE and Bayesian estimation procedures; (3) we introduce a subfamily of the more general panel data models (dynamic network logistic regression)—which allows for vertex dynamics—and expand standard MLE procedures to include Bayesian estimation; (4) through simulation and empirical examples we explore the effect of different prior specifications on both parameter estimation/hypothesis tests and predictive adequacy; (5) finally, we conclude with a summary and discussion of our findings.

1.2 Statistical Models for Social Network Data

The literature on statistical models for network analysis has grown substantially over the last two decades (for a brief review see Butts, 2008b). Further, the literature on dynamic networks has expanded extensively in this last decade – a good overview can be found in Almquist and Butts (2014). In this chapter we use a combination of commonly used statistical and graph theoretic notation. First, we briefly introduce necessary notation and literature for the current state of the art in network panel data models, then we review these panel data models in their general form, including their Bayesian representation. Last, we discuss a specific model family (DNR) which reduces to an easily employed regression-like structure, and formalize it to the Bayesian context.

1.2.1 Network Data and Nomenclature

For purposes of this chapter, we will focus on networks (social or otherwise) that can be represented in terms of dichotomous (i.e., unvalued) ties among pairs of discrete entities. [For more general discussion of network representation, see, e.g., Wasserman and Faust (1994); Butts (2009).] We represent the set of potentially interacting entities via a *vertex set* (V), with the set of interacting pairs (or ordered pairs, for directed relationships) represented by an *edge set* (E). In combination, these two sets are referred to as a *graph*, $G = (V, E)$. (Here, we will use the term "graph" generically to refer to either directed or undirected structures, except as indicated otherwise.) Networks may be static, e.g., representing relationships at a single time point or aggregated over a period of time, or dynamic, e.g., representing relationships appearing and disappearing in continuous time or relationship status at particular discrete intervals.

For many purposes, it is useful to represent a graph in terms of its *adjacency matrix*: for a graph G of order $N = |V|$, the adjacency matrix $Y \in \{0,1\}^{N \times N}$ is a matrix of indicator variables such that $Y_{ij} = 1$ iff the ith vertex of G is adjacent (i.e., sends a tie to) the jth vertex of G. Following convention in the social network (but not graph theoretic) literature, we will refer to N as the *size* of G.

The above extends naturally to the case of dynamic networks in discrete time. Let us consider the time series $\ldots, G_{t-1}, G_t, G_{t+1}, \ldots$, where $G_t = (V_t, E_t)$ represents the state of a system of interest at time t. This corresponds in turn to the adjacency matrix series $\ldots, Y_{\cdot\cdot t-1}, Y_{\cdot\cdot t}, Y_{\cdot\cdot t+1}, \ldots$, with $N_t = |V_t|$ being the size of the network at time t and $Y_{\cdot\cdot t} \in \{0,1\}^{N_t \times N_t}$ such that $Y_{ijt} = 1$ iff the ith vertex of G_t is adjacent to the jth vertex of G_t at time t. As this notation implies, the vertex set of an evolving network is not necessarily fixed; we shall be particularly interested here in the case in which V_t is drawn from some larger risk set, such that vertices may enter and leave the network over time.

1.2.2 Exponential Family Random Graph Models

When modeling social or other networks, it is often helpful to represent their distributions via random graphs in discrete exponential family form. Graph distributions expressed in this way are called *exponential family random graph models* or ERGMs. Holland and Leinhardt (1981) are generally credited with the first explicit use of statistical exponential families to represent random graph models for social networks, with important extensions by Frank and Strauss (1986) and subsequent elaboration by Wasserman and Pattison (1996), Pattison and Wasserman (1999), Pattison and Robins (2002), Snijders et al. (2006), Butts (2007), and others. The power of this framework lies in the extensive body of inferential, computational, and stochastic process theory [borrowed from the general theory of discrete exponential families, see, e.g., Barndorff-Nielsen (1978); Brown (1986)] that can be brought to bear on models specified in its terms.

We begin with the "static" case in which we have a single random graph, G, with support $\mathcal{G}$. It is convenient to model G via its adjacency matrix Y, with $\mathcal{Y}$ representing the associated support (i.e., the set of adjacency matrices corresponding to all elements in $\mathcal{G}$). In ERGM form, we express the pmf of Y as follows:

$$\Pr(Y = y \mid S, \theta, X) = \frac{\exp\left(\theta^T S(y, X)\right)}{\sum_{y' \in \mathcal{Y}} \exp\left(\theta^T S(y', X)\right)} \mathbb{I}_{\mathcal{Y}}(y), \tag{1.1}$$

where $S : \mathcal{Y}, \mathcal{X} \to \mathbb{R}^s$ is a vector of sufficient statistics, $\theta \in \mathbb{R}^s$ is a vector of natural parameters, $X \in \mathcal{X}$ is a collection of covariates, and $\mathbb{I}_{\mathcal{Y}}$ is the indicator function (i.e., 1 if its argument is in the support of $\mathcal{Y}$, 0 otherwise).[1] If $|\mathcal{G}|$ is finite, then the pmf for any G can obviously be written with finite-dimensional S, θ (e.g., by letting S be a vector of indicator variables for elements of $\mathcal{Y}$); this is not necessarily true in the more general case, although a representation with S, θ of countable dimension still exists. In practice, it is generally assumed that S is of low dimension, or that at least that the vector of natural parameters can be mapped to a low-dimensional vector of "curved" parameters [see, e.g., Hunter and Handcock (2006)].

While the extreme generality of this framework has made it attractive, model selection and parameter estimation are often difficult due to the normalizing factor ($\kappa(\theta, S, X) = \sum_{y' \in \mathcal{Y}} \exp(\theta^T S(y', X))$) in the denominator of equation (1.1). This normalizing factor is analytically intractable and difficult to compute, except in special cases such as the Bernoulli and dyad-multinomial random graph families (Holland and Leinhardt, 1981); the first applications of this family (stemming from Holland and Leinhardt's seminal 1981 paper) focused on these special cases. Later, Frank and Strauss (1986) introduced a more general estimation procedure based on cumulant methods, but this proved too unstable for practical use. This, in turn, led to an emphasis on approximate inference using maximum pseudo-likelihood (MPLE) estimation (Besag, 1974), as popularized in this application by Strauss and Ikeda (1990) and later Wasserman and Pattison (1996). Although MPLE coincides with maximum likelihood estimation (MLE) in the limiting case of edgewise independence, the former was found to be a poor approximation to the MLE in many practical settings, thus leading to a consensus against its general use [see, e.g., Besag (2001) and van Duijn et al. (2009)]. The late 1990s saw the development of effective Markov chain Monte Carlo strategies for simulating draws from ERG models (Anderson et al., 1999; Snijders, 2002) which led to the current focus on MLE methods based either on first order method of moments (which coincides with MLE for regular ERGMs) or on importance sampling (Geyer and Thompson, 1992).[2]

Theoretical developments in the ERGM literature have arguably lagged inferential and computational advances, although this has become an increasingly active area of research. A major concern of the theoretical literature on ERGMs is the problem of *degeneracy*, defined differently by different authors but generally involving an inappropriately large concentration of probability mass on a small set of (generally unrealistic) structures. This issue was

[1] $\mathbb{I}_{\mathcal{Y}}$ can also be interpreted here as the counting measure on $\mathcal{Y}$. Although this is adequate for all cases considered here, see Krivitsky (2012) for the importance of reference measures in valued ERGM settings.

[2] Algorithms for parameter estimation and model selection using these approaches are implemented in a number of software packages (see, e.g., Snijders et al., 2007; Handcock et al., 2003; Wang et al., 2009), and empirical applications are increasingly common (e.g., Goodreau et al., 2009; Robins and Pattison, 2001, etc.).

recognized as early as Strauss (1986), who showed asymptotic concentration of probability mass on graphs of high density for models based on triangle statistics. [This motivated the use of *local* triangulation by Strauss and Ikeda (1990), a recommendation that went unheeded in later work.] More general treatments of the degeneracy problem can be found in Handcock (2003), Schweinberger (2011), and Chatterjee and Diaconis (2011). Butts (2011) introduced analytical methods that can be used to bound the behavior of general ERGMs by Bernoulli graphs (i.e., ERGMs with independent edge variables), and used these to show sufficient conditions for ERGMs to avoid certain forms of degeneracy as $N \to \infty$. One area of relatively rich theoretical development in the ERGM literature has been the derivation of sufficient statistics from first principles (particularly dependence conditions). Following the early work of Frank and Strauss (1986), many papers in this area employ Hammersley-Clifford constructions (Besag, 1974) in which initially posited axioms for conditional dependence among edge variables (usually based on substantive theory) are used to generate sets of statistics sufficient to represent all pmfs with the posited dependence structure. Examples of such work for single-graph ERGMs include Wasserman and Pattison (1996), Pattison and Robins (2002), and Snijders et al. (2006), with multi-relational examples including Pattison and Wasserman (1999) and Koehly and Pattison (2005). Snijders (2010) has showed that statistics based on certain forms of dependence allow for models that allow conditional marginalization across components (i.e., graph components are conditionally independent); this suggests statistics that may be appropriate for social processes in which edges can only influence each other "through" the network itself, and provides insight into circumstances which facilitate inference for population network parameters from data sampled at the component level (see also Shalizi and Rinaldo, 2013). An alternative way to motivate model statistics is via generative models that treat the observed network as arising from a stochastic choice process. Examples of such developments include Snijders (2001) and Almquist and Butts (2013).

1.2.2.1 Bayesian Inference for ERGM Parameters Given the likelihood of equation (1.1), Bayesian inference follows in the usual fashion by application of Bayes' Theorem, i.e.,

$$
\begin{aligned}
p(\theta|Y = y, S, X) &= \frac{ERG(y|\theta, S, X)p(\theta|S, X)}{\int_{\mathbb{R}^s} ERG(y|\theta', S, X)p(\theta'|S, X)d\theta'} \\
&\propto ERG(y|\theta, S, X)p(\theta|S, X),
\end{aligned}
$$

where $p(\theta|Y = y, S, X)$ is the posterior density of θ given the observed state of Y, statistic vector S, and covariate set X, $p(\theta|S, X)$ is the corresponding prior density of theta on $\mathbb{R}^s$, and $ERG(y|\theta, S, X)$ represents the ERGM likelihood for $\Pr(Y = y|\theta, S, X)$ from equation (1.1). In the case of ERGMs belonging to regular exponential families (e.g., non-curved), we can immediately obtain a conjugate prior for θ using known properties of exponential families:

$$p(\theta|S,X) = \frac{\exp\left(\theta^T\phi - \nu \ln \kappa(\theta,S,X)\right)}{\int_{\mathbb{R}^s} \exp\left(\theta'^T\phi - \nu \ln \kappa(\theta',S,X)\right) d\theta'},$$

where $\phi \in \mathbb{R}^s$ and $\nu > 0$ are hyperparameters and κ is the ERGM normalizing factor (as defined above). Note that ϕ and ν have natural interpretations in terms of "prior pseudo-data" and "prior pseudo-sample size," as is clear from the joint posterior:

$$\begin{aligned}
p(\theta|Y=y,S,X) &\propto ERG(y|\theta,S,X)\frac{\exp\left(\theta^T\phi - \nu \ln \kappa(\theta,S,X)\right)}{\int_{\mathbb{R}^s} \exp\left(\theta'^T\phi - \nu \ln \kappa(\theta',S,X)\right) d\theta'} \\
&= \frac{\exp\left(\theta^T S(y,X)\right)}{\kappa(\theta,S,X)} \frac{\exp\left(\theta^T\phi - \nu \ln \kappa(\theta,S,X)\right)}{\int_{\mathbb{R}^s} \exp\left(\theta'^T\phi - \nu \ln \kappa(\theta',S,X)\right) d\theta'} \\
&= \frac{\exp\left(\theta^T\left(\phi + S(y,X)\right) - (\nu+1)\ln \kappa(\theta,S,X)\right)}{\int_{\mathbb{R}^s} \exp\left(\theta'^T\phi - \nu \ln \kappa(\theta',S,X)\right) d\theta'} \\
&\propto \frac{\exp\left(\theta^T\left(\phi + S(y,X)\right) - (\nu+1)\ln \kappa(\theta,S,X)\right)}{\int_{\mathbb{R}^s} \exp\left(\theta'^T\left(\phi + S(y,X)\right) - (\nu+1)\ln \kappa(\theta',S,X)\right) d\theta'},
\end{aligned} \tag{1.2}$$

with equation (1.2) giving the (re)normalized form.

Despite the attractiveness of the conjugate prior, it is less helpful than it might be due to the intractability of the ERGM normalizing factor. While standard MCMC methods (e.g., the Metropolis-Hastings algorithm) can often manage intractable normalizing constants of a posterior density when the posterior density in question is known up to a constant, the kernel of equation (1.2) also involves the (usually intractable) normalizing factor κ from the ERGM likelihood. Such posteriors have been described as "doubly intractable" (Murray et al., 2012), and pose significant computational challenges in practice. In the more general case for which $p(\theta)$ does not necessarily include κ (i.e., non-conjugate priors), MCMC or related approaches must generally deal with posterior odds of the form

$$\begin{aligned}
\frac{p(\theta|Y=y,S,X)}{p(\theta'|Y=y,S,X)} &= \frac{ERG(y|\theta,S,X)}{ERG(y|\theta',S,X)} \frac{p(\theta|S,X)}{p(\theta'|S,X)} \\
&= \exp\left(S(y,X)^T(\theta-\theta')\right) \frac{\kappa(\theta',S,X)}{\kappa(\theta,S,X)} \frac{p(\theta|S,X)}{p(\theta'|S,X)},
\end{aligned}$$

which still require evaluation of normalizing factor ratios at each step. Provided that the prior ratio can be easily calculated, the complexity of this calculation is no worse than the associated ratios required for likelihood maximization, and indeed MAP can be performed in such cases using MCMC-MLE methods (see e.g. Hunter et al., 2008, 2012, for the MLE case) via the addition of prior odds as a penalty function. Approaches to direct posterior simulation in this regime include the use of exchange algorithms (Caimo and Friel, 2011) and other approximate MCMC methods (see Hunter et al., 2012, for a review). To date these latter methods have proven too computationally expensive for routine use, but the area is one of active research.

An alternative (if less generically satisfying) approach to the problem arises by observing that there are some classes of models for which κ is directly computable, and hence for which Bayesian analysis is more readily performed. An early example of this work is that of Wong (1987), who provided a fully Bayesian treatment of the p_1 family of Holland and Leinhardt (1981). Because the likelihood for this family factors as a product of categorical pmfs (the four edge variable states associated with each dyad), κ is easily calculated and Bayesian inference is greatly simplified. This intuition was subsequently elaborated by van Duijn et al. (2004), who used it as a basis for a much richer family of effects. Although we are focused here on models in ERGM form, it should also be noted that many latent variable models for networks can be viewed as positing that Y is drawn from an ERGM with strong conditional independence properties (leading to a tractable normalizing factor), given a (possibly very complex) set of latent covariates on which a prior structure is placed. Models such as those of Hoff et al. (2002), Handcock et al. (2007), Nowicki and Snijders (2001) and Airoldi et al. (2008) can be viewed in this light. While the simultaneous dependence in cross-sectional data tends to limit the utility of simplified ERGMs (or to require a shifting of computational burden into a complexly specified parameter structure), this problem is sometimes reduced in dynamic data due to the ability to condition on past observations (i.e., replacing simultaneous dependence in the present with dependence on the past) (Almquist and Butts, 2014). It is to this setting that we now turn.

1.2.3 Temporal Models for Network Data

Temporal models for social network data can be generally classified into two broad categories: (1) continuous time models; and (2) panel data models. Here we will focus only on panel data models – for examples of models for continuous time interaction data see Butts (2008a), DuBois, Butts, McFarland, and Smyth (2013), and DuBois, Butts, and Smyth (2013). Current theory and software are focused on statistical inference for panel data models based on four general approaches. The first is the family of *actor oriented models*, which assumes an underlying continuous-time model of network dynamics, where each observed event represents a single actor altering his or her outgoing links to optimize a function based on sufficient statistics (for details, see Snijders, 1996; Snijders and Van Duijn, 1997; Snijders, 2001, 2005). The second is the family of *latent dynamic structure models*, which treat network dynamics as emerging from a simple network process influenced by the evolution of set of latent covariates; for example, see Sarkar and Moore (2005), Sarkar et al. (2007), and Foulds et al. (2011). The third is the family of *temporal exponential family random graph models* (TERGMs), which attempt to directly parameterize the joint pmf of a graph sequence using discrete exponential families (Hanneke and Xing, 2007a; Hanneke et al., 2010; Hanneke and Xing, 2007b; Cranmer and Desmarais, 2011; Desmarais and Cranmer, 2011, 2012; Almquist and Butts, 2012, 2013, 2014). Finally, the fourth approach is the *separable temporal ERGM* family (or STERGM), which assumes each panel observation is a cross-sectional observation from a latent continuous time process in which edges evolve via two separable processes of edge formation and edge dissolution (Krivitsky and Handcock, 2010). Here, we focus on the TERGM case.

TERGMs can be viewed as the natural analog of time series (e.g., VAR) models for the random graph case. Typically, we assume a time series of adjacency matrices $\ldots, Y_{t-1}, Y_t, \ldots$ and parameterize the conditional pmf of $Y_t|Y_{t-1}, Y_{t-2}, \ldots$ in ERGM form. As with classical time series models, it is typical to introduce a temporal Markov assumption of limited dependence on past states; specifically, we assume the existence of some $k \geq 0$ such that that Y_t is independent of $Y_{t-k-1}, Y_{t-k-2}, \ldots$ given $Y_{t-1}, \ldots, Y_{t-k} \equiv Y_{t-k}^{t-1}$. Under this assumption, the standard TERGM likelihood for a single observation is written as

$$\Pr(Y_t = y_t \mid Y_{t-k}^{t-1} = y_{t-k}^{t-1}, \theta, S, X_t) = \frac{\exp\left(\theta^T S(y_t, y_{t-k}^{t-1}, X_t)\right)}{\sum_{y' \in \mathcal{Y}} \exp\left(\theta^T S(y'_t, y_{t-k}^{t-1}, X_t)\right)} \mathbb{I}_{\mathcal{Y}}(y_t). \tag{1.3}$$

As before, S is an s-vector of real-valued sufficient statistics, but for the TERGMs $S : \mathcal{Y}^{k+1}, \mathcal{X} \mapsto \mathbb{R}^s$ (i.e., each function may involve observations at the k time points prior to t instead of a single graph). Otherwise, nothing is intrinsically different from the cross-sectional case. (In particular, note that from the point of view of Y_t, y_{t-k}^{t-1} is a fully observed covariate. This is useful for the development that follows.) The denominator of (1.3) is again intractable in the general case, as it is for ERGMs.

For a complete TERGM series, the joint likelihood of the sequence $Y_1, \ldots, Y_t$ is given by $\prod_{i=1}^{t} TERG(y_i|Y_{i-k}^{i-1} = y_{i-k}^{i-1}, \theta, S, X_i)$, where $TERG$ refers to the single-observation TERGM likelihood of equation (1.3). MCMC-based maximum likelihood estimation for θ is feasible for very short series, but becomes costly as sequence length grows. Cranmer and Desmarais (2011) propose estimation via MPLE combined with a bootstrapping procedure to estimate standard errors as a computationally cheaper alternative. Alternately, scalable estimation is greatly simplified for TERGMs with no simultaneous dependence terms; i.e., models such that Y_{ijt} is conditionally independent of Y_{klt} given Y_{t-k}^{t-1} for all distinct $(i, j), (k, l)$. The TERGM likelihood for such models reduces to a product of Bernoulli graph pmfs, and hence the corresponding inference problem is equivalent to (dynamic) logistic regression. Although by no means novel, these conditional Bernoulli families have recently been advocated by Almquist and Butts (2014) as viable alternatives for network time series in which the time period between observations is on or faster than the time scale of network evolution, or whether it is for other reasons possible to capture much of the simultaneous dependence among edges by conditioning on the past history of the network. Almquist and Butts (2014) also show how this family can be easily extended to incorporate endogenous vertex dynamics (a feature not currently treated in other dynamic network families). In the remainder of this chapter, we focus on this case, with a particular emphasis on Bayesian inference for joint vertex/edge dynamics.

1.2.3.1 TERGM with Vertex Dynamics The TERG model in Section 1.2.3 can be further extended to handle vertex dynamics by employing a separable parameterization between the vertex set and edge set as proposed by Almquist and Butts (2014). Here we take the vertex set V_t to arise at each point in time from a fixed support of possible vertex sets, $\mathcal{V}$, with the associated pmf parameterized via a discrete exponential family. Y_t then arises from an ERG distribution conditional on V_t. To clarify notation, let $Z_t = (V_t, Y_t)$ be a representation for graph G_t, and as before let Z_a^b be the network time series $Z_a, \ldots, Z_b$. The pmf for a single observation under vertex dynamics is then

$$\begin{aligned}
\Pr(Z_t = z_t \mid Z_{t-k}^{t-1} = z_{t-k}^{t-1}, \psi, \theta, W, S, X_t)&\\
&= \Pr(V_t = v_t \mid Z_{t-k}^{t-1} = z_{t-k}^{t-1}, \psi, W, X_t)\\
&\times \Pr(Y_t = y_t \mid V_t = v_t, Z_{t-k}^{t-1} = z_{t-k}^{t-1}, \theta, S, X_t)\\
= \frac{\exp\left(\psi^T W(v_t, z_{t-k}^{t-1}, X_t)\right)}{\sum_{v' \in \mathcal{V}} \exp\left(\psi^T W(v'_t, z_{t-k}^{t-1}, X_t)\right)} &\times \frac{\exp\left(\theta^T S(y_t, v_t, z_{t-k}^{t-1}, X_t)\right)}{\sum_{y' \in \mathcal{Y}_{v_t}} \exp\left(\theta^T S(y'_t, v_t, z_{t-k}^{t-1}, X_t)\right)},
\end{aligned} \tag{1.4}$$

where $\mathcal{Y}_{v_t}$ is the set of possible adjacency matrices compatible with vertex set v_t, W is a w-vector of sufficient statistics on the vertex set, and ψ is a w-vector of vertex set parameters. The joint TERGM likelihood for a time series is then the product of the likelihoods for each observation. We refer to the conditional likelihood of a single observation in equation (1.4) as $TERGV$ (i.e., temporal exponential family random graph with vertex processes) in the discussion that follows.

The likelihood of equation (1.4) is inferentially "separable" in the sense that it factorizes into terms respectively dealing with ψ (and the vertex set) and with θ (and the edge set). These may be estimated separately, even when both depend on the same data (i.e., the edge history and vertex history may both enter into S and W). On the other hand, inferential separability does not imply *predictive* separability: the vertex model will strongly impact the edge structure of graphs drawn from the model, and in some cases vice versa. [See Almquist and Butts (2014) for a discussion.]

1.2.3.2 Bayesian Estimation of TERGMs As before, Bayesian inference for the full TERGM family (with vertex dynamics) is based on the posterior distribution of θ, ψ given $Z_1, \ldots, Z_t$:

$$p(\psi, \theta | Z_1^t = z_1^t, S, W, X) \propto p(\psi, \theta | S, W, X) \times \prod_{i=1}^{t} TERGV(z_i | Z_{i-k}^{i-1} = z_{i-k}^{i-1}, \psi, \theta, W, S, X_i).$$

It is frequently reasonable to treat the parameters of the edge and vertex processes as *a priori* independent given X. In that case, the above factors as

$$\begin{aligned} p(\psi, \theta | Z_1^t = z_1^t, S, W, X) &\propto p(\psi | W, X) p(\theta | S, X) \\ &\quad \times \prod_{i=1}^{t} TERGV(z_i | Z_{i-k}^{i-1} = z_{i-k}^{i-1}, \psi, \theta, W, S, X_i) \\ &= p(\psi | W, X) \frac{\exp\left(\psi^T W(v_t, z_{t-k}^{t-1}, X_t)\right)}{\sum_{v' \in \mathcal{V}} \exp\left(\psi^T W(v_t', z_{t-k}^{t-1}, X_t)\right)} \\ &\quad \times p(\theta | S, X) \frac{\exp\left(\theta^T S(y_t, v_t, z_{t-k}^{t-1}, X_t)\right)}{\sum_{y' \in \mathcal{Y}_{v_t}} \exp\left(\theta^T S(y_t', v_t, z_{t-k}^{t-1}, X_t)\right)}, \end{aligned}$$

which implies that the joint posterior itself factors as

$$p(\psi | Z_1^t = z_1^t, W, X) \propto p(\psi | W, X) \frac{\exp\left(\psi^T W(v_t, z_{t-k}^{t-1}, X_t)\right)}{\sum_{v' \in \mathcal{V}} \exp\left(\psi^T W(v_t', z_{t-k}^{t-1}, X_t)\right)} \quad \text{and}$$

$$p(\theta | Z_1^t = z_1^t, S, X) \propto p(\theta | S, X) \frac{\exp\left(\theta^T S(y_t, v_t, z_{t-k}^{t-1}, X_t)\right)}{\sum_{y' \in \mathcal{Y}_{v_t}} \exp\left(\theta^T S(y_t', v_t, z_{t-k}^{t-1}, X_t)\right)}. \quad (1.5)$$

This is a manifestation of the inferential separability remarked on previously. Although ψ and θ are jointly dependent on both the covariates and the observed data, the two may be analyzed independently. In the special case where no vertex dynamics are present, or where such dynamics are exogenous, the joint posterior simplifies to that of equation (1.5).

As with the ERGM case, posterior estimation for TERGMs inherits the normalizing factor problem (exacerbated in the case of vertex dynamics by the presence of two distinct exponential families, each with a normalizing factor!). Because of these technical complications there has been very little work in applying Bayesian analysis to the more general TERGM framework.[3] In the special case in which all observations in the present are independent conditional on the past, however, the normalizing factor becomes tractable and analysis is greatly simplified. As noted above, the similarity of the resulting inference problem to logistic regression (and the direct analogy with existing

[3] There has been some work into modeling the generative process of TERGM models through Bayesian networks (not to be confused with Bayesian models *of* networks) by Guo et al. (2007).

network regression models) has led to these being dubbed "dynamic network logistic regression" (DNR) families (Almquist and Butts, 2014). In the following section we will discuss Bayesian estimation in the DNR case, with and without vertex dynamics.

1.3 Dynamic Network Logistic Regression with Vertex Dynamics

Dynamic Network Logistic Regression with vertex dynamics was introduced by Almquist and Butts (2014), who modeled $\Pr(V_t = v_t \mid Z_{t-k}, X_t)$ and $\Pr(Y_t = y_t \mid V_t, Z_{t-k}, X_t)$ as separable processes whose elementwise realizations are independent conditional on covariates and past history. Under the necessary conditional independence, homogeneity, and temporal Markov assumptions one can derive the likelihood function for DNR from the general $TERGV$ likelihood, with the vertex likelihood given by

$$\begin{aligned}\Pr(V_t = v_t \mid Z_{t-k}^{t-1} = z_{t-k}^{t-1}, \psi, W, X_t)\\ = \prod_{i=1}^{n} B\left(\mathbb{I}(v^{(i)} \in v_t) \left| \mathrm{logit}^{-1}\left(\psi^T W(i, z_{i-k}^{t-1}, X_t)\right)\right.\right)\end{aligned}$$

and the edge likelihood by

$$\begin{aligned}&\Pr(Y_t = y_t \mid V_t = v_t, Z_{t-k}^{t-1} = z_{t-k}^{t-1}, \theta, S, X_t)\\ &= \prod_{(v^{(i)}, v^{(j)}) \in v_t \times v_t}^{n} B\left(Y_{ijt} = y_{ijt} \left| \mathrm{logit}^{-1}\left(\theta^T S(i, j, v_t, z_{t-k}^{t-1}, X_t)\right)\right.\right), \quad (1.6)\end{aligned}$$

where B is understood to be the Bernoulli pmf, $\mathbb{I}$ is the indicator function, and $v^{(i)}$ indicates the ith vertex from a known total risk set $V^{\max}$. (Thus, the support $\mathcal{V}$ of V_t is the power set of $V^{\max}$.) The analogy of this model family with logistic regression is clear from the form of the joint likelihood, which is equivalent to a (relatively complex) logistic regression of indicator variables for edge and vertex set memberships on a set of statistics associated with the network history and/or covariates. In the special case when V_t is exogenously varying, the joint likelihood of the data reduces to the edge process in equation (1.6); when it is fixed, the likelihood reduces to the "classic" dynamic network logistic regression model. Model specification, maximum likelihood based inference, and adequacy checking for this family are discussed in Almquist and Butts (2013, 2014).

1.3.1 Bayesian Inference for DNR Parameters

Because the DNR family reduces to a logistic regression structure, Bayesian inference is nominally straightforward. However, choice of prior structure for

DNR families has not been explored to date. Justifiably or otherwise, researchers typically seek to employ a default prior specification if they do not have a strong rationale for endorsing a specific prior. There is an extensive literature on noninformative, default, and reference prior distributions within the Bayesian statistical field (see Jeffreys, 1998; Hartigan, 1964; Bernardo, 1979; Spiegelhalter and Smith, 1982; Yang and Berger, 1994; Kass and Wasserman, 1996). More recent work has continued the traditions of both research on informative prior distributions using application-specific information and on minimally informative priors (often motivated by invariance principles) (for a review see, Gelman et al., 2008). One increasingly widely used approach to evaluating default priors (particularly in the machine learning literature) is the use of *predictive assessment*, i.e., examination of the extent to which a given prior structure reliably leads to accurate predictions on test data for a given body of training data. While arguably less principled than priors derived from other considerations, priors found to give good predictive performance on past data may be attractive on pragmatic grounds; by turns, such priors can also be justified more substantively as representing distributions *compatible with past observations* on similar data, and hence plausible at least as a crude starting point. Likewise, priors that consistently lead to poor predictive performance on test data should be suspect, whatever the principles used to construct them. The balance of this chapter is thus concerned with the predictive evaluation of various candidate priors in the context of DNR models.

While both MCMC and MAP are feasible for inference in DNR families, our focus here will be on posterior simulation via MCMC. In addition to giving us a more complete view of the posterior distribution, posterior simulation is particularly well-adapted to predictive model adequacy checking (Gelman et al., 2004). Specifically, simulation of future observations conditional on a point estimate (e.g., the posterior mean or mode) can greatly underestimate the uncertainty associated with the posterior distribution, and by extension can fail to reveal the benefits to be gained by, e.g., prior specifications that reign in extreme parameter values without greatly changing the central tendency of the posterior distribution.

Given the above, there are many reasonable choices of prior specifications for DNR families that may be applicable in one or another context. Given that our focus here is on evaluating simple, default priors, we will focus our attention on four prior specifications suggested as default priors for logistic regression in the Bayesian statistical literature (e.g., Gelman et al., 2008). Our core questions are as follows. First, what are the inferential consequences of employing these reference priors versus maximum likelihood estimation for DNR families in typical social network settings? Second, to what extent do various reasonable default priors lead to differences in either point estimation or posterior uncertainty in such settings? Finally, what differences (if any) does selection of one or another default prior make to prediction, in the specific sense of forecasting properties of an evolving network? If, in typical settings, inferential and predictive outcomes are fairly insensitive to choice of prior, then selection based on computational or other factors may be a reasonable practice. If, by contrast, we find substantial differences in inferential and/or predictive performance among default priors, then these choices must be scrutinized far more carefully. The balance of this chapter is intended as a first step towards assessing these questions.

Although simulation and inference for ERGMs is in general a highly specialized art (see, e.g., Snijders, 2002; Hunter et al., 2008; Wang et al., 2009), the logistic form of DNR families facilitates parameter estimation (if not network simulation) using more standardized tools and techniques. Examples of off-the-shelf toolkits suitable for posterior simulation in the cases studied here include Winbugs, JAGS or MCMCpack in `R` (Spiegelhalter et al., 2003; Plummer, 2003; Martin et al., 2011). Here, we employ the Metropolis-Hastings algorithm as implemented Martin et al. (2011) for posterior simulation, with sufficient statistics computed in the same manner as Almquist and Butts (2014, 2013) via custom `statnet`-based tools (Handcock et al., 2003). The latter were also used for posterior predictive simulation of graph-theoretic quantities. MCMC convergence was assessed using both Geweke's convergence diagnostic (Geweke et al., 1991) and Raftery and Lewis's diagnostic (Raftery and Lewis, 1992).

1.3.2 Bayesian Estimation of DNR with Vertex Dynamics

As noted above, inferential separability of the edge and vertex processes in the DNR context allows both to be treated as independent logistic regression problems (so long as the associated parameters are *a priori* independent). As this implies, there is no reason that the same prior structure must be used for both; for simplicity and practicality, however we here consider the case where we assume both the edge and vertex parameters have the same prior distributions.

As noted above, we here consider a number of typical priors for logistic regression (which have been recommended in the literature) for use as priors in DNR with and without vertex dynamics. The baseline point of comparison for all Bayesian results will be the ML estimate (and its sampling distribution), reflecting the dominant practice within the ERGM literature. In addition to this baseline, we consider five prior specifications within three general classes: (1) an improper uniform prior (i.e., the fully-Bayesian counterpart to maximum likelihood estimation); (2-3) independent normally distributed priors (one centered at 0, and one offset to inflate prior density); and (4-5) weakly informative t families of proper priors recommended by Gelman et al. (2008) (a Cauchy distribution centered at 0 with a scale parameter of 2.5, and a t_7 prior centered at 0). Empirical experiments on the effects of these priors on the analysis and interpretation can be found in Section 1.4.

1.4 Empirical Examples and Simulation Analysis

Given the six specifications described above (including the MLE), we seek to evaluate the practical consequences of these priors for Bayesian inference in typical settings. To that end, we consider a comparative analysis of inference under our proposed priors for two empirical cases. The first is a dynamic network of citations among bloggers during the 2004 US presidential election ("*Blog*"), and the second is a dynamic network of face-to-face communication ties among windsurfers on a southern California beach ("*Beach*"). Both networks are typical of social network data sets in that each involves multiple, complex mechanisms of interaction as well as actor-level heterogeneity. These networks have also been studied in the literature, making them useful reference cases for our present analysis.

1.4.1 Blog Data

The first data set considered in this chapter is a dynamic inter- and intra-group blog citation network collected by Butts and Cross (2009) and previously analyzed with DNR in Almquist and Butts (2013). This dynamic network consists of interactions among all blogs credentialed by the Democratic National Committee (DNC) or Republican National Committee (RNC) for their respective 2004 conventions. The vertex set is comprised of actors representing 34 DNC and 14 RNC credentialed blogs, as well as the one blog credentialed by both groups, and is static over the observation period. The edge set is comprised of directed hyperlinks on the front pages of the aforementioned blogs as observed over a 121 day period. The data was obtained by automatically querying the main page of each blog at six hour intervals starting at midnight, Pacific time. The period of study begins on 7/22/04 (shortly before the DNC convention) and ends 11/19/04 (shortly after the Presidential election), leading to a total of 484 time points. There exists an edge from blog i to blog j at time t iff a link to blog j appears on the main page of i at time t. Specifically, this data is represented as an adjacency array, Y, such that $Y_{ijt} = 1$ if i cites (i.e., links to) j at time t, and 0 otherwise.

1.4.2 Beach Data

The second data set considered in this chapter was originally collected by Freeman et al. (1988). It is a dynamic interpersonal communication network of windsurfers in the late 1980s which Freeman et al. (1988) subsequently analyzed in aggregate. These data have been used in a number of influential articles (see Cornwell, 2009; Hummon and Doreian, 2003; Zeggelink et al., 1996, among others), and later analyzed dynamically using DNR with vertex dynamics in Almquist and Butts (2014). The network was originally collected daily (aggregated over a morning and an afternoon observation period) for 31 days from August 28, 1986 to September 27, 1986.[4] The data set consists of 95 individuals which were originally classified ethnographically by Freeman et al. (1988) into *"regulars"* ($N = 54$)—frequent attendees who were well-integrated into the social life of the beach community—and *"irregulars"* ($N = 41$). Further, the *regulars* were categorized by the researchers into two groups, *group 1* ($N = 22$) and *group 2* ($N = 21$), with 11 individuals not classified as belonging to either group 1 or group 2. The number of windsurfers appearing on the beach ranged from 3 to 37 individuals over the observation period. The number of communication ties per day ranging from 0 to 96 in the network.

1.4.3 Case Analysis: Static Vertex Set

Given a collection of alternative priors, what are the consequences of choosing one or another for inference in the case of a DNR family with a fixed vertex set? To assess this, we consider parameter estimation and prediction on the Blog data. While the entire time series consists of 484 time points, we here restrict ourselves to a much smaller series of 32 time points, in order to form a reasonable comparison with the Beach data. As noted below, the likelihood specification (i.e., sufficient statistics or effects) for models employed here is based on prior work by Almquist and Butts (2013); we hold the effects constant across all models to isolate the impact of prior specifications.

Because V_t is fixed here, we need only consider priors for the edge parameters (θ). Posterior distributions based on these models are compared the standard ML estimate derived from iterative weighted least squares methods (McCullagh and Nelder, 1999), noting that the ML estimate can also be thought of as a MAP estimator under an improper uniform prior. The posteriors evaluated are as follows: (1) an improper uniform prior (i.e., a fully Bayesian analog to the MLE); (2) independent $N(0, .1)$ priors on each parameter (a simple default choice with relatively strong shrinkage towards 0); (3) independent $N(5, 1)$ priors on each parameter (included as a non-normative test to evaluate the consequences of poor prior specification); (4) independent standard t_7 priors on each parameter, emulating the example used by Gelman et al. (2008); and (5) independent scaled Cauchy priors $C(0, 2.5)$ as recommended by Gelman et al. (2008) for logistic regression in another context. All Bayesian DNR models were be estimated via posterior simulation by MCMC, thus allowing for improved predictive modeling. This is discussed in the following section.

Our specification of the Blog network model follows Almquist and Butts (2013); for expository purposes, we have simplified the original model slightly by removing less important effects. Those retained fall into two main classes: (1) density effects, e.g., DNC/RNC mixing and daily seasonality; and (2)

[4]Unfortunately, one day (September 21st) is missing due to a race on a different beach, which precluded data collection. Thus, complete data is available for 30 days during the observation period.

lagged network effects, e.g., interaction or preferential attachment effects (Wasserman and Faust, 1994). In Table 1.1 we consider first mixing effects for DNC and RNC internal ties (respectively), followed by terms for interaction between DNC and RNC blogs; next we consider daily effects, *Tuesday* to *Sunday* effects with *Monday* as the reference category; a lag effect (Y_{t-1}); and a lagged indegree effect $\log(Deg(Y_{t-1}))$, which can be thought of as preferential attachment effect. The table is laid out to illustrate the differences and similarities of the parameter estimates for different minimally informative priors on estimation. Table 1.1 demonstrates that most reasonable minimally informative priors produce results comparable to each other and the MLE, as suggested by Gelman et al. (2008). It is worth pointing out that the prior *can* have a large effect on the parameter estimate (and thus the resulting predictions, see Section 1.4.3.1): the $N(5,1)$ parameter is noticeably farther from the other estimates (although the bias is not always in a positive direction, e.g., for the lag and lagged degree terms). That said, it is perhaps reassuring that even a fairly strong prior still leads to generally comparable estimates for a data set of only moderate size (44 nodes over 32 time points). In general, the MLE and posterior mean estimates for the standard default priors differ by an amount that is approximately an order of magnitude smaller than the statistical uncertainty associated with the estimates themselves; for the "extreme" $N(5,1)$ prior, the differences are on the scale of the posterior standard deviation. Likewise, the posterior standard deviations themselves are generally quite similar to each other, and to the standard error of the MLE (although, again, the "extreme" prior deviates somewhat).

To further illustrate the similarities and differences among estimates, we consider Figure 1.1 which contains the 95% posterior marginals for each parameter (where the ML case shows the asymptotic Gaussian approximation to the sampling distribution of the MLE). Here we see again that the typical minimally informative priors result in relatively similar posterior distributions, both in the central mass and in the tails; all are likewise quite similar to the MLE. The deviations associated with the "extreme" prior are clearer here, and demonstrate visually the fact that (for models of this sort) a bias in the prior may not manifest within the posterior in a clear way. Specifically, for this case the inflated prior mean on the baseline interaction rate parameters tends to lead to a compensating deflation of the lagged terms; such effects suggest that it is important to be careful when using highly informative priors, since influence on one parameter may propagate to the posterior marginals for other parameters in a non-obvious manner.

The fact that all reasonable prior specifications lead to posterior marginals that are close to each other and to the sampling distribution of the MLE suggests that we are effectively operating within the regime in which the posterior

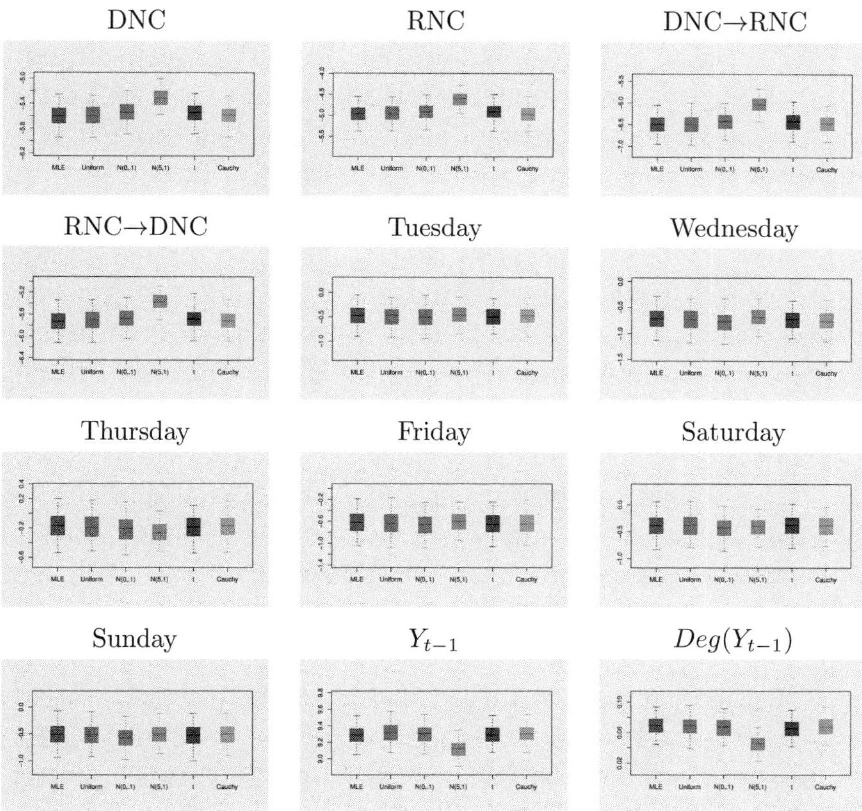

Figure 1.1: Posterior marginals for the five prior specifications discussed in Section 1.4.3; MLE sampling distribution based on asymptotic Gaussian approximation

Table 1.1: Parameter estimates: MLE and posterior means under the five prior specifications discussed in Section 1.4.3 (standard errors and posterior standard deviations are given in parentheses).

	MLE	U	$N(0,.1)$	$N(5,1)$	t_7	$C(0,2.5)$
DNC	−5.598	−5.599	−5.543	−5.308	−5.547	−5.589
	(0.176)	(0.184)	(0.176)	(0.152)	(0.164)	(0.164)
RNC	−4.959	−4.956	−4.919	−4.61	−4.917	−4.966
	(0.208)	(0.219)	(0.217)	(0.178)	(0.201)	(0.216)
DNC→RNC	−6.486	−6.494	−6.432	−6.027	−6.425	−6.474
	(0.225)	(0.243)	(0.221)	(0.196)	(0.238)	(0.218)
RNC→DNC	−5.728	−5.711	−5.672	−5.37	−5.681	−5.714
	(0.199)	(0.212)	(0.196)	(0.164)	(0.198)	(0.194)
Y_{t-1}	9.284	9.317	9.302	9.121	9.296	9.302
	(0.119)	(0.129)	(0.122)	(0.108)	(0.117)	(0.119)
$Deg(Y_{i-1})$	0.069	0.068	0.067	0.045	0.065	0.068
	(0.013)	(0.013)	(0.013)	(0.011)	(0.013)	(0.013)
Monday	–	–	–	–	–	–
Tuesday	−0.469	−0.495	−0.509	−0.461	−0.501	−0.482
	(0.213)	(0.219)	(0.219)	(0.197)	(0.198)	(0.205)
Wednesday	−0.717	−0.728	−0.781	−0.683	−0.734	−0.745
	(0.214)	(0.222)	(0.221)	(0.193)	(0.198)	(0.202)
Thursday	−0.167	−0.187	−0.216	−0.257	−0.197	−0.179
	(0.184)	(0.182)	(0.183)	(0.147)	(0.174)	(0.169)
Friday	−0.619	−0.636	−0.667	−0.601	−0.646	−0.635
	(0.219)	(0.232)	(0.222)	(0.186)	(0.216)	(0.219)
Saturday	−0.388	−0.388	−0.428	−0.412	−0.388	−0.393
	(0.216)	(0.231)	(0.206)	(0.183)	(0.211)	(0.208)
Sunday	−0.51	−0.517	−0.571	−0.505	−0.529	−0.514
	(0.217)	(0.213)	(0.210)	(0.195)	(0.217)	(0.210)

is close to its asymptotic Gaussian limit (Gelman et al., 2004). Although this phenomenon is data dependent, our case suggests that, for models of this kind, data sets of moderate size may prove adequate for the asymptotic approximation to hold. As such, any long-tailed, symmetric prior distribution is likely to work well from an inferential point of view, and choice of distribution can be reasonably made for computational or other reasons.

1.4.3.1 One-step Prediction for Model Assessment Inference is vital for hypothesis testing, model understanding, and the like; however, one can obtain parameters that are qualitatively reasonable from models that nevertheless predict poorly. In this section we will focus on one type of predictive model assessment procedure, described by Almquist and Butts (2014) in the DNR context as *inhomogeneous Bernoulli prediction*. The algorithm as suggested by Almquist and Butts (2014) is as follows: for each time point t we predict $Y_t|Y_{t-k}^{t-1} = y_{t-k}^{t-1}$ n times (in other words, we take n draws from the posterior predictive distribution of the network at time t given the previous k observed time steps) to predict the edge structure. We then summarize the resulting network via a suite of Graph Level Indices (GLI; Anderson et al., 1999), yielding a GLI distribution for each time point.

Here we consider two important graph level indices: density (Wasserman and Faust, 1994) and the fraction of triads that form 3-cliques (i.e., triangles) (Wasserman and Faust, 1994). The first of these is an extremely fundamental GLI that describes the fraction of all possible ties that are present in the network; the second is a simple index related to clustering. For purposes of assessment, we examine the one-step inhomogeneous Bernoulli predictions of graph density and 3-clique formation rates (respectively) under the ML and Bayesian posterior predictive distributions. The results are summarized in Figures 1.2 and 1.3. An important advantage of using posterior predictive distributions in a temporal modeling context (versus point estimates) is that this allows for a realistic propagation of posterior uncertainty in parameter estimates into the resulting predictive distribution for prediction in temporal modeling is that it allows for more realistic prediction through sampling of the prior distribution. We can see this quite clearly in Figures 1.2 and 1.3, where the prediction intervals for the MCMC Bayesian estimates are typically much larger than those of the MLE (blue dotted lines). Further, the we see that the "extreme" $N(5, 1)$ prior has surprisingly minimal effect on the prediction estimates and prediction intervals. Overall, it appears that the increased variance in the model is a more realistic portrayal of the data as wider prediction interval often covers the observed data, while the "pure" ML estimated prediction often does not.

1.4.4 Bayesian DNR with Vertex Dynamics

To experiment with the effects of different minimally informative priors on models with vertex dynamics, we again look to an empirical case; here we begin by considering the beach data set collected by Freeman et al. (1988), which is comprised of windsurfers engaging in interpersonal communication on a beach in South California in early fall of 1986. This is a temporally

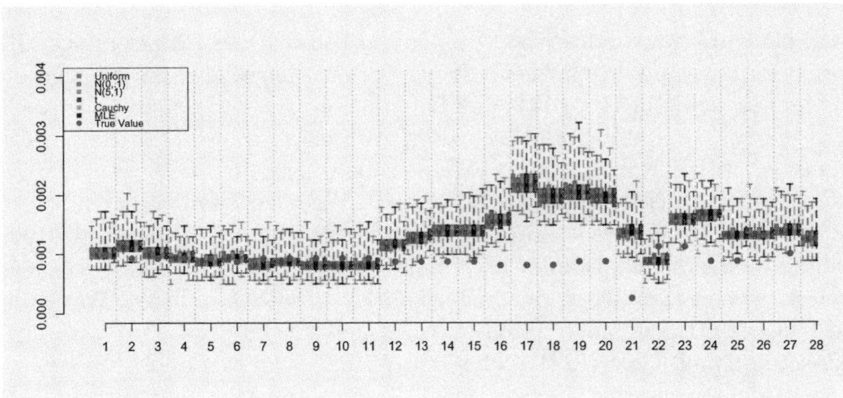

Figure 1.2: One-step prediction of graph density; red dots indicate observed values.

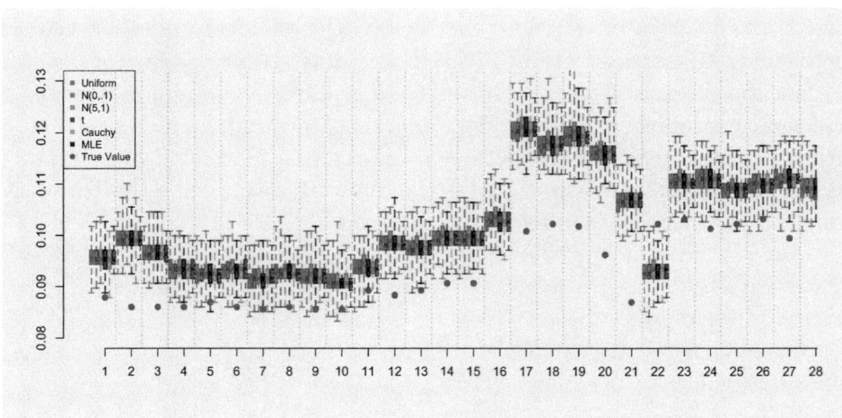

Figure 1.3: One-step prediction of fraction of triads forming 3-cliques; red dots indicate observed values.

evolving network that includes endogenous vertex dynamics, i.e., change in windsurfers who show up over time. Thus, our priors here involve two classes of parameters: θ (the edge parameters), and ψ (the vertex parameters). For the present study, we use the same priors as in the Blog case (along with the MLE), with the same prior specification employed in each instance for θ and for ψ.

Our specification of the likelihood for the Beach data is a simplified version of the model employed by Almquist and Butts (2014). Because the model includes vertex dynamics, note that effects must be specified for both the edge set (given the vertices present each day) and the vertex set (given the past history up to the day in question). We include two main classes of effects: (1) density effects, e.g., weekday/weekend effects and log of the vertex set size[5]; and (2) lagged network effects, e.g., interaction or appearance on the day before and the number of interactions one engaged in the day before (i.e., degree effects). Degree effects may be particularly important in this context as they capture general tendency towards engagement in the network the day before. In Table 1.2 we consider *weekend* and *weekday* effects for both the vertex and edge set dynamics (in a beach setting we expect this to be particularly important as it captures the natural rhythm of activity of the work week in the United States), a $\log(n_t)$ effect for the edge set, a lag effect for both the edge (Y_{t-1}) and vertex set (V_{t-1}), and last a lagged degree effect for both the edge $\log(Deg(Y_{t-1}))$ and vertex set $Deg(V_{t-1})$. The table is laid out to illustrate the differences and similarities of the parameter estimates for different minimally informative priors on estimation (as well as the "extreme" $N(5,1)$ prior as a point of comparison).

Table 1.2 demonstrates that (as with the Blog model), we obtain comparable point estimates using either the MLE or standard default priors, with variation across prior specifications generally at or below the level of statistical uncertainty associated with the estimates themselves. As before, the "extreme" prior has a noteworthy effect, although for many parameters the posterior means given this prior are quite close to those obtained via other specifications. The primary exceptions in this regard are the density related effects, which are more weakly estimated and subject to considerable influence by a strongly informative prior. On the whole, however, our results suggest that reasonable default priors would lead to qualitatively (and quantitatively similar conclusions) for the processes shaping the Beach network.

Turning to the posterior marginal distributions, Figure 1.4 shows results that closely parallel the above (and our findings for the Blog data). The posteriors under reasonable default priors closely mirror each other and the MLE

[5]This helps to stabilize mean degree in the presence of changing vertex set sizes; see Krivitsky et al. (2011) and Butts (2011).

Table 1.2: Posterior means and MLE estimates for the Beach data under the prior specifications discussed in Section 1.4.4 (posterior standard deviations and MLE standard errors given in parentheses).

Edge Parameters						
	MLE	U	$N(0,.1)$	$N(5,1)$	t_7	$C(0,2.5)$
Weekend (E)	−0.966	−0.932	−0.922	0.240	−0.687	−0.875
(s.e./p.s.d.)	(0.379)	(0.390)	(0.375)	(0.318)	(0.337)	(0.367)
Weekday (E)	−1.679	−1.635	−1.635	−0.318	−1.365	−1.583
(s.e./p.s.d.)	(0.428)	(0.434)	(0.423)	(0.356)	(0.380)	(0.415)
$\log(n_t)$	−0.096	−0.108	−0.110	−0.491	−0.188	−0.126
(s.e./p.s.d.)	(0.125)	(0.127)	(0.123)	(0.105)	(0.111)	(0.121)
Y_{t-1}	0.995	0.989	0.991	1.072	0.956	0.981
(s.e./p.s.d.)	(0.174)	(0.176)	(0.175)	(0.173)	(0.171)	(0.175)
$Deg(Y_{t-1})$	0.037	0.036	0.037	0.026	0.036	0.037
(s.e./p.s.d.)	(0.015)	(0.014)	(0.015)	(0.014)	(0.014)	(0.014)
Vertex Parameters						
	MLE	U	$N(0,.1)$	$N(5,1)$	t_7	$C(0,2.5)$
Weekday (V)	−1.858	−1.86	−1.858	−1.828	−1.846	−1.858
(s.e./p.s.d.)	(0.076)	(0.073)	(0.077)	(0.075)	(0.076)	(0.076)
Weekend (V)	−1.935	−1.93	−1.937	−1.850	−1.910	−1.930
(s.e./p.s.d.)	(0.128)	(0.129)	(0.128)	(0.122)	(0.126)	(0.128)
V_{t-1}	0.737	0.725	0.727	0.768	0.699	0.723
(s.e./p.s.d.)	(0.230)	(0.232)	(0.235)	(0.221)	(0.223)	(0.231)
$Deg(V_{t-1})$	0.596	0.598	0.602	0.57	0.604	0.603
(s.e./p.s.d.)	(0.144)	(0.142)	(0.146)	(0.141)	(0.139)	(0.146)

sampling distribution, suggesting that we are once again within the Gaussian asymptotic regime. The use of an extreme informative prior does clearly disrupt inference, with a stronger effect on some parameters than others; as before, the sign of the effect on a given parameter may not reflect the direction of prior bias, due to relationships among the parameters within the likelihood.

1.4.4.1 One-step Prediction for Model Assessment For the Blog models, we saw that while the MLE was inferentially comparable to the posterior distributions obtained by reasonable default priors, the resulting predictions (MLE-based simulations versus the full posterior predictive) were quite different.

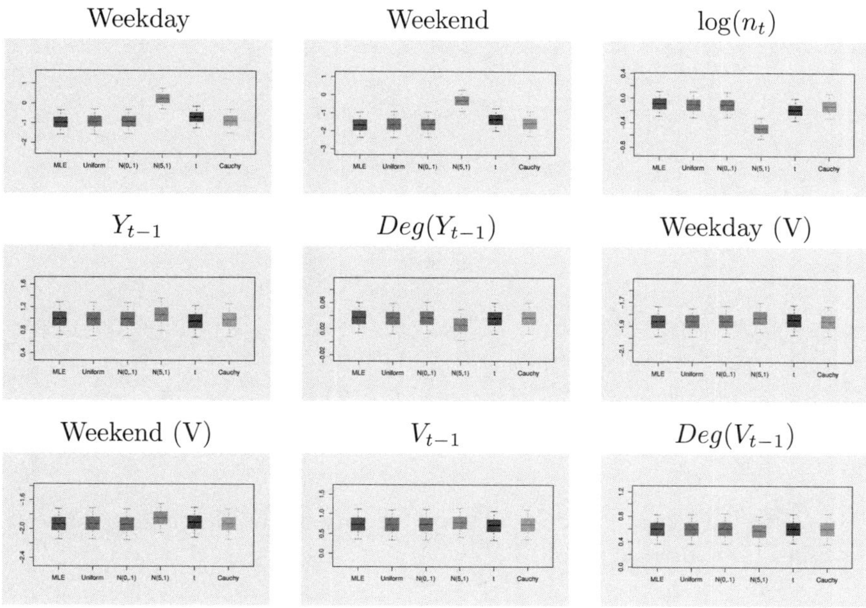

Figure 1.4: Posterior marginals and MLE sampling distributions for the Beach data.

Here, we repeat this analysis for the Beach models, bearing in mind that we now sample vertices as well as edges. (I.e., at each step we first draw the vertex set from the associated model, next drawing the edge set conditional on the realization of the vertex set.) As before, we employ the density and fraction of triads forming 3-cliques in the next time period as our outcome measures.

The results of our simulation experiment are summarized in Figures 1.5 and 1.6. Broadly, the results are similar to those of the Blog data: incorporating posterior uncertainty into our predictions leads to much wider prediction intervals, and these are often necessary to capture the observed data. In general, predicting clique structure is hard for this model (it has no simultaneous dependence, and many other potential predictors have been removed), but the uncertainty associated with the model predictions is clearer for the Bayesian models than for the MLE. There is not an obvious "winner" from the default priors, although the extreme tails of their prediction intervals do vary somewhat from time point to time point. In general, the differences between them appear small and unsystematic.

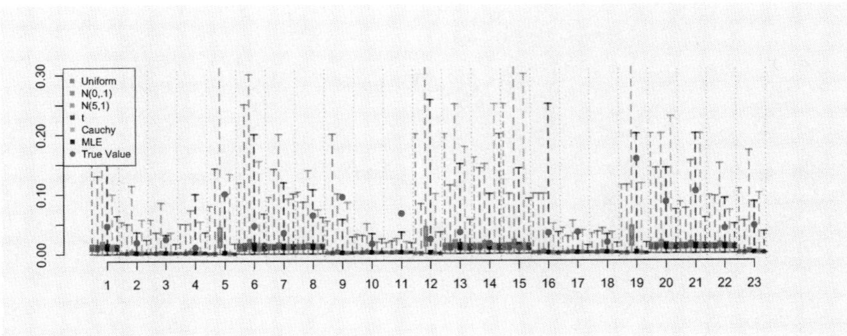

Figure 1.5: One-step prediction of graph density on the Beach data for all five Bayesian DNR models and MLE.

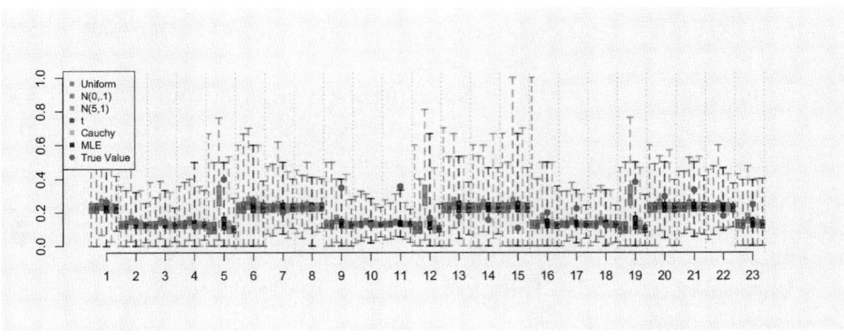

Figure 1.6: One-step prediction of graph 3-cliques on the Beach data for all five Bayesian DNR models and MLE.

1.5 Discussion

In our test cases for Bayesian DNR with and without vertex dynamics, we see that several standard minimally informative priors produce both inferential and predictive results that are comparable to each other and—in the case of inference—to maximum likelihood estimation. Differences between point estimates for the default priors tested here are typically smaller (often by an order of magnitude) than the statistical uncertainty associated with the estimates themselves, and any are hence likely to lead to very similar interpretation in practice. Since these results were achieved with two very different types of dynamic social networks of relatively moderate size, we are led to the preliminary conclusion that selection of default priors for DNR families is unlikely to strongly impact results in typical settings (and the choice of prior can thus be made on computational or other grounds).

It is noteworthy that there is a general advantage for the use of Bayesian posterior predictive distributions versus predictive distributions based on the MLE, in terms of reduced overconfidence. This is a general and well-known phenomenon, and not particular to network models; however, it is perhaps worth reinforcing the point that predictions should incorporate uncertainty regarding parameters, and the Bayesian approach greatly facilitates this practice.

While we saw that inference in our test cases was generally fairly robust to prior specification, it is of course possible to "break" a data set by employing a sufficiently informative prior. Here, we observed that use of an "extreme" prior biased in a single direction strongly affected some (but not all) parameters, and that the direction of bias in the posterior did not necessarily correspond to the direction of bias in the prior (due to correlations among the associated statistics). As would be expected, parameters that are more poorly estimated are more subject to influence from the prior, and these (along with others closely related to them via the likelihood) are the ones that are most vulnerable to poor prior specifications. As in other contexts, our results suggest that use of strongly informative priors in DNR must be undertaken with caution, and in particular with an awareness of the degree to which parameters are entangled via the likelihood. Robustness tests are strongly recommended.

The strong concordance between posteriors under various choices of prior with each other and the sampling distribution of the MLE strongly suggests that our test cases place us in the asymptotic Gaussian regime, an encouraging development given that our networks are of relatively modest size. That said, it is important to bear in mind the large number of degrees of freedom inherent in dynamic network data. In general, data size for such problems grows as $\mathcal{O}(N^2T)$, where N is the vertex set size and T is the number of time points. For graphs of even moderate size—and even for small numbers of time points—this can easily result in an extremely large number of edge variables. Of course, the asymptotic limit depends on more than simply the number of degrees of freedom (the sparsity of the data is also important), but this heuristic provides a reasonable intuition for why the Gaussian approximation is likely to work well here. This is a property that is potentially exploitable for e.g. approximate Bayesian computation in large N,T settings.

1.6 Conclusion

In this chapter, we have reviewed the problem of modeling static and dynamic networks in exponential family form. As we have shown, Bayesian analysis of both problems in the general case is made difficult via the central role of the incomputable ERGM normalizing factor, which enters into the likelihood (and sometimes the prior) in a manner that makes traditional MCMC-based sampling schemes slow and/or impractical. For some model families, however, this problem does not apply (due to the presence of a tractable normalizing factor); such families are especially useful in the case of dynamic network modeling, where conditioning on the past can in some cases allow us to model edges as conditionally independent in the present. For such cases, the temporal ERGM form reduces to a simple product of inhomogeneous Bernoulli graph likelihoods, dubbed "dynamic network regression" because of the similarity of the resulting model to logistic regression on time series data. Because of the simplicity of this family, and its similarity to logistic regression, it would be desirable to be able to employ standard "default" priors for analysis in routine settings. Our experiments with two different data sets suggest that this is a reasonable approach: alternative default priors lead to very similar conclusions, with all being similar to inferences resulting from maximum likelihood estimation. One reason for this concordance is that even fairly modest dynamic network data sets supply enough data degrees of freedom to—for DNR families—place the posterior within the asymptotic Gaussian regime. While it is possible to obtain poor results by selection of an especially inappropriate prior, reasonable choices thus lead to reasonable outcomes. Given this, and given the advantages of the Bayesian approach for problems such as prediction, there seems little reason not to recommend this as a standard technique for analyzing network dynamics with TERGM DNR families.

Acknowledgements: This work was supported in part by Office of Naval Research (ONR) award # N00014-08-1-1015, National Science Foundation (NSF) awards # BCS-0827027 and # SES-1260798 and National Institute of Health (NIH)/ National Institute of Child Health & Human Development (NICHD) award # 1R01HD068395-01.

Bibliography

Airoldi, E. M., D. M. Blei, S. E. Fienberg, and E. P. Xing (2008, June). Mixed membership stochastic blockmodels. *Journal of Machine Learning Research 9*, 1981–2014.

Almquist, Z. W. and C. T. Butts (2012). Evolving context: Evidence from temporal change in organizational collaboration over the course of the 2005 katrina disaster. Working paper, University of California, Irvine.

Almquist, Z. W. and C. T. Butts (2013). Dynamic network logistic regression: A logistic choice analysis of inter- and intra-group blog citation dynamics in the 2004 US presidential election. *Political Analysis 21*(4), 430–448.

Almquist, Z. W. and C. T. Butts (2014). Logistic network regression for scalable analysis of networks with joint edge/vertex dynamics. *Sociological Methodology* (forthcoming).

Anderson, B. S., C. Butts, and K. Carley (1999). The interaction of size and density with graph-level indices. *Social Networks 21*(3), 239–267.

Anderson, C. J., S. Wasserman, and B. Crouch (1999). A p* primer: Logit models for social networks. *Social Networks 21*(1), 37–66.

Barndorff-Nielsen, O. (1978). *Information and Exponential Families in Statistical Theory.* New York: John Wiley and Sons.

Bernardo, J. M. (1979). Reference posterior distributions for Bayesian inference. *Journal of the Royal Statistical Society. Series B (Methodological)*, 113–147.

Besag, J. (1974). Spatial interactions and the statistical analysis of lattice systems. *Journal of the Royal Statistical Society. Series B (Methodological) 36*(2), 192–236.

Besag, J. (2001). Markov chain Monte Carlo for statistical inference. *Center for Statistics and the Social Sciences Working Papers*(9).

Brown, L. D. (1986). *Fundamentals of Statistical Exponential Families: with Applications in Statistical Decision Theory.* Hayward, CA: Institute of Mathematical Statistics.

Butts, C. T. (2007). Permutation models for relational data. *Sociological Methodology 37*(1), 257–281.

Butts, C. T. (2008a). A relational event framework for social action. *Sciological Methodology 38*(1), 155–200.

Butts, C. T. (2008b). Social networks: A methodological introduction. *Asian Journal of Social Psychology 11*(1), 13–41.

Butts, C. T. (2009). Revisiting the foundations of network analysis. *Science 325*, 414–416.

Butts, C. T. (2011). Bernoulli graph bounds for general random graphs. *Sociological Methodology 41*, 299–345.

Butts, C. T. and B. R. Cross (2009). Change and external events in computer-mediated citation networks: English language weblogs and the 2004 U.S. electoral cycle. *The Journal of Social Structure 10*(3), 1–29.

Caimo, A. and N. Friel (2011). Bayesian inference for exponential random graph models. *Social Networks 33*(1), 41–55.

Chatterjee, S. and P. Diaconis (2011). Estimating and understanding exponential random graph models. arXiv:1102.2650.

Cornwell, B. (2009). Good health and the bridging of structural holes. *Social Networks 31*(1), 92–103.

Cranmer, S. J. and B. A. Desmarais (2011). Inferential network analysis with exponential random graph models. *Political Analysis 19*(1), 66–86.

Desmarais, B. and S. Cranmer (2012). Statistical mechanics of networks: Estimation and uncertainty. *Physica A: Statistical Mechanics and its Applications 391*(4), 1865–1876.

Desmarais, B. A. and S. J. Cranmer (2011). Forecasting the locational dynamics of transnational terrorism: A network analytic approach. In *Proceedings of the European Intelligence and Security Informatics Conference (EISIC)*. IEEE Computer Society.

DuBois, C., C. T. Butts, D. McFarland, and P. Smyth (2013). Hierarchical models for relational event sequences. *Journal of Mathematical Psychology*, 57(6), 297–309.

DuBois, C., C. T. Butts, and P. Smyth (2013). Stochastic blockmodeling of relational event dynamics. In *Proceedings of the 16th International Conference on Artificial Intelligence and Statistics*, (AISTATS).

Foulds, J. R., C. DuBois, A. U. Asuncion, C. T. Butts, and P. Smyth (2011). A dynamic relational infinite feature model for longitudinal social networks. *Journal of Machine Learning Research - Proceedings Track 15*, 287–295.

Frank, O. and D. Strauss (1986). Markov graphs. *Journal of the American Statistical Association 81*(395), 832–842.

Freeman, L. C., S. C. Freeman, and A. G. Michaelson (1988). On human social intelligence. *Journal of Social Biological Structure 11*, 415–425.

Gelman, A., J. B. Carlin, H. S. Stern, and D. B. Rubin (2004). *Bayesian Data Analysis* (2nd ed.). New York, NY: Chapman & Hall/CRC.

Gelman, A., A. Jakulin, M. G. Pittau, and Y.-S. Su (2008). A weakly informative default prior distribution for logistic and other regression models. *The Annals of Applied Statistics*, 1360–1383.

Geweke, J. et al. (1991). *Evaluating the accuracy of sampling-based approaches to the calculation of posterior moments*. Federal Reserve Bank of Minneapolis, Research Department.

Geyer, C. J. and E. A. Thompson (1992). Constrained monte carlo maximum likelihood for dependent data. *Journal of the Royal Statistical Society. Series B(Methodological) 54*(3), 657–699.

Goodreau, S. M., J. A. Kitts, and M. Morris (2009). Birds of a feather, or friend of a friend?: Using exponential random graph models to investigate adolescent social networks. *Demography 46*(1), 103–125.

Guo, F., S. Hanneke, W. Fu, and E. P. Xing (2007). Recovering temporally rewiring networks: A model-based approach. In *Proceedings of the 24th International Conference on Machine Learning*, pp. 321–328. ACM.

Handcock, M. S. (2003). Statistical models for social networks: Inference and degeneracy. In R. Breiger, K. M. Carley, and P. Pattison (Eds.), *Dynamic Social Network Modeling and Analysis*, pp. 229–240. Washington, DC: National Academies Press.

Handcock, M. S., D. R. Hunter, C. T. Butts, S. M. Goodreau, and M. Morris (2003). `statnet`: A suite of R packages for the statistical modeling of social networks. Electronic.

Handcock, M. S., A. E. Raftery, and J. M. Tantrum (2007). Model based clustering for social networks. *Journal of the Royal Statistical Society, Series A 170*, 301–354.

Hanneke, S., W. Fu, and E. P. Xing (2010). Discrete temporal models of social networks. *Electronic Journal of Statistics 4*, 585–605.

Hanneke, S. and E. P. Xing (2007a). Discrete temporal models of social networks. In *Statistical Network Analysis: Models, Issues, and New Directions*, Volume 4503 of *Lecture Notes in Computer Science*, pp. 115–125. Berlin, Germany: Springer.

Hanneke, S. and E. P. Xing (2007b). Statistical network analysis: Models, issues, and new directions: ICML 2006 workshop on statistical network analysis, Pittsburgh, PA, USA, June 29, 2006, revised selected papers. *TBA 4503*, 115–125.

Hartigan, J. (1964). Invariant prior distributions. *The Annals of Mathematical Statistics 35*(2), 836–845.

Hoff, P. D., A. E. Raftery, and M. S. Handcock (2002). Latent space approaches to social network analysis. *Journal of the American Statistical Association 97*(460), 1090–1098.

Holland, P. W. and S. Leinhardt (1981). An exponential family of probability distributions for directed graphs. *Journal of the American Statistical Association 76*(373), 33–50.

Hummon, N. and P. Doreian (2003). Some dynamics of social balance processes: bringing Heider back into balance theory. *Social Networks 25*(1), 17–49.

Hunter, D. R. and M. S. Handcock (2006). Inference in curved exponential family models for networks. *Journal of Computational and Graphical Statistics 15*, 565–583.

Hunter, D. R., M. S. Handcock, C. T. Butts, S. M. Goodreau, and M. Morris (2008). ergm: A package to fit, simulate and diagnose exponential-family models for networks. *Journal of Statistical Software 24*(3).

Hunter, D. R., P. N. Krivitsky, and M. Schweinberger (2012). Computational statistical methods for social network models. *Journal of Computational and Graphical Statistics 21*(4), 856–882.

Jeffreys, H. (1998). *The Theory of Probability.* Oxford University Press.

Kass, R. E. and L. Wasserman (1996). The selection of prior distributions by formal rules. *Journal of the American Statistical Association 91*(435), 1343–1370.

Koehly, L. M. and P. Pattison (2005). Random graph models for social networks: Multiple relations or multiple raters. In P. J. Carrington, J. Scott, and S. Wasserman (Eds.), *Models and Methods in Social Network Analysis*, Chapter 9, pp. 162–191. Cambridge: Cambridge University Press.

Krivitsky, P. N. (2012). Exponential-family random graph models for valued networks. *Electronic Journal of Statistics 6*, 1100–1128.

Krivitsky, P. N. and M. S. Handcock (2010). A separable model for dynamic networks. Technical Report arXiv:1011.1937, arXive e-print.

Krivitsky, P. N., M. S. Handcock, and M. Morris (2011). Adjusting for network size and composition effects in exponential-family random graph models. *Statistical Methodology 8*, 319–339.

Martin, A. D., K. M. Quinn, and J. H. Park (2011). MCMCpack: Markov chain Monte Carlo in R. *Journal of Statistical Software 42*(9), 1–21.

McCullagh, P. and J. A. Nelder (1999). *Generalized Linear Models* (2nd ed.). New York: Chapman & Hall/CRC.

Murray, I., Z. Ghahramani, and D. MacKay (2012). MCMC for doubly-intractable distributions. arXiv preprint, arXiv:1206.6848.

Nowicki, K. and T. A. B. Snijders (2001). Estimation and prediction for stochastic blockstructures. *Journal of the American Statistical Association 96*, 1077–1087.

Pattison, P. E. and G. L. Robins (2002). Neighborhood-based models for social networks. *Sociological Methodology 32*, 301–337.

Pattison, P. E. and S. Wasserman (1999). Logit models and logistic regressions for social networks: II. Multivariate relations. *British Journal of Mathematical and Statistical Psychology 52*, 169–193.

Plummer, M. (2003). Jags: A program for analysis of Bayesian graphical models using gibbs sampling. In *Proceedings of the 3rd International Workshop on Distributed Statistical Computing (DSC 2003). March*, pp. 20–22.

Raftery, A. and S. Lewis (1992). One long run with diagnostics: Implementation strategies for Markov chain Monte Carlo. *Statistical Science 7*, 493–497.

Robins, G. and P. Pattison (2001). Random graph models for temporal processes in social networks. *Journal of Mathematical Sociology 25*(1), 5–41.

Sarkar, P. and A. W. Moore (2005, December). Dynamic social network analysis using latent space models. *SIGKDD Explor. Newsl. 7*(2), 31–40.

Sarkar, P., S. M. Siddiqi, and G. J. Gordon (2007). A latent space approach to dynamic embedding of co-occurrence data. In M. Meila and X. Shen (Eds.), *Proceedings of the Eleventh International Conference on Artificial Intelligence and Statistics (AISTATS 2007)*.

Schweinberger, M. (2011). Instability, sensitivity, and degeneracy of discrete exponential families. *Journal of the American Statistical Association 106*(496), 1361–1370.

Shalizi, C. R. and A. Rinaldo (2013). Consistency under sampling of exponential random graph models. *The Annals of Statistics 41*(2), 508–535.

Snijders, T. A. (1996). Stochastic actor-oriented models for network change. *Journal of Mathematical Sociology 21*(1–2), 149–172.

Snijders, T. A. (2001). The statistical evaluation of social network dynamics. In M. Sobel and M. B. Boston (Eds.), *Sociological Methodology*, pp. 361–395. London: Basil Blackwell.

Snijders, T. A. (2002). Markov chain Monte Carlo estimation of exponential random graph models. *Journal of Social Structure 3*(2).

Snijders, T. A. (2005). Models for longitudinal network data. In P. Carrington, J. Scott, and S. Wasserman (Eds.), *Models and Methods in Social Network Analysis*, Chapter 11. New York: Cambridge University Press.

Snijders, T. A., C. E. Steglich, M. Schweinberger, and M. Huisman (2007). *Manual for SIENA Version 3.1*. University of Groningen: ICS Department of Sociology; University of Oxford: Department of Statistics.

Snijders, T. A. and M. Van Duijn (1997). Simulation for statistical inference in dynamic network models. In R. Conte, R. Hegselmann, and P. Terna (Eds.), *Simulating Social Phenomena*, pp. 493–512. Berlin: Springer.

Snijders, T. A. B. (2010). Conditional marginalization for exponential random graph models. *Journal of Mathematical Sociology 34*, 239–252.

Snijders, T. A. B., P. E. Pattison, G. L. Robins, and M. S. Handcock (2006). New specifications for exponential random graph models. *Sociological Methodology 36*.

Spiegelhalter, D., A. Thomas, N. Best, and D. Lunn (2003). Winbugs user manual. *Cambridge: MRC Biostatistics Unit.*

Spiegelhalter, D. J. and A. F. Smith (1982). Bayes factors for linear and log-linear models with vague prior information. *Journal of the Royal Statistical Society. Series B (Methodological)*, 377–387.

Strauss, D. (1986). On a general class of models for interaction. *SIAM Review 28*(4), 513–527.

Strauss, D. and M. Ikeda (1990). Psuedolikelihood estimation for social networks. *Journal of the American Statistical Association 85*(409), 204–212.

van Duijn, M. A. J., K. Gile, and M. S. Handcock (2009). Comparison of maximum pseudo likelihood and maximum likelihood estimation of exponential family random graph models. *Social Networks 31*(1), 52–62.

van Duijn, M. A. J., T. A. B. Snijders, and B. H. Zijlstra (2004). p_2: A random effects model with covariates for directed graphs. *Statistica Neerlandica 58*, 234–254.

Wang, P., G. Robins, and P. Pattison (2009, September). Pnet: Program for the simulation and estimation of exponential random graph (p*) models. `http://www.sna.unimelb.edu.au/pnet/pnet.html`.

Wasserman, S. and K. Faust (1994). *Social Network Analysis: Methods and Applications.* New York: Cambridge University Press.

Wasserman, S. and P. Pattison (1996). Logit models and logistic regressions for social networks: I. An introduction to Markov graphs and p^*. *Psychometrika 61*(3), 401–425.

Wong, G. Y. (1987). Bayesian models for directed graphs. *Journal of the American Statistical Association 82*(397), 140–148.

Yang, R. and J. O. Berger (1994). Estimation of a covariance matrix using the reference prior. *The Annals of Statistics*, 1195–1211.

Zeggelink, E. P., F. N. Stokman, and G. G. van der Bunt (1996). The emergence of groups in the evolution of friendship networks. *Journal of Mathematical Sociology 21*(1-2), 29–55.

CHAPTER 2

ETHNIC MINORITY RULE AND CIVIL WAR: A BAYESIAN DYNAMIC MULTILEVEL ANALYSIS

Xun Pang

Tsinghua University, Beijing, China

This chapter applies a Bayesian dynamic multilevel model to analyze the effect of ethnic minority rule (EMR) on civil war onset. EMR's effect on civil war is unlikely to be constant across countries with different political, economic, and social backgrounds. Without considering the variation of EMR's effect, theoretical and empirical conclusions would be overgeneralized and misleading. To analyze the varying-effect of EMR and to improve statistical predictions of civil war, this chapter adopts a Bayesian dynamic multilevel model to control for multiple confounders caused by the time-series cross-section structure of the civil war data. The empirical results suggest that the relationship between civil war and EMR is moderated by several background factors including political instability and governance quality. Modeling the stochastic process of the errors improves forecasting, which indicates that information in the errors, which has often been neglected, is valuable for predicting civil war.

2.1 Introduction: Ethnic Minority Rule and Civil War

Since the end of the Cold War, civil war has been one of the most studied research topics in political science. Civil wars have caused immense human suffering, both from direct casualties and the long-term effects of disease, dislocation, and lasting grievances. Various political and economic theories have been developed to identify the causes of civil war (Cederman et al., 2013; Sambanis, 2002; Cederman et al., 2011; Miguel et al., 2004; Fearon, 2004; Fearon and Laitin, 2003b; Collier and Hoeffler, 2002). To many political scientists, one of the most surprising empirical findings is that ethnicity does not have a clear or robust effect on civil war, despite strong nationalist theories predicting a direct causal relationship between the two (Huntington, 1968; Russett, 1964; Scott, 1976; Deutsch, 1985, 1953; Anderson, 1983; Horowitz, 1985; Huntington, 1996). To further scrutinize the effect of ethnicity, recent studies on civil war take a statist perspective and focus on ethnic political dominance rather than ethnic diversity when analyzing the effect of ethnicity on nationalist grievance against the state (Cederman and Girardin, 2007; Fearon et al., 2007). Compared to the general question about the relationship between ethnicity and civil war, these recent studies asked a more specific question about whether EMR increases the risk of civil war.

However, EMR is not straightforward to measure because its meaning may vary in different political and social contexts. The measure constructed by Cederman and Girardin (2007) (henceforth CG), rely on the ethnicities of politicians in important governmental positions. This measure is subject to aggregation difficulties. Fearon et al. (2007) (henceforth FKL) simply use the government leader's ethnicity, which is also problematic because the leader's ethnic group is not necessarily the dominant group in politics, such as African-Americans in the United States under the Obama government. Hence, it is unrealistic to expect that EMR, coded with either criterion, has the same meaning, generates the same degree of nationalist grievance, or provides the same rebellion opportunities in different countries with varying political, economic, and social backgrounds. Observed or unobserved background factors are likely to alter the relationship between EMR and civil war onset. This poses a challenge for large N studies, because, without considering those variations, heterogeneity bias is likely to be present in quantitative analyses.

This chapter focuses on explaining the context-dependent effect of EMR on civil war onset. The statistical model applied to handle the methodological challenges is a Bayesian generalized linear multilevel model. In addition to the varying effect of EMR altered by background factors, empirical analysis on civil war has to control for other sources of confounding. Because the civil war data are longitudinal and rich in structure, there are multiple sources of possible confounding, including serial correlation and unobserved heterogeneity in both the time and spatial dimensions. In order to control for serial correlation and analyze the dynamics in the error term, the model also specifies a pth-order autoregressive error process. The Bayes factor is used to compare different model specifications and to determine the lag order *ex post*. The empirical results suggest strong positive autocorrelation lasting for multiple time periods and a salient variation of EMR's effect on civil war onset across countries. Political instability is the most important background factor which amplifies EMR's effect on civil war onset. Male secondary school enrollment, which proxies for governance quality and the strength of the government, is likely to reduce the effect of EMR on civil war risk. No evidence is found to support the importance of ethnic diversity for the relationship between EMR and civil war onset. The robustness of these findings is checked with alternative measures of EMR. The generalized linear multilevel model with an autoregressive error process (henceforth GLMM-AR(p) model) improves the reliability of statistical inferences and the performance of within-sample forecasting.

2.2 EMR: Grievance and Opportunities of Rebellion

Following the greed-grievance framework of Collier and Hoeffler (2004) and Collier et al. (2004), motives (mainly grievance) and opportunities are two dimensions for analyzing civil war, and the two are entangled. Rebellion is not solely driven by grievance or anger, but also based on the rational cost-benefit analysis. Minority political dominance may cause nationalist resentment *and* perception of greater opportunities to take over power (Wimmer, 2002; Gellner, 1983). The leader's ethnicity, as a measure of EMR, is straightforward and easy to code. It also has the advantage of being time-varying and being able to include much more country-years in the analysis (161 countries in 55 years). KFL investigated the relationship between EMR and civil war onset, and their empirical results do not suggest any statistically significant connection between the two. Checking the samples case by case, FKL recognized that using the leader's ethnicity to measure EMR had its own problems. First, in some countries, the ethnic group that the government leader comes from does not necessarily mean that the group is politically dominant. For example, the Britain Prime Minister Douglas-Home (1963) was Scottish, Franco in Spain was Galician, Stalin in the USSR was Georgian, several General Secretaries in Czechoslovakia were Slovak, and the Presidency in Switzerland is rotating among ethnic groups. This is to say, the ethnicity of the leader does not necessarily measure the primary explanatory variable—the dominance of the leader's ethnic group for some (democratic) institutional or (imperial) policy reasons. In this situation the nationalist grievance mechanism is not relevant (Collier et al., 2004). Recognizing this type of confounding, they recoded these cases and got similar empirical results.

However, there is a more general problem that FKL mentioned in their paper but did not provide a solution for. The meaning of EMR may vary from country to country. Furthermore, it is highly likely that some observable and unobservable factors affect the relationship between a leader's ethnicity and civil war. There are various intervening factors that can alter the mechanism by which EMR affects civil war risk, causing the variation of this relationship across countries. One example of such factors may be political instability. In an unstable regime, a country may be more vulnerable to civil war under ethnic minority rule than under ethnic plurality rule (EPR) (Collier et al., 2004; Hegre et al., 2001). The effect of EMR may also depend on the economy. When the economy is good, the opportunity cost of rebellion is high because the expected gain from being in the labor market is higher than when the economy is bad (Miguel et al., 2004; Grossman, 1995; Hirschleifer, 1995; Collier and Hoeffler, 2004). Likewise, ethnically more diverse societies may mean a more tolerant attitude towards EMR (Sambanis, 2002; Collier and Hoeffler, 2004). In addition, the effect of EMR on civil war may depend on regime type. In democracies, EMR may not necessarily incur more civil war than EPR because the leader is elected and has legitimacy, but in non-democracies, the legitimacy of the leader is more questionable and more easily challenged by other groups (Dahl, 1989; Gasiorowski, 1995; Acemoglu and Robinson, 2003). In addition, international and historical factors can change the direction or alter the degree that EMR affects civil war risk. For example, external shocks could temporally make the country array around the flag, and domestic nationalist grievances are dwarfed by grievances against external entities or by external threats to the survival of the country as a whole. Another example may be historical memory: if a country has experienced conflicts (not necessarily civil war) between the minority group the leader belongs to and the plurality group, or, in other words, if there is unresolved historical hostility between the two groups, it is more likely that EMR will increase civil war risk.

2.3 Bayesian GLMM-AR(p) Model

This chapter applies multilevel modeling to incorporate the intervening effects of background factors when analyzing the relationship between EMR and civil war. Multilevel analysis has been widely used in longitudinal data analysis (more generally longitudinal analysis) in many disciplines, including political science. It has been well developed in both the conventional and Bayesian frameworks (Hagenaars, 1990; Singer and Willett, 2003; Yang et al., 2004; Skrondal and Rabe-Hesketh, 2004; Molenberghs and Verbeke, 2005; Zhu, 2012). One particularly important characteristic of multilevel modeling for the current study is that the predictors can enter the model at the correct level, and variations within and between groups can be modeled for both theoretical and methodological reasons (Gill, 2007, p. 395). In addition, several slow-moving or time-invariant predictors are important in the study of civil war, but collinearity is a big concern for the fixed-effect panel model (Shor et al., 2007). In contrast, multilevel modeling avoids this pitfall by placing those variables into the group-level regression (Gelman and Hill, 2006; Gelman, 2006, pp. 246–247).

In addition to the variation of EMR's effect, there is likely to be other unobserved heterogeneity of the sample countries. If this heterogeneity is correlated with any of the regressors, the estimator is biased. Also, theories of international relations emphasize the importance of the international system. Disturbances in the international system form common observed or unobserved shocks on countries which may affect civil war risk. Again, endogeneity arises when those omitted common shocks are not strictly exogenous. Furthermore, serial correlation is not only a source of inefficiency but can also be a further cause of endogeneity because several important covariates are conventionally lagged to avoid endogeneity (such as per capita income, population, and democracy). In longitudinal analysis, lagged values should be used with caution and in conjunction with serial correlation correction; if the residual in the previous period is correlated with the residual in the current period, i.e., $\text{Cov}(\xi_t, \xi_{t-1}) \neq 0$, and if some of the covariates are lagged because they may correlated with the residual, namely, $\text{Cov}(x_{t-1}, \xi_{t-1}) \neq 0$, we have $\text{Cov}(x_{t-1}, \xi_t) \neq 0$, resulting in endogeneity.

Pang Pang (2010) proposes a Bayesian generalized linear multilevel model with pth-order autoregressive errors (GLMM-AR(p)). This model can handle the methodological challenges caused by varying coefficients, serial correlation and dichotomous outcome variables. Before applying the GLMM-AR(p) model to analyze civil war data, it is necessary to describe the model specification and Markov chain Monte Carlo (MCMC) strategies for model estimation and comparison.

2.3.1 General Model Specification

For $i = 1, 2, \cdots, N$ and $t = 1, 2, \cdots, T$, the general model specification proposed by Pang Pang (2010) can be expressed as follows:

$$
\begin{aligned}
y_{it} &= \mathbb{I}(z_{it} > 0), && (2.1)\\
z_{it} &= \mathbf{x}'_{1it}\beta_1 + \mathbf{w}'_{it}\beta_{2i} + \mathbf{s}'_{it}\beta_{3t} + \xi_{it}, && (2.2)\\
\beta_{2i} &= \mathbf{A}_i\beta_2 + \mathbf{b}_i, && (2.3)\\
\beta_{3t} &= \mathbf{F}_t\beta_3 + \mathbf{c}_t, && (2.4)\\
\xi_{it} &= \rho_1\xi_{i,t-1} + ... + \rho_p\xi_{i,t-p} + \epsilon_{it}, && (2.5)
\end{aligned}
$$

where $\mathbb{I}(\cdot)$ is an indicator function. The $NT \times K_1$ matrix $\mathbf{x}_{1it}$, $NT \times K_2$ matrix $\mathbf{w}_{it}$, and the $NT \times K_3$ matrix $\mathbf{s}_{it}$ contain explanatory variables with constant effects (β_1), unit-specific effects (β_{2i}), and time-specific effects (β_{3t}) on y, respectively. The two design matrices $\mathbf{A}_i$ and $\mathbf{F}_t$ are group-level coefficient matrices, and $(\rho_1, ..., \rho_p)$ are autoregressive coefficients which satisfy the stationarity conditions. The reduced form of the model will be used for model estimation. Denote $K = K_1 + K_2 + K_3$, and the $NT \times K$ matrix

$\mathbf{x}_{it} = (\mathbf{x}_{1,it}, \mathbf{w}'_{it}\mathbf{A}_i, \mathbf{s}'_{it}\mathbf{F}_t)$ and the K-dimensional vector $\beta = (\beta_1, \beta_2, \beta_3)$, then the reduced form of the latent model can be written as follows

$$z_{it} = \mathbf{x}'_{it}\beta + \mathbf{w}'_{it}\mathbf{b}_i + \mathbf{s}'_{it}\mathbf{c}_t + \xi_{it} \tag{2.6}$$

$$\xi_{it} = \rho_1\xi_{i,t-1} + \ldots + \rho_p\xi_{i,t-p} + \epsilon_{it}. \tag{2.7}$$

2.3.2 Parameter Estimation

The probit multilevel model with autoregressive errors is difficult to estimate using non-Bayesian approaches. Pang (2010) proposed an MCMC algorithm to estimate the parameters and choose the lag order. In the Bayesian framework, models with different explanatory variables, lag orders, and specifications can be compared based on the Bayes factor. Moreover, the Bayesian method has other advantages over conventional methods. The Bayesian method is more efficient than point-estimation methods in that it averages over the uncertainty in multilevel modeling (Shor et al., 2007; Gelman and Hill, 2006, Chapter 18). In addition, since the civil war data are not balanced, for unbalanced data the sampling distribution for the elements of the variance-covariance matrix is skewed. The Bayesian estimator does not suffer from this pitfall because inferences are not conditional on the specific point estimates of the covariance matrix (Raudenbush and Bryk, 2002; Shor et al., 2007).

Priors are necessary to complete a Bayesian model, which assume the following distributive forms:

$$\beta \sim \mathcal{N}_K(\beta_0, \mathbf{B}_0), \qquad \epsilon_{it} \sim \mathcal{N}(0,1), \qquad \{\mathbf{b}_i\} \sim \mathcal{N}_{K_2}(\mathbf{0}, \mathbf{D}),$$
$$\mathbf{D}^{-1} \sim \mathcal{W}_{K_2}(\nu_0, \mathbf{D}_0), \qquad \{\mathbf{c}_t\} \sim \mathcal{N}_3(\mathbf{0}, \mathbf{E}), \qquad \mathbf{E}^{-1} \sim \mathcal{W}_{K_3}(\eta_0, \mathbf{E}_0),$$
$$\rho \sim \mathcal{U}_p(\rho : \rho \in S_\rho),$$

where $\mathcal{N}_Q(a, A)$ is a Q-dimensional multivariate normal distribution with mean a and variance-covariance matrix A, $\mathcal{W}(b, B)$ denotes a Wishart distribution with b degrees of freedom and scale matrix B, and $\mathcal{U}_p$ is a p-dimensional Uniform distribution in the space of S (in this chapter, it is the stationarity space of the autoregressive coefficients[1]). In the data analysis in the next section, diffuse proper priors are used. In addition, simulations are conducted on the prior distributions to ensure that they are in reasonable spaces for the substantive questions. Sensitivity analysis is performed by changing the expectation and variance of the priors to be two times bigger or one-half smaller. The empirical posteriors are not sensitive to different prior choices.

Because the variance-covariance matrix of the error $\xi_i = \{\xi_{i1}, \ldots, \xi_{iT}\}$ is not diagonal, MCMC updating of the latent responses $\mathbf{z}_t$ requires sampling

[1] An AR process is stationary if all the characteristic roots of the polynomial are outside the unit circle. For different autoregressive processes the stationarity space is different.

from a T-dimensional multivariate normal distribution. The GHK simulator is a widely-applied approach to sample from multivariate normal distributions, but it is computationally expensive when the dimension is high (Geweke, 1991). This chapter adopts an auxiliary approach to orthogonalize the variance-covariance matrix and sample $\mathbf{z}$s independently. Chib and Jeliazkov (2006) introduce an auxiliary parameter $\mathbf{u} \sim \mathcal{N}(\mathbf{0}, \mathbf{I})$ to orthogonalize the serially correlated errors with Cholesky Decomposition. They applied this approach to simplify sampling from a nonparameteric function. But their approach can actually also improves the efficiency of sampling from the posterior of $\mathbf{z}$. More specifically, the approach works as follows: the variance-covariance matrix $\boldsymbol{\Sigma}_{\xi_i}$ can be decomposed into two parts $\boldsymbol{\Sigma}_{\xi_i} = \boldsymbol{\Omega}_i + \kappa \mathbf{I}_i$, where $\boldsymbol{\Omega}_i$ is a symmetric positive definite matrix, κ is a positive constant, and $\mathbf{I}$ is the identity matrix. Following Chib and Jeliazkov (2006), I choose $\lambda_i/2$ for κ_i, where λ_i is the smallest eigenvalue of $\boldsymbol{\Omega}_i$. This choice makes the algorithm numerically stable, and then $\boldsymbol{\Omega}_i$ is further decomposed as $\mathbf{V}_i'\mathbf{V}_i$, in which $\mathbf{V}_i'$ is the lower triangular matrix produced by the Cholesky decomposition. Hence, the covariance matrix can be rewritten as $\boldsymbol{\Sigma}_{\xi_i} = \mathbf{V}_i'\mathbf{V}_i + \kappa \mathbf{I}_T$. Now, the model in equation (2.6) can be written as

$$\mathbf{z}_i = \mathbf{x}_i'\beta + \mathbf{w}_i'\mathbf{b}_i + \mathbf{s}_i'\mathbf{c}_T + \mathbf{V}_i'\mathbf{u}_i + \epsilon_i, \tag{2.8}$$

where $\epsilon_i \sim N(0, \kappa_i \mathbf{I}_T)$, and ϵ_i and $\mathbf{u}_i$ are mutually independent. Define $(q_{i1}, q_{i2}, ..., q_{iT}) \equiv \mathbf{q}_i \equiv \mathbf{V}_i'\mathbf{u}_i$, and the full MCMC algorithm is as follows:

1. Sample $\beta, \{\mathbf{b}_i\}, \{\mathbf{u}_i\}$ from their joint posterior distribution

$$\pi(\beta, \{\mathbf{b}_i\}, \{\mathbf{u}_i\}) = \pi(\{\mathbf{u}_i\}|\beta, \{\mathbf{b}_i\})\pi(\{\mathbf{b}_i\}|\beta)\pi(\beta)$$

   - $\beta|\{\mathbf{b}_i\}, \{\mathbf{u}_i\} \sim N_K(\bar{\beta}, \mathbf{B}_1)$, where $\mathbf{B}_1 = (\mathbf{B}_0^{-1} + \sum_{i=1}^N \mathbf{x}_i'\mathbf{H}_i^{-1}\mathbf{x}_i)^{-1}$, $\bar{\beta} = \mathbf{B}_1\left(\mathbf{B}_0^{-1}\beta_0 + \sum_{i=1}^N \mathbf{x}_i'\mathbf{H}_i^{-1}(\mathbf{z}_i - \mathbf{s}_i'\mathbf{c}_T)\right)$, and $\mathbf{H}_i = (\boldsymbol{\Sigma}_i + \mathbf{w}_i'\mathbf{D}\mathbf{w}_i)$;
   - $\mathbf{b}_i|\beta \sim N_{K_2}(\bar{\mathbf{b}}_i, \mathbf{D}_{1i})$, where $\mathbf{D}_{1i} = (\mathbf{D}^{-1} + \mathbf{w}_i'(\boldsymbol{\Sigma}_i)^{-1}\mathbf{w}_i)^{-1}$ and $\bar{\mathbf{b}}_i = \mathbf{D}_{1i}\mathbf{w}_i'(\boldsymbol{\Sigma}_i)^{-1}(\mathbf{z}_i - \mathbf{x}_i'\beta - \mathbf{s}_i'\mathbf{c}_{T_i})$;
   - $\mathbf{u}_i \sim N(\bar{\mathbf{u}}_i, \mathbf{U}_i)$, where $\mathbf{U}_i = (\mathbf{I}_T + \mathbf{V}_i\mathbf{V}_i'/\kappa)^{-1}$, and $\bar{\mathbf{u}}_i = \mathbf{U}_i\mathbf{V}_i(\mathbf{z}_i - \mathbf{x}_i'\beta - \mathbf{w}_i'\mathbf{b}_i - \mathbf{s}_i'\mathbf{c}_{T_i})/\kappa_i$;

2. $\{\mathbf{c}_t\} \sim N_{K_3}(\bar{\mathbf{c}}_t, \mathbf{E}_{1i})$, where $\mathbf{E}_{1i} = (\mathbf{E}^{-1} + \mathbf{s}_i'(\kappa_{N_t}\mathbf{I}_N)^{-1}\mathbf{s}_i)^{-1}$ and $\bar{\mathbf{c}}_t = \mathbf{E}_{1i}\mathbf{s}_i'(\kappa_{N_t}\mathbf{I}_N)^{-1}(\mathbf{z}_i - \mathbf{x}_i'\beta - \mathbf{s}_t'\mathbf{b}_{N_t} - q_t)$;

3. $z_{it} \sim \mathcal{TN}_s(\mathbf{x}_{it}'\beta + \mathbf{w}_{it}'\mathbf{b}_i + \mathbf{s}_{it}'\mathbf{c}_t + q_{it}, \kappa_i^2)$, where $\mathcal{TN}_s(a, b)$ is a truncated normal distribution with mean a, variance b, and region of truncation s. The truncation region is determined by the observed outcome y_{it}: if $y_{it} = 0$, then $s = (-\infty, 0]$; if $y_{it} = 1$, then $s = (0, \infty)$.

4. $\mathbf{D}^{-1} \sim W_{K_2}(\nu_1, \mathbf{D}_1)$ and $\mathbf{E}^{-1}|\{\mathbf{c}_t\} \sim W_{K_3}(\eta_1, \mathbf{E}_1)$, where $\nu_1 = \nu_0 + N$, $\mathbf{D}_1 = (\mathbf{D}_0^{-1} + \sum_{i=1}^N \mathbf{b}_i\mathbf{b}_i')^{-1}$, $\eta_1 = \eta_0 + T$, and $\mathbf{E}_1 = (\mathbf{E}_0^{-1} + \sum_{i=1}^T \mathbf{c}_t\mathbf{c}_t')^{-1}$;

5. $\rho \sim \Psi(\rho) \times N(\hat{\rho}, \mathbf{P})$, following Albert and Chib (1993), I use a Metropolis-Hasting algorithm to update ρ by using the tailored kernel $N(\hat{\rho}, \mathbf{P})$,

where N_k denotes a k-dimensional multivariate normal distribution, W_k a k-dimensional Wishart distribution, and $\mathbf{\Psi}$ a non-standard distribution.

2.3.3 Bayesian Model Comparison

For model comparison and lag-order selection, Pang (2010) adopts the marginal likelihood approach of Chib (1995) and Chib and Jeliazkov (2001) to compute the Bayes factor. The marginal likelihood is the normalizing constant in Bayes' formula, and can be expressed as follows:

$$m(y) = \frac{f(y|\theta)\pi(\theta)}{\pi(\theta|y)}. \tag{2.9}$$

Fixing values of θ at θ^*, the marginal likelihood on logarithm scale can be computed in the following way:

$$\ln \hat{m}(y) = \ln \hat{f}(y|\theta^*) + \ln \hat{\pi}(\theta^*) - \ln \hat{\pi}(\theta^*|y). \tag{2.10}$$

The symbol a^* means an arbitrarily chosen fixed value of the posterior distribution of a random variable a. In practice, this fixed valued is often the mean or the median of the posterior. The prior ordinates are easy to compute. Here I only present the algorithm for the likelihood ordinate and the posterior ordinate.

For the likelihood ordinate, denote by θ all the parameters, except the auxiliary parameter $\mathbf{u}$, and apply the following formula:

$$\begin{aligned}\hat{f}(\mathbf{y}|\theta^*) = & \frac{1}{M}\sum_{m=1}^{M}\prod_{i=1}^{N}\prod_{t=1}^{T}\Phi\left(\frac{\mathbf{x}'_{it}\beta^* + \mathbf{w}'_{it}\mathbf{b}^*_i + \mathbf{s}'_i\mathbf{c}^*_t + q_{it}^{(m)}}{\sqrt{\kappa_i^{(m)}}}\right)^{y_{it}} \\ & \times\left(1 - \Phi\left(\frac{\mathbf{x}'_{it}\beta^* + \mathbf{w}'_{it}\mathbf{b}^*_i + \mathbf{s}'_i\mathbf{c}^*_t + q_{it}^{(m)}}{\sqrt{\kappa_i^{(m)}}}\right)\right)^{1-y_{it}}.\end{aligned} \tag{2.11}$$

For the posterior ordinate, the approach to approximate it is as follows:

$$\begin{aligned}\hat{\pi}(\beta^*, \mathbf{b}^*, \mathbf{D}^*, \rho^*, \mathbf{E}^*|\mathbf{y}) = & \hat{\pi}(\rho^*|\mathbf{y})\hat{\pi}(\mathbf{c}^*|\rho^*, \mathbf{y})\hat{\pi}(\mathbf{E}^*|\mathbf{c}^*, \rho^*, \mathbf{y})\hat{\pi}(\mathbf{b}^*|\mathbf{E}^*, \mathbf{c}^*, \rho^*, \mathbf{y}) \\ & \times \hat{\pi}(\mathbf{D}^*|\mathbf{b}^*, \mathbf{E}^*, \mathbf{c}^*, \rho^*, \mathbf{y})\hat{\pi}(\beta^*|\mathbf{D}^*, \mathbf{b}^*, \mathbf{E}^*, \mathbf{c}^*, \rho^*, \mathbf{y})\end{aligned}$$

1. For the ordinate $\hat{\pi}(\rho^*|\mathbf{y})$, denote all parameters except ρ and $\mathbf{u}$ as ψ:

$$\hat{\pi}(\rho^*|\mathbf{y}) = \frac{K\sum_{i=1}^{N}\left(\alpha(\rho^{(j)}, \rho^*|\mathbf{y}, \psi^{(j)}, \mathbf{u}^{(j)})q(\rho^{(j)}, \rho^*|\mathbf{y}, \psi^{(j)}, \mathbf{u}^{(j)})\right)}{J\sum_{k=1}^{K}\left(\alpha(\rho^*, \rho^{(k)}|\mathbf{y}, \psi^{(k)}, \mathbf{u}^{(k)})\right)}. \tag{2.12}$$

The numerator is the sample expectation with respect to $\pi(\psi, \mathbf{u}|\mathbf{y})$, and the MCMC output can be directly used to integrate those parameters in the conditional part. The denominator is the sample expectation with respect to the conditional product measure $\pi(\psi, \mathbf{u}|\mathbf{y})q(\rho^*, \rho|\mathbf{y}, \psi, \mathbf{u})$. Here, one reduced run is needed: fix ρ at ρ^*, do a reduced run to get ψ and $\rho^{(k)}$ in each iteration by using $\psi^{(k)}$, and then plug all the draws of the parameters and augmented data into the denominator, and compute the quantity.

2. $\hat{\pi}(\mathbf{c}^*|\rho^*, \mathbf{y})$: directly use the reduced run conducted above;

3. $\hat{\pi}(\mathbf{E}^*|\mathbf{c}^*, \rho^*, \mathbf{y}) = \hat{\pi}(\mathbf{E}^*|\mathbf{c}^*, \mathbf{y})$: no reduced run is required;

4. $\hat{\pi}(\mathbf{b}^*|\mathbf{E}^*, \mathbf{c}^*, \rho^*, \mathbf{y}) = \prod_{i=1}^{N} \hat{\pi}(\mathbf{b}_i^*|\mathbf{c}^*, \rho^*, \mathbf{y}_i)$: Conduct a reduced run by fixing $\mathbf{E}, \mathbf{c}, \rho$, and keep the output of $\beta, \mathbf{D}, \mathbf{z}$ together with the fixed values to compute this quantity;

5. $\hat{\pi}(\mathbf{D}^*|\mathbf{b}^*, \mathbf{E}^*, \mathbf{c}^*, \rho^*, \mathbf{y}) = \hat{\pi}(\mathbf{D}^*|\mathbf{b}^*, \mathbf{y})$: no reduced run is required here;

6. $\hat{\pi}(\beta^*|\mathbf{D}^*, \mathbf{b}^*, \mathbf{E}^*, \mathbf{c}^*, \rho^*, \mathbf{y}) = \hat{\pi}(\beta^*|\mathbf{b}^*, \mathbf{c}^*, \rho^*, \mathbf{y})$: conduct a reduced run by fixing $\mathbf{D}, \mathbf{b}, \mathbf{E}, \mathbf{c}, \rho$ and keep the output of $\mathbf{z}$ together with the fixed values of $\mathbf{b}^*, \mathbf{c}^*, \rho^*$ to compute this quantity.

2.4 Variables, Model, and Data

This section applies the GLMM-AR(p) model to analyze the civil war data. The following is the core civil war model specification:

$$
\begin{aligned}
\texttt{Onset}_{it} &= \mathbf{1}(\texttt{Latent}_{it} > 0), && (2.13)\\
\texttt{Latent}_{it} &= \beta_t + b_{1i} + \beta_{2,i} * \texttt{EMR} + \beta_{11}\texttt{ongwar} + \beta_{12} * \texttt{gdpenl} \\
&\quad + \beta_{13} * \texttt{lpopl} + \beta_{14} * \texttt{lmtnest} + \beta_{15} * \texttt{ncotig} \\
&\quad + \beta_{16} * \texttt{oil} + \beta_{17} * \texttt{newstate} + \beta_{18} * \texttt{instab} \\
&\quad + \beta_{18} * \texttt{polity2} + \beta_{19} * \texttt{ancol} + \beta_{210} * \texttt{ethfrac} + \xi_{it}, && (2.14)\\
\beta_{2,i} &= \beta_{21} + \beta_{22} * \texttt{minstab} + \beta_{23} * \texttt{methfrac} \\
&\quad + \beta_{24} * \texttt{second} + b_{2i}, && (2.15)\\
\beta_t &= \beta_0 + c_t, && (2.16)\\
\xi_{it} &= \rho_1 \xi_{i,t-1} + ... + \rho_p \xi_{i,t-p} + u_{it}, && (2.17)
\end{aligned}
$$

The meanings of the covariate symbols and the within-country and within-year variations of the included variables are summarized in Table 2.1, and the coding criteria of the variables are reported in Fearon and Laitin (2003a,b) and FKL. The individual-level regression, equation (2.14), follows the core specification in Fearon and Laitin (2003b), CG, and FKL. Unlike FKL, I include *Democracy* (`polity2`) in the model because CG found that democracy merited inclusion, while FKL did not test it. I use country-specific and year-specific intercepts to control for unobserved heterogeneity in the time and spatial dimensions. Note that this specification cannot be allowed in the fixed-effect panel model because of the slow-moving and time-invariant variables, such as *Mountainous* (`lmtnest`) and *Ethno-linguistic fractionalization* (`ethfrac`). The variable, *Secondary school enrollment*, has no variation across time for almost all countries, and its variation among countries does not change over time either. This suggests that using its country averages in the group-level regression does not discard much information.

In the country-level regression, equation (2.15), the variation of EMR's effect, $\beta_{2,i}$, is explained by both observed (the country-level regressors) and unobserved background factors (the error term, b_{2i}). *Democracy* is not included as a country-level covariate for two reasons: first, it is a time-varying variable and the average of this variable poorly describes the background; second, preliminary model specifications with democracy averaged within countries have worse model quality than the present one (according to the log-marginal likelihoods), and the posterior distribution of its effect is estimated to be centered at zero and with a small error band. This does not necessarily mean that democracy is not an important intervening factor. The null effect could be attributed to the fact that the average does not capture the background. Therefore, the general regime or institutional factors are left as unobserved in the error term. *Political instability* is coded by Fearon and Laitin as "a dummy variable indicating whether the country had a three-or-greater change on the Polity IV regime index in any of the three years prior to the country-year in question" Fearon and Laitin (2003b), which is labeled by them as "political instability." Since whether a regime is stable or not should not be judged by the presence or absence of transformation. Instead, the frequency of regime change during a certain period of time reflects political or regime instability. Therefore, the average is a good measure of instability of a country during the sample years.

Male secondary school enrollment is included in the model to proxy for the background factor of government strength and governance quality. *Per capita income* and *male secondary school enrollment* are highly correlated ($\rho \approx 0.8$). Since the latter is almost time-invariant and the former is a time-varying variable, it is better to use *secondary school enrollment* in the group-level regression as a proxy for economic development. Furthermore, secondary school enrollment directly reflects the capability and willingness of the government to provide public goods to society and measures, more indirectly, governance quality. The variable of *Ethno-linguistic fractionalization* is a direct measure of the ethnic background of a country, and it is almost time-invariant; the country average of this variable is ready to be used in the group-level regression. Unobserved background factors are included in the group-level error term.

Table 2.1: Within-group variation of variables.

Variable	Symbol	Within-country variation			Within-year variation		
		Min.	Mean	Max.	Min.	Mean	Max.
War outbreak	`onset`	0.00	0.07	0.35	0.00	0.11	0.24
Ongoing war$_{t-1}$	`ongwar`	0.00	0.15	0.51	0.00	0.32	0.42
Income per capita$_{t-1}$	`gdpenl`	0.02	1.26	17.50	1.81	4.03	6.55
Log(population)$_{t-1}$	`lpopl`	0.00	0.25	0.71	1.36	1.45	1.53
Log(% Mountainous)	`lmtnest`	0.00	0.00	0.00	1.27	1.39	1.45
Noncontiguous state	`ncontig`	0.00	0.00	0.50	0.36	0.38	0.43
Oil producer	`oil`	0.00	0.06	0.50	0.26	0.33	0.39
New state	`nwstate`	0.00	0.17	0.49	0.00	0.13	0.40
Political instability	`instab`	0.00	0.26	0.53	0.00	0.35	0.45
Anocracy$_{t-1}$[a]	`anocl`	0.00	0.26	0.53	0.31	0.42	0.50
Ethno-linguistic fractionalization	`ethfrac`	0.00	0.00	0.13	0.26	0.28	0.30
Ethnicity minority leader	`minldr1`	0.00	0.11	0.51	0.44	0.47	0.49
Democracy$_{t-1}$	`polity2l`	0.00	2.88	9.23	6.73	7.39	7.74
Secondary school enrollment	`second`	0.00	0.00	0.04	0.11	0.11	0.11

[a]Anocracy is a mixture regime of a democracy with autocratic features.

2.5 Empirical Results and Interpretation

The GLMM-AR(p) model is estimated with increasing lag orders starting with $p = 0$. The log-marginal likelihood of each model is computed and compared. Model searching stops when increasing the lag order decreases the log-marginal likelihood. The more sophisticated GLMM-AR(p) model is also compared with two commonly-used alternative specifications, i.e., the completely pooled probit model (denoted as "PROBIT" in Figure 2.1) and the probit multi-level model without year-specific random effects or serial correlation (denoted as "GLMM-CL1" in Figure 2.1[2]).

The simple probit model does not consider the variation of EMR's effect and serves as a replication of FKL's model. The GLMM-CL1 model can be regarded as a special case of the GLMM-AR(p) model with two additional assumptions: the time-specific random effects have no variation, and the errors are uncorrelated. The GLMM-CL1 model with serially correlated errors was estimated, but the error process was likely to be nonstationary (with an AR(1) and AR(2) process, the first ten thousand draws of the autoregressive coefficients were very close to the unit circle and the MCMC simulation halted because no legitimate proposals could be drawn within the stationary space. Therefore, only the AR(0) GLMM-CL1 model could be successfully estimated by ignoring the dynamics). The suspected nonstationary error process makes inferences based on the GLMM-CL1 model unreliable, but for model comparison purposes, I still summarize the MCMC output of the static GLMM-CL1 model in Figure 2.1. However, adding the year-specific effects makes the same specification stationary. This is not surprising, because if time-specific effects are correlated with each other, putting them into the error term will exaggerate serial correlation and create spurious dynamics (Heckman, 1981). Alternatively, if time-specific effects are correlated with other regressors, omitted time-dimension heterogeneity changes the relationship between the dynamic process of civil war and the dynamic processes the explanatory variables represent, which affects cointegration.

Bayes factors are computed for all those models, and they indicate that the model with a second order error autoregressive process is the best one (see Figure 2.1 for the marginal likelihood of each model). An insurance run of the GLMM-AR(4) is conducted, which finds that its quality decreases compared to the AR(2) and AR(3) model. The number of iterations for each model is 500,000 with a burn-in stage of 50,000 iterations. The chains of random-effect and low-level parameters mix much more slowly, which slow down the whole process of convergence, since partial convergence should not be regarded as convergence (Gill, 2008). Because the total number of quantities to be updated is more than 12,800 (including the latent responses), it is unrealistic to

[2]This model is equivalent to a fixed effect panel model regarding the country-level regression as interaction terms and merging the group-level error term $b_{2i}$with the random intercept (b_{1i}).

conduct formal diagnostics to check convergence for all of them. Also, the latent responses and the auxiliary parameters are not substantively interesting. Therefore, I did multiple convergence diagnostics by using the `coda` package in `R` on all β, some of $\mathbf{b}_i$ and $\mathbf{c}_t$s. For the augmented data and the auxiliary parameter $\mathbf{u}$, I randomly chose 100 and conducted convergence diagnostics. No evidence of non-convergence is found in those diagnostics. Figure 2.1 presents all six models with 95% credible intervals and the posteriors of the fixed parameters (at all levels). The random effects (the random intercepts of b_{1i} and β_t, and the random effects of the ethnic minority leadership $\beta_{2,i}$) are plotted in Figure 2.2 based on the GLMM-AR(2) model which has the best goodness-of-fit.

The posteriors based on the probit model are very similar to the results in FKL, though the scales are different because of the different link functions (FKL used the logistic link). With a big sample size, the point estimates of the Bayesian model (posterior mean) are very similar to the maximum likelihood estimates if the same link function were used. Unlike in FKL, I include two additional variables (*democracy* and *ethno-linguistic fractionalization*) in the model. Without modeling serial correlation and heterogeneity among countries, the estimator tends to be biased toward 0 and has much smaller error bands compared to posteriors based on the other five models. For some of the variables, it yields very different substantive results. Larger populations increase civil war onset propensity with high certainty, but after controlling for heterogeneity, the direction of its effect reverses; modeling serial correlation further, its effect becomes unclear. Similar changes have been observed in the existing studies: the effect of population in the fixed-effect panel model is often different from that in completely pooled models (Fearon and Laitin, 2003b; Collier et al., 2004; Fearon et al., 2007). It suggests that population could be correlated with some omitted country-specific characteristics that also affect civil war. Those characteristics might include the territory, economic structure, climate or other geographical features, and so on. It is often argued that a more populous country incurs higher civil war risk because large population size enlarges the pool of potential recruits (Fearon and Laitin, 2003b), but once confounding factors are controlled for, population does not have a certain effect. The probit model also finds that non-contiguous countries are likely to have higher risk of civil war but the effect is uncertain. Models controlling for heterogeneity in either time or spatial dimension suggest that it has a positive effect with a high level of certainty. The completely pooled model also exaggerates the effect of *new state status*. For the primary explanatory variable, `EMR`, the results confirm FKL and suggest that EMR may increase the likelihood of civil war onset, but this effect only has a low level of certainty.

After controlling for heterogeneity, the other five models all find that EMR, on average, is more likely to decrease civil war risk, but this statement can only be made with a low level of credibility (a mere 60%). Among the five multilevel models, the only difference between GLMM-CL1 and GLMM-AR(0) is that the latter considers the heterogeneity across years. The estimated time-dimension hetergeneity is mild. The lower-right graph in Figure 2.2 shows the year-level variation (the year-specific intercept) based on the GLMM-AR(2) model. We can see that there is year-to-year difference but the variation is not large. The Bayes factor ($\log_{10}$) of the GLMM-AR(0) versus GLMM-CL1 is 2.66, and decisively supports adding the year-level regression (Greenberg, 2007). The improvement on model fit by modeling the year-level heterogeneity is also reflected by the forecasting performance, which is discussed in detail in the next section. Between the two models, the posteriors of the fixed effect parameters are very similar. Nonetheless, for several variables, the GLMM-CL1 model generates posteriors with the tendency of being biased toward 0. The GLMM-AR(p) models with $p > 0$ find positive serial correlation which is not weak (the posterior mean of ρ_1 is 0.5 based on the AR(1) model) and probably lasts for more than one time period (posterior means of ρ_1 and ρ_2 are both about 0.3 based on the AR(2) model). Serial correlation correction and dynamic modeling greatly improve model quality, and the log-marginal likelihood increases by more than 20 (see Figure 2.2). Compared to the probit and GLMM-CL1 models, modeling serial correlation yields three key differences: first, the error bounds are systematically bigger than those of the GLMM model without modeling serial correlation, confirming the statistical theory that ignoring positive serial correlation results in narrower error bands (Gourieroux et al., 1984; Poirier and Ruud, 1988). Second, the effect of ongoing war in the previous time period is almost twice as big as in the models without considering serial correlation. Apparently, serial correlation causes biased estimation of this variable's effect: ongoing war at time $t-1$ is very likely to be correlated with ϵ_{t-1} because the omitted factors which affect civil war onset are also likely to be correlated with ongoing civil war. If ϵ_{t-1} is correlated with ϵ_t, there is endogeneity arising from the correlation between ongoing civil war and the error term. Without modeling serial correlation, the dynamics captured by the variable of ongoing civil war are misleading. Further evidence is that, based on the estimated autoregressive coefficients, the propensity for civil war is positively correlated. This is a different mechanism than that suggested by using the variable of previous ongoing war since their signs are opposite. Third, within-sample prediction is improved dramatically by making good use of the information contained in the dynamic error process, which is analyzed in more detail in the next section.

All of the multilevel models detect salient heterogeneity of EMR's effect on civil war onset across countries. Figure 2.2 shows the heterogeneity estimated by the GLMM-AR(2),[3] which indicates considerable heterogeneity across countries. The effect of EMR is more likely to be negative for about two thirds of the sample countries. The GLMM-AR(2) model finds that for

[3]The cross-country heterogeneities estimated by all the multilevel models are very similar to each other, and therefore only the random-effect posteriors based on the GLMM-AR(2) model are presented in the chapter.

Figure 2.1: Posterior summaries with 95% credibility intervals (six models).

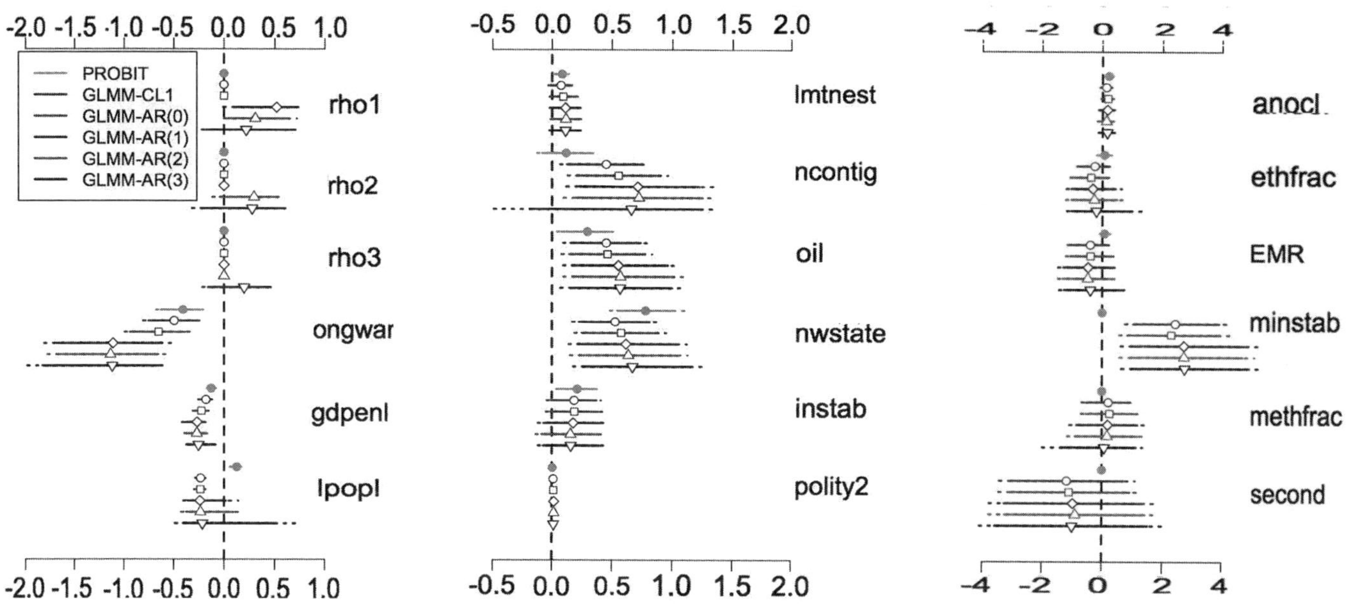

The log-marginal likelihoods of the six models — PROBIT: 531.783; GLMM-CL1: −524.291; GLMM-AR(0): −518.174; GLMM-AR(1): −476.712; GLMM-AR(2): −470.546; and GLMM-AR(3): −494.009

Figure 2.2: Random intercepts and random effects.

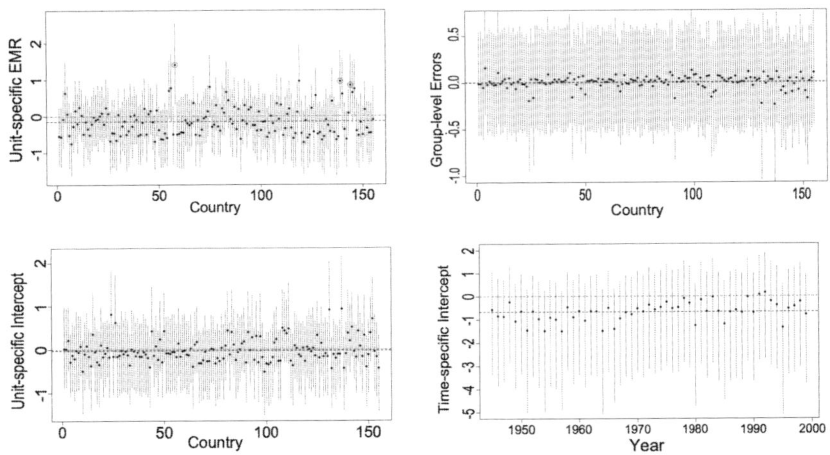

three countries, i.e., Azerbaijan, Pakistan, and Thailand, ethnic minority rule increases the likelihood of civil war onset with a 95% credible level. However, the effect of EMR is at a low level of credibility for most sample countries, and its directions vary widely across countries. In Figure 2.1, the three posteriors of the group-level regressors partially explain this variation. The effect of political instability is highly robust to various model specifications, and suggests that in a country which experiences frequent regime change, ethnic minority rule is more vulnerable and creates higher risk of civil war than in politically stable countries. Compared with the posteriors of the fixed-effect parameters, the error band of political instability is much larger, which can be explained by the much smaller sample size than at the individual level. The variable of *male secondary school enrollment* is likely to decrease the danger of civil war increased by EMR, but it is not very certain (at a 70% credibility level). *Ethno-linguistic fractionalization* seems not to have an effect on this relationship. The heterogeneity of EMR's effect across countries can be compared with the relatively homogeneous country-level errors presented in the upper-right graph in Figure 2.1. This comparison suggests that the group-level regressors explain a large proportion of the variation of EMR's effect on civil war onset. The lower-left graph displays the country-specific intercept, which shows that, aside from the hetergoneous effect of EMR on civil war, countries still demonstrate differences in other respects and these differences are relevant to civil war onset.

2.6 Civil War: Prediction

Assessing the goodness-of-fit of a statistical model is necessary to confidently draw causal inferences based on the estimates (Hoeting et al., 1999). The Bayes factor is a comprehensive criterion for this purpose, but within-sample and out-of-sample predictions are not only important alternative ways to assess model quality but also valuable in themselves. Especially for civil war studies, which have important policy implications, better predictions based on statistical models can serve as warning systems and provide valuable information to policy makers. The GLMM-AR(p) model, which better identifies the causal relationships and makes better use of the information in the data (the dynamic process of the errors), performs better in forecasting.[4] This section compares the forecast performance of the following four models: the probit model, the GLMM-CL1, the GLMM-AR(p) model without serial correlation consideration, and the GLMM-AR(p) with an appropriate lag order (the second order) of the autoregressive errors.

2.6.1 Predictive Probabilities of Civil War

The general within-sample predicting performance of the four models is presented in Figure 2.3. To avoid setting arbitrary or *post hoc* thresholds for classifications of civil war or non-civil war, the numeric predictive probabilities of civil war are directly investigated in such a way that the distributions of predictive probabilities of the civil war group and the non-civil war group are presented side by side. If the densities of the two groups are well separated (the density of the civil war group is plotted on the right side and that of the non-civil war group on the left side in each figure), forecasts based on those predictive probabilities will make few mistakes in both classifications ($y = 0$ and $y = 1$), which suggests that the model performs well in forecasting. As shown in Figure 2.3, the pooled probit model does a bad job in differentiating the civil war cases from the non-civil war ones, and the density kernels of the two groups overlap most of the time. The GLMM-CL1 model, by modeling the variation of EMR's effect and other country-level heterogeneity, separates the two groups better than the pooled model, but there is still a large overlapping area. The GLMM-AR(0) model further considers heterogeneity in the time periods and reduces the overlap compared to the GLMM-CL1 model. It is not a good model either, though it is better than the GLMM-CL1 model. The GLMM-AR(2) model, which models the dynamics and makes good use of the information contained in the error term, classifies the two groups well: the two density kernels are almost completely separated from each other and only a very small part at each tail is connected with the other.

[4]The relationship between better study of causal relationships and better and more stable forecasting is articulated in King and Zeng (2001).

Table 2.2: Within-sample predictions of civil war onset.

Best Predictions					*Worst Predictions*									
Probs. (Civil War Observed)					Probs. (Civil War Observed)					Probs. (Civil War NOT Observed)				
Country	Year	AR2	AR0	CL1	Country	Year	AR2	AR0	CL1	Country	Year	AR2	AR0	CL1
Angola	1975	0.686	0.394	0.338	The U.K.	1969	0.326	0.023	0.028	Indonesia	1952	0.299	0.391	0.338
Indonesia	1950	0.674	0.475	0.431	Sri Lanka	1987	0.346	0.051	0.051	Indonesia	1949	0.274	0.408	0.431
Burma	1948	0.627	0.248	0.199	Argentina	1973	0.365	0.047	0.041	Azerbaijan	1991	0.271	0.343	0.247
Phlippines	1946	0.624	0.303	0.297	Iran	1978	0.366	0.021	0.021	Taijikistan	1991	0.241	0.297	0.209
Congo(DR)	1960	0.623	0.310	0.301	Cyprus	1974	0.367	0.061	0.071	Russia	1998	0.236	0.176	0.127
Congo(DR)	1996	0.623	0.219	0.190	China	1950	0.375	0.067	0.073	Burundi	1991	0.226	0.218	0.132
Azerbaijan	1992	0.612	0.359	0.246	India	1982	0.376	0.067	0.072	Burundi	1992	0.223	0.219	0.129
Rwanda	1962	0.611	0.265	0.257	Colombia	1963	0.377	0.053	0.061	Indonesia	1955	0.214	0.266	0.269
Indonesia	1958	0.601	0.329	0.265	Argentina	1955	0.381	0.075	0.077	Pakistan	1970	0.210	0.230	0.213
Nicaragua	1981	0.600	0.200	0.135	Jordan	1970	0.395	0.053	0.059	Turkey	1983	0.206	0.154	0.078
Pakistan	1971	0.591	0.202	0.199	Bangladesh	1976	0.397	0.083	0.072	Georgia	1991	0.204	0.233	0.149
Tajikistan	1992	0.582	0.283	0.197	Paraguay	1947	0.401	0.058	0.081	Pakistan	1948	0.198	0.390	0.321
Indonesia	1965	0.580	0.297	0.241	S. Africa	1983	0.409	0.049	0.026	Indonesia	1963	0.197	0.251	0.243
Haiti	1991	0.577	0.259	0.182	Chad	1994	0.412	0.103	0.109	Rwanda	1989	0.191	0.172	0.099
Colombia	1948	0.570	0.238	0.157	Rwanda	1990	0.414	0.080	0.100	Indonesia	1962	0.189	0.312	0.296

Both "best" and "worst" predictions are sorted based on the GLMM-AR(2) model. Also, because the best 15 predictions of the observation of non-civil war from all three models are $\Pr(y = 1|\mathbf{x}, \theta) = 0.000$, these probabilities are not listed in the table.

Figure 2.3: Comparison of within-sample predicted probabilities.

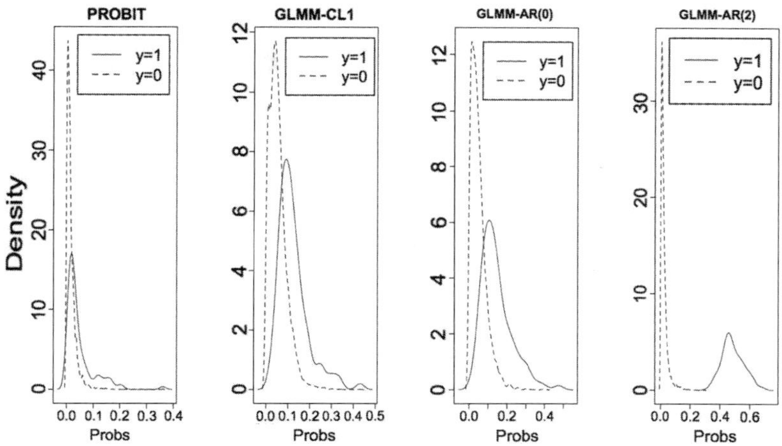

Predictive Prob. of 1s and 0s

Table 2.2 shows the 15 best and worst predictions based on the GLMM-AR(2) model and compares them with those based on the GLMM-AR(0) and GLMM-CL1 models. Because the probit model is the worst one, it is omitted here due to space limitation. For the war cases, even the worst predictions of the GLMM-AR(2) are much better than the other two models, and most of the worst predictions of non-civil war cases are also better. Figure 2.4 presents predictions of eight selected countries in the sample years. The selected countries experience most to fewest civil war onset events (from 6 episodes to 0). All those figures and tables demonstrate that the GLMM-AR(2) performs much better in statistical forecasting because it takes into account two-dimensional heterogeneity and the dynamic process of the error term. For the three models with a static error process, the predicted probabilities demonstrate an AR(2) or AR(1) process, which suggests that their poor forecast performance should be largely attributed to ignoring the dynamics.

2.6.2 Receiver-Operating Characteristic Curve

The second way of evaluating and comparing the predicting performance of the competing models is to use the *Receiver-Operating Characteristic (ROC) curve*, as in King and Zeng (2001). The idea of using the ROC curve to evaluate predicting performance of models is simple: given a level of correct classification of one group (say, $y = 1$), the model performs better if the rate of correct classification of the other group (say, $y = 0$) is higher based on this model. Graphically, the curve dominating other curves represents the best model among the competing ones. This approach has the advantage

Figure 2.4: Random intercepts and random effects.

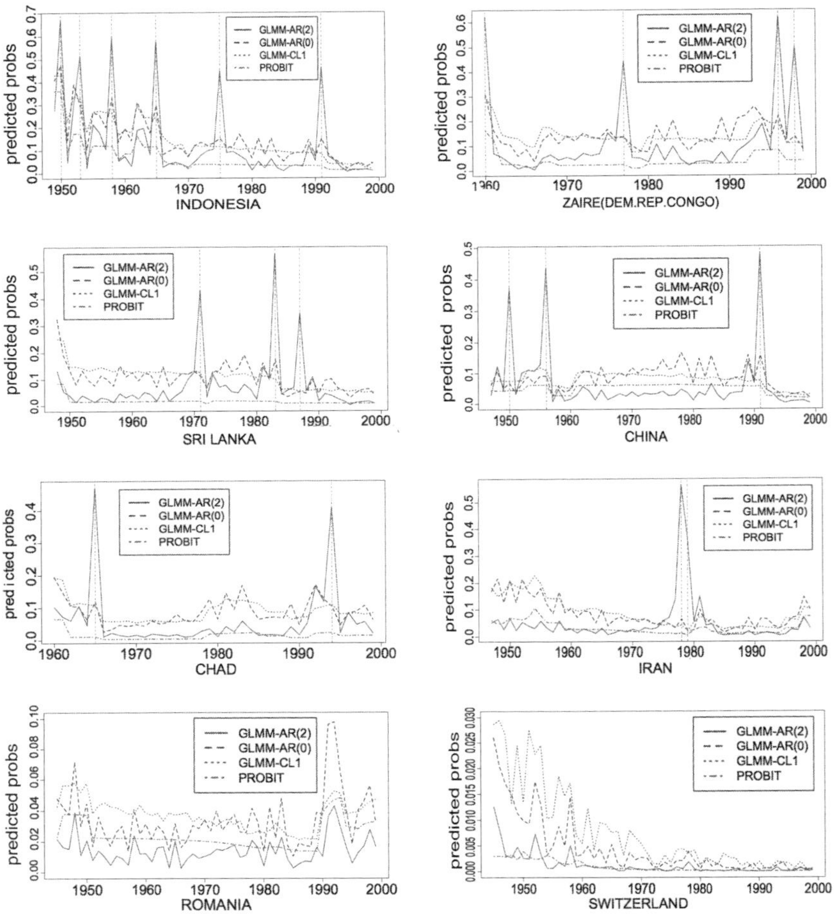

Gray vertical lines in the graphs indicate episodes of civil war onset.

Figure 2.5: Receiver-operating characteristic (ROC) curves.

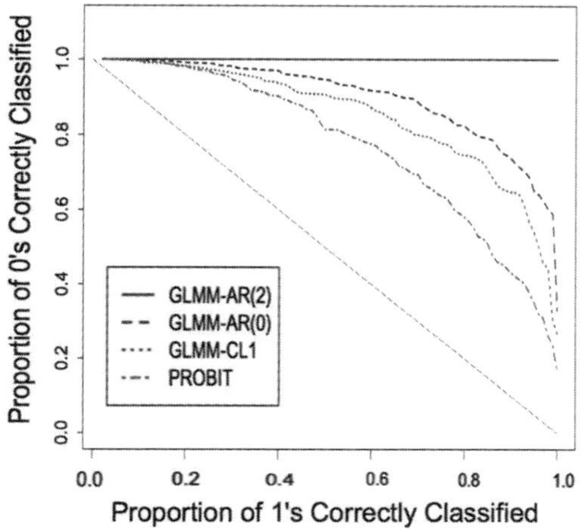

of avoiding assignment of any fixed threshold for classification and assessing model performance based on any specific cutoff value. In Figure 2.5, the diagonal line is the reference line, indicating the extreme situation that the predictive probabilities of the two groups are completely mixed. The ROC curve based on the completely pooled probit model is the lowest one. The GLMM-CL1 model improves forecasting by modeling heterogeneity among countries and its curve globally dominates the one of the pooled probit model. The GLMM-AR(0) model has an ROC curve globally above the one of the GLMM-CL1, but is dominated everywhere by the curve of the GLMM-AR(2) model. The ROC curve of the GLMM-AR(2) model is almost a horizontal line, indicating that, since the densities of the predictive probabilities of the 1 and 0 groups are well separated, there is barely any trade-off between the two types of classification.[5]

By taking multivariate heterogeneity and higher order dynamics seriously, the empirical findings of parameter estimation and within-sample prediction both suggest that simpler models are likely to be misspecified in fundamental

[5]Note that the close-to-horizontal ROC curve does not suggest that the GLMM-AR(2) model will correctly classify all the country-years. This is because the threshold of classification should be set *ex ante* based on the risk assessment and pre-specified loss function, rather than being chosen after the pattern of predictive probabilities are observed. For instance, if setting a *post hoc* threshold of 0.32, then the GLMM-AR(2) model makes no misclassification of either civil war or non-civil war; but we may set an *ex ante* threshold of 0.60 if we think misclassification of civil war is much less expensive than mistakenly identifying non-civil war as civil war, then most of the observed civil wars will be misclassified as non-civil wars even based on the GLMM-AR(2) model.

ways in longitudinal analysis, as was similarly found in Chib and Jeliazkov (2006).

2.7 Robustness Checking: Alternative Measures of EMR

To check the robustness of the empirical findings, different model specifications were used: this included some other variables, such as *Democracy*, in the country-group regression and the *Cold war* dummy in the year-level regression, excluding *ethno-linguistic fractionalization* from the individual-level regression, assigning different priors, and so on (the details are not reported here). None of them cause important changes, and the cold war dummy does not have explanatory power. A more important way of checking the robustness of the empirical findings is to use alternative measures of EMR, as FKL did. This section re-estimates the GLMM-AR(2) model with alternative measures of EMR in order to check the robustness of the empirical findings discussed above.

FKL presented two other alternative measures of EMR: for the first alternative, the previous measure of EMR is modified by examining the data case by case and recoding those country-years as plurality group dominance when leader's ethnicity does not mean the dominance of their minority ethnic groups; the second alternative measure is to further recode white in the same ethnic group as mestizo in Latin America based on the first alternative measure. This changes the proportion of EMR in the sample substantially: the percentage of minority leader country-years decreases from 32.20% to 25.12%. I estimate the GLMM-AR(2) model with the two alternative measures and compare them with the results based on the previous measure. The summary of posterior means and 95% credible intervals is presented in Table 2.3. The posteriors are symmetric with their means very close to medians; therefore, I do not report the medians in the table. The posteriors of the parameters in the individual-level regression are very similar among the three models, and the dynamics of the error process is almost the same in the three models. The average effect of EMR demonstrates differences with different measures, but overall, its effect is uncertain in all models. Even though the average effect of EMR is different, the degree that political instability alters EMR's effect on civil war onset is robust and has similar posteriors across models. For the other two explanatory variables, *average ethno-linguistic fractionalization* and *secondary school enrollment*, they seem to be more likely to decrease EMR's effect on civil war onset based on the two alternative measures than the measure adopted in the previous section. Figure 2.6 also shows that the country-specific effect of EMR on civil war onset, as well as its variation across countries, are similar to those in Figure 2.2, only with slightly bigger variations. The log-marginal likelihoods of the three models are reported in Table 2.3; they support the original measure over the two alternatives.

Table 2.3: Robustness check: Alternative measures of EMR.

	Measure I		Measure II		Measure III	
Covariate	Mean	Credible interval	Mean	Credible interval	Mean	Credible interval
ρ_1	0.32	(0.00, 0.72)	0.34	(0.06, 0.65)	0.32	(0.01, 0.65)
ρ_2	0.28	(−0.11, 0.57)	0.32	(−0.07, 0.62)	0.31	(−0.01, 0.57)
`ongwar`	−1.15	(−1.78, −0.58)	−1.25	(−1.97, −0.66)	−1.23	(−2.26, −0.60)
`gdpenl`	−0.27	(−0.39, −0.15)	−0.27	(−0.42, −0.15)	−0.28	(−0.44, −0.14)
`lpopl`	−0.20	(−0.43, 0.19)	−0.22	(−0.64, 0.32)	−0.21	(−0.70, 0.47)
`lmtnest`	0.12	(−0.01, 0.27)	0.12	(−0.02, 0.28)	0.11	(−0.04, 0.28)
`ncontig`	0.72	(0.10, 1.37)	0.78	(0.02, 1.60)	0.76	(−0.12, 1.67)
`oil`	0.58	(0.09, 1.11)	0.58	(0.08, 1.14)	0.59	(0.09, 1.18)
`nwstate`	0.64	(0.15, 1.13)	0.65	(0.15, 1.17)	0.65	(0.15, 1.21)
`instab`	0.15	(−0.14, 0.43)	0.15	(−0.14, 0.45)	0.15	(−0.13, 0.43)
`polity2`	0.02	(−0.01, 0.04)	0.02	(−0.01, 0.04)	0.02	(−0.01, 0.04)
`anocl`	0.13	(−0.17, 0.42)	0.12	(−0.18, 0.42)	0.13	(−0.20, 0.43)
`ethrac`	−0.27	(−1.21, 0.68)	−0.22	(−1.28, 0.71)	−0.28	(−1.40, 0.72)
`EMR`	−0.49	(−1.47, 0.48)	−0.14	(−1.41, 1.08)	0.09	(−1.25, 1.47)
`minstab`	2.76	(0.58, 5.09)	2.92	(0.49, 5.72)	2.47	(0.01, 5.32)
`methfrac`	0.17	(−1.15, 1.43)	−0.08	(−1.47, 1.35)	−0.17	(−1.75, 1.27)
`second`	−0.90	(−3.78, 1.86)	−1.65	(−5.11, 1.71)	−1.74	(−5.21, 1.70)
Marg. Lik.	−470.546		−476.891		−481.754	

Figure 2.6: Random intercepts and random effects (alternative measures).

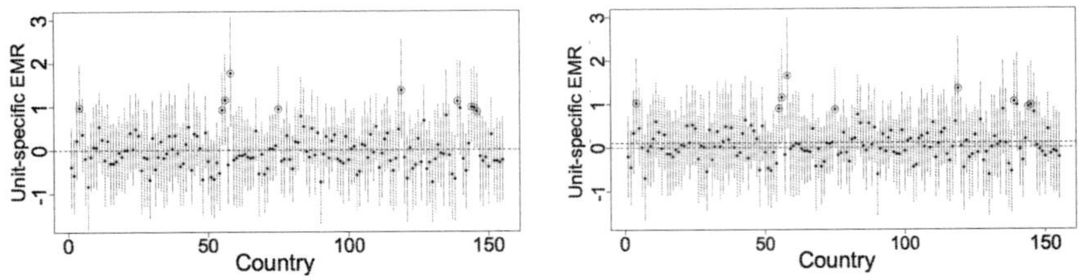

Measure II of EMR

Measure III of EMR

2.8 Conclusion

This chapter revisits the relationship between minority ethnic dominance on civil war onset raised by recent studies in the civil war literature. With the complicated causal chain between the two variables, this relationship is altered by multiple observed and unobserved background factors and varies from country to country. This chapter applies a Bayesian model for empirical analyses, which not only directly model the varying degree of EMR's effect on civil war onset, but also carefully controls for multiple sources of confounding caused by the longitudinal data structure of the civil war data set. The major findings include the heterogeneity of EMR's effect on civil war onset across countries, and countries with an unstable political regime are more prone to civil war when they are under ethnic minority rule than are politically stable countries. This finding is robust across different model specifications and alternative measures of EMR. Finally, modeling the dynamic process of the errors and differentiating dynamics from heterogeneity in both year and country dimensions dramatically improve statistical forecasts.

Acknowledgements: This research project is supported by the Humanities and Social Science Research Fund, China's Ministry of Education (Grant # 13YJCGJW009).

Bibliography

Acemoglu, D. and J. A. Robinson (2003). *Political Origins of Dictatorship and Democracy.* Massachusetts Institute of Technology.

Anderson, B. (1983). *Imagined Communities.* London: Verso.

Cederman, L. E. and L. Girardin (2007). Beyond fractionalization: Mapping ethnicity onto nationalist insurgencies. *American Political Science Review 101*(1), 173–185.

Cederman, L. E., K. S. Gleditsch, I. Salehyana, and J. Wucherpfennig (2013). Transborder ethnic kin and civil war. *International Organization 67*(2), 389–410.

Cederman, L. E., N. B. Weidmann, and K. S. Gleditsch (2011). Horizontal inequalities and ethnonationalist civil war: A global comparison. *American Political Science Review 105*(3), 478–495.

Chib, S. (1993). Bayes regression with autoregressive errors: A Gibbs sampling approach. *Journal of Econometrics 58*(3), 275–294.

Chib, S. (1995). Marginal likelihood from the Gibbs output. *Journal of the American Statistical Association 90*(432), 1313–1321.

Chib, S. and I. Jeliazkov (2001). Marginal likelihood from the Metropolis-Hastings output. *Journal of the American Statistical Association 96*(453), 270–281.

Chib, S. and I. Jeliazkov (2006). Inference in semiparametric dynamic models for binary longitudinal data. *Journal of the American Statistical Association 101*(474), 685–700.

Collier, P. and A. Hoeffler (2002). On the incidence of civil war in Africa. *Journal of Conflict Resolution 46*(1), 13–28.

Collier, P. and A. Hoeffler (2004). Greed and grievance in civil war. *Oxford Economic papers 56*(4), 563–595.

Collier, P., P. H. P., and M. Soderbom (2004). On the duration of civil war. *Journal of Peace Research 41*(3), 253–273.

Dahl, R. A. (1989). *Democracy and Its Critics.* New Haven: Yale University Press.

Deutsch, E. N. M. (1985). Income inequality, regime repressiveness, and political violence. *American Sociology Review 50*(1), 47–61.

Deutsch, K. W. (1953). *Nationalism and Social Communication.* Cambridge, MA: MIT Press.

Fearon, J. D. (2004). Why do some civil wars last so much longer than others. *Journal of Peace Research 41*(3), 275–301.

Fearon, J. D., K. Kasara, and D. D. Laitin (2007). Ethnic minority rule and civil war onset. *American Political Science Review 101*(1), 187–193.

Fearon, J. D. and D. D. Laitin (2003a). Additional tables for ethnicity, insurgency, and civil war. Supplementary Materials.

Fearon, J. D. and D. D. Laitin (2003b). Ethnicity, insurgency, and civil war. *American Political Science Review 97*(1), 75–90.

Gasiorowski, M. J. (1995). Economic crisis and political regime change: An event history analysis. *The American Political Science Review 89*(4), 882–897.

Gellner, E. (1983). *Nations and Nationalism.* Ithaca, NY: Cornell University Press.

Gelman, A. (2006). Multilevel (hierarchical) modeling: What it can and cannot do. *Technometrics 48*, 432–435.

Gelman, A. and J. Hill (2006). *Data Analysis Using Regression and Multilevel/Hierarchical Models.* New York: Cambridge University Press.

Geweke, J. (1991). Efficient simulation from the multivariate normal and student-t distributions subject to linear constraints. In E. M. Keramidas (Ed.), *Computing Science and Statistics: Proceedings of the Twenty Third Symposium on the Interface*, pp. 571–578. Fairfax, VA: Interface Foundation of North America.

Gill, J. (2007). *Bayesian Methods: A Social and Behavioral Sciences Approach* (2 ed.). Boca Raton, FL: Chapman & Hall/CRC.

Gill, J. (2008). Is partial-dimension convergence a problem for inferences from MCMC algorithms? *Political Analysis 16*(2), 153–178.

Gourieroux, C., A. Monfort, and A. Trognon (1984). Estimation and test in probit models with serial correlation. In J. P. Florens, M. Mouchart, J. P. Raoult, and L. Simar (Eds.), *Alternative Approaches to Time Series Analysis.* Brussels: Publications des Facultes Universitaires Saint-Louis.

Greenberg, E. (2007). *Introduction to Bayesian Econometrics.* Cambridge: New York: Cambridge University Press.

Grossman, H. I. (1995). Insurrections. In K. Hartley and T. Sandler (Eds.), *Handbook of Defense Economics*, pp. 191–212. Amsterdam: Elsevier.

Hagenaars, J. A. (1990). *Categorical Longitudinal Data: Log-Linear Analysis of Panel, Trend and Cohort Data.* London: Sage.

Heckman, J. (1981). Heterogeneity and state dependence. In S. Rosen (Ed.), *Labor Markets*, pp. 91–131. Chicago: University of Chicago Press.

Hegre, H., T. Ellingsen, S. Gates, and N. P. Gleditsch (2001). Toward a democratic civil peace? Democracy, political change, and civil war, 1861-1992. *American Political Science Review 95*(1), 33–48.

Hirschleifer, J. (1995). Theorizing about conflict. In K. Hartley and T. Sandler (Eds.), *Handbook of Defense Economics*, pp. 165–192. Elsevier, Amsterdam: Vol. 1.

Hoeting, J. A., D. Madigan, A. E. Raftery, and C. T. Volinsky (1999). Bayesian model averaging: A tutorial (with discussion). *Statistical Science 14*(4), 382–401.

Horowitz, D. L. (1985). *Ethinic Groups in Conflict.* Berkeley: University of California Press.

Huntington, S. P. (1968). *Social Order in Changing Societies.* New Haven CT: Yale University Press.

Huntington, S. P. (1996). *The Clash of Civilizations and the Remaking of World Order.* New York: Simon and Shuster.

King, G. and L. Zeng (2001). Improving forecasts of state failure. *World Politics 53*(4), 623–658.

Miguel, E., S. Satyanath, and E. Sergenti (2004). Economic shocks and civil conflict: An instrumental variables approach. *Journal of Political Economy 112*(4), 725–753.

Molenberghs, G. and G. Verbeke (2005). *Models for Discrete Longitudinal Data.* New York: Springer.

Pang, X. (2010). Modeling heterogeneity and serial correlation in binary tscs data: A Bayesian multilevel model with AR(p) errors. *Political Analysis 18*(4), 470–498.

Poirier, D. J. and P. A. Ruud (1988). Probit with dependent observations. *The Review of Economic Studies 55*(4), 593–614.

Raudenbush, S. W. and A. S. Bryk (2002). *Hierarchical Linear Models: Applications and Data Analysis Methods.* Newbury Park, CA: Sage Publications.

Russett, B. M. (1964). Inequality and instability: The relations between land tenure to politics. *World Politics 16*(3), 442–54.

Sambanis, N. (2002). A review of recent advances and future directions in the quantitative literature on civil war. *Defence and Peace Economics 13*(3), 215–243.

Scott, J. C. (1976). *The Moral Economy of the Peasant.* New Haven CT: Yale University Press.

Shor, B., J. Bafumi, L. Keele, and D. Park (2007). A Bayesian multilevel modeling approach to time-series cross-sectional data. *Political Analysis 15*(2), 165–181.

Singer, J. D. and J. B. Willett (2003). *Applied Longitudinal Data Analysis: Modelling Change and Event Occurrence.* New York: Oxford University Press.

Skrondal, A. and S. Rabe-Hesketh (2004). *Generalized Latent Variable Modeling: Multilevel, Longitudinal, and Structural Equation Models.* New York: Chapman & Hall/CRC.

Wimmer, A. (2002). *Nationalist Exclusion and Ethnic Conflict: Shadow of Modernity.* Cambridge: Cambridge University Press.

Yang, Y., W. Fu, and K. C. Land (2004). A methodological comparison of age-period-cohort models: The intrinsic estimator and conventional generalized linear models. *Sociological Methodology 34*(1), 75–110.

Zhu, L. (2012). Panel data analysis in public administration: Substantive and statistical considerations. *Journal of Public Administration Research and Theory*, forthcoming.

CHAPTER 3

BAYESIAN ANALYSIS OF TREATMENT EFFECT MODELS

MINGLIANG LI[1] AND JUSTIN L. TOBIAS[2]

[1]SUNY-Buffalo, USA.
[2]Purdue University, USA.

This chapter reviews Bayesian approaches and Markov chain Monte Carlo (MCMC) methods for estimating treatment-response models. We begin by reviewing the standard continuous outcome with continuous treatment specification under normality and then move on to discuss procedures for handling limited dependent treatment variables and outcomes within this framework. We also discuss methods for relaxing the standard "Gaussian" assumptions commonly made in textbook treatments of this class of problems and commonly seen in empirical applications. In so doing, we discuss issues of model comparison in finite mixture models and conclude with references to some recent work on this topic, including instrument imperfection.

3.1 Introduction

In many fields in Economics—labor, public, industrial organization, international trade and development, to name but a few—the identification and estimation of *causal effects* or *treatment effects* from observational data is a central issue. In these instances the researcher is primarily interested in estimating the effect of some treatment variable, say x, on an outcome, say y, but is concerned that a variety of unobserved factors (i.e., *unobserved confounders*) may be responsible for much of the observed correlation or partial correlation between these variables. To provide a specific and well-studied example, simple regressions of hourly wages (or their logarithm) on education may not provide a good estimate of the causal effect of education on hourly earnings since unobserved ability, motivation, and personality might simultaneously influence both educational attainment and earnings in the labor market.

The dominant technique in the applied literature for estimating causal effects with observational data is to employ *instrumental variables* (IV) or *two-stage least squares (2SLS).* When such methods are employed, the researcher identifies a set of variables which have a conditional impact on the treatment variable x but, given x, are believed to have no other direct influence on the conditional mean of y. Such variables, or *instruments*, serve to both identify the model and can be exploited to consistently estimate the model parameters.

The IV exercise in practice is both an art and a science - conditioned on a valid instrument, or a set of them, one can derive the asymptotic distribution of the IV estimator, test for overidentifying restrictions, investigate inferential consequences when instruments are weak - the usual fare of econometricians and statisticians. Coming up with compelling instruments in actual applied work, however, involves a great deal of imagination and creative skill on the part of the researcher, and the success of such studies depends in no small part on that researcher's ability to form a compelling argument that the instrument in question is plausibly excludable.

While IV/2SLS and GMM in nonlinear settings remain the most common approaches in literature, a comparably small but growing Bayesian literature also exists for the estimation of treatment-response models and identification of causal effects. It is our goal in this chapter to review some of these methods in a variety of different settings and to discuss estimation of the model parameters via Markov chain Monte Carlo (MCMC) methods.

The outline of this chapter is as follows. The following section begins with a standard Gaussian linear treatment-response model and discusses identification and MCMC implementation in that context. Section 3.3 extends these ideas to nonlinear settings. Here we spend some time discussing the popular Roy model (Roy, 1951; Heckman and Honoré, 1990) and present a general treatment that nests a variety of nonlinear treatment-response specifications. Section 3.4 discusses several departures from these cases, including taking up the case of non-Gaussian errors, model comparison and selection, and issues of instrument imperfection. An illustrative application is presented in Section 3.5, and the chapter concludes with a summary in Section 3.6.

3.2 Linear Treatment Response Models Under Normality

To fix ideas, we begin by considering the following treatment-response model for a continuous outcome y and continuous treatment variable x (throughout using the convention of boldface to denote vectors and bold capitals to denote matrices):

$$y_i = \beta_{y,0} + x_i\beta_{y,x} + \boldsymbol{w}_i\boldsymbol{\beta}_{y,w} + \epsilon_i \tag{3.1}$$

$$x_i = \beta_{x,0} + \boldsymbol{z}_i\boldsymbol{\beta}_{x,z} + u_i, \tag{3.2}$$

where

$$\left[\begin{array}{c}\epsilon_i \\ u_i\end{array}\right] \Bigg| \boldsymbol{W}, \boldsymbol{Z} \stackrel{iid}{\sim} \mathcal{N}\left[\left(\begin{array}{c}0 \\ 0\end{array}\right), \left(\begin{array}{cc}\sigma_\epsilon^2 & \sigma_{\epsilon u} \\ \sigma_{\epsilon u} & \sigma_u^2\end{array}\right)\right] \equiv \mathcal{N}(\boldsymbol{0}, \boldsymbol{\Sigma}), \quad i = 1, 2, \ldots, n. \tag{3.3}$$

The system in equations (3.1) and (3.2), and the type of model we consider throughout this chapter, is *triangular* – meaning that x enters as a right-hand side variable in (3.1), yet y does not enter as a right-hand side variable in (3.2). Thus, the structure here seems appropriate for, say, joint modeling of post-schooling wages (y) and schooling (x), but perhaps not for something like health and income, since in the latter case, we might more naturally consider a fully simultaneous system where health directly affects income, and, conversely, income directly affects health. We also take up the case where there is a single endogenous variable x in (3.1), although the case of several endogenous variables represents a straightforward extension.

Equation (3.3) assumes that the errors are i.i.d. bivariate normal. This is a common assumption, and serves as a useful starting point, although it may be potentially inappropriate in many applications. We will take up the issue of relaxing this assumption in Section 3.4. The primary parameter of interest in virtually all cases is the causal effect $\beta_{y,x}$, and unobserved confounders that simultaneously associate with u (and thus x) and ϵ (and thus y) are captured through the covariance parameter $\sigma_{\epsilon u}$, which is suspected as to be non-zero. Finally, $\boldsymbol{W}$ and $\boldsymbol{Z}$, both assumed exogenous, are $n \times k_w$ and $n \times k_z$ matrices (respectively) constructed from the $\boldsymbol{w}_i$ and $\boldsymbol{z}_i$ vectors as follows:

$$\boldsymbol{W} \equiv \left[\begin{array}{c}\boldsymbol{w}_1 \\ \boldsymbol{w}_2 \\ \vdots \\ \boldsymbol{w}_n\end{array}\right], \quad \boldsymbol{Z} \equiv \left[\begin{array}{c}\boldsymbol{z}_1 \\ \boldsymbol{z}_2 \\ \vdots \\ \boldsymbol{z}_n\end{array}\right].$$

In empirical work, the elements of $\boldsymbol{w}_i$ and $\boldsymbol{x}_i$ will likely have some degree of overlap, meaning that many elements of $\boldsymbol{w}$ will also be contained in $\boldsymbol{z}$. For example, if y represents earnings and x represents educational attainment, we might believe that variables such as gender, race and ethnicity, test scores, family characteristics, etc., should play a role in both equations. As shown below, however, we will require the appearance of at least one column (or variable) in $\boldsymbol{Z}$ that is not contained in $\boldsymbol{W}$ - these will be our *instruments* which will serve to identify the model parameters and provide a means for parameter estimation.

3.2.1 Instruments and Identification

To see why such an exclusion restriction (or instrument) is necessary in the absence of any additional model structure, let us begin by letting $\boldsymbol{\theta}$ denote all the parameters of the model. We can decompose the bivariate error distribution into the product of a conditional times a marginal:

$$p(\epsilon_i, u_i|\boldsymbol{\theta}) = p(\epsilon_i|u_i, \boldsymbol{\theta})p(u_i|\boldsymbol{\theta}). \tag{3.4}$$

Noting that the Jacobian of the transformation from (ϵ_i, u_i) to (y_i, x_i) is unity (given the triangularity of the model), we obtain

$$\begin{aligned} p(y_i, x_i|\boldsymbol{\theta}) &= \phi(y_i|\mu_{y_i|x}, \sigma^2_{y|x}) \\ &\quad \times\phi(x_i|\beta_{x,0} + \boldsymbol{z}_i\boldsymbol{\beta}_{x,z}, \sigma^2_u), \end{aligned} \tag{3.5}$$

where

$$\mu_{y_i|x} \equiv \beta_{y,0} + x_i\beta_{y,x} + \boldsymbol{w}_i\boldsymbol{\beta}_{y,w} + \frac{\sigma_{\epsilon u}}{\sigma^2_u}(x_i - \beta_{x,0} - \boldsymbol{z}_i\boldsymbol{\beta}_{x,z}), \tag{3.6}$$

$$\sigma^2_{y|x} \equiv \sigma^2_\epsilon(1-\rho^2_{\epsilon u}), \tag{3.7}$$

$$\rho_{\epsilon u} \equiv \sigma_{\epsilon u}/[\sigma_\epsilon \sigma_u], \tag{3.8}$$

and $\phi(s|\mu_s, \sigma^2_s)$ is simply the notation denoting a normal density function for the random variable s with mean μ_s and variance σ^2_s.

It is useful to pause and discuss identification in the context of this system of equations. To this end, let us first consider the case where the set of exogenous covariates are exactly common to both equations in the sense that $\boldsymbol{z}_i = \boldsymbol{w}_i$. In this instance, (3.5) can be written as

$$\begin{aligned} p(y_i, x_i|\boldsymbol{\theta}) &= \phi\left(y_i|\psi_0 + x_i\psi_1 + \boldsymbol{w}_i\boldsymbol{\psi}_2, \sigma^2_{y|x}\right) \\ &\quad \times\phi\left(x_i|\beta_{x,0} + \boldsymbol{z}_i\boldsymbol{\beta}_{x,z}, \sigma^2_u\right), \end{aligned} \tag{3.9}$$

where

$$\psi_0 = \left[\beta_{y,0} - \beta_{x,0}\frac{\sigma_{\epsilon u}}{\sigma^2_u}\right], \quad \psi_1 = \left[\beta_{y,x} + \frac{\sigma_{\epsilon u}}{\sigma^2_u}\right], \quad \boldsymbol{\psi}_2 = \left[\boldsymbol{\beta}_{y,w} - \boldsymbol{\beta}_{x,z}\frac{\sigma_{\epsilon u}}{\sigma^2_u}\right]. \tag{3.10}$$

Some quick accounting, then, shows that the likelihood is completely determined by 7 (blocks of) parameters:

$$\beta_{x,0},\ \boldsymbol{\beta}_{x,z},\ \sigma_u^2,\ \psi_0,\ \psi_1,\ \boldsymbol{\psi}_2,\ \text{and}\ \sigma_{y|x}^2, \tag{3.11}$$

whereas we seek to recover 8 "structural" parameters in $\boldsymbol{\theta}$:

$$\beta_{y,0},\quad \beta_{y,x},\quad \boldsymbol{\beta}_{y,w},\quad \beta_{x,0},\quad \boldsymbol{\beta}_{x,z},\quad \sigma_u^2,\quad \sigma_\epsilon^2,\ \text{and}\ \sigma_{\epsilon u}. \tag{3.12}$$

To summarize, the quantities in equation (3.11) are identified by the likelihood and consistently estimable whereas the full set of structural parameters in equation (3.12) is not identifiable. Importantly, observe that the "causal effect" $\beta_{y,x}$ - the object that garners most attention in practice - is among the parameters that are not identifiable when the set of covariates appearing in equations (3.1) and (3.2) are the same. What is identifiable in this case is $\psi_1 = \beta_{y,x} + \sigma_{\epsilon u}\sigma_u^{-2}$, the "total partial effect" of x on y, which combines the desired causal effect $\beta_{y,x}$ with an effect arising from unobserved confounding.

The mapping between the identified quantities and the structural parameters of interest also reveals the important role that instruments can play in parameter identification. To see this, consider the case where there is at least one variable in $\boldsymbol{z}_i$ that is not contained in $\boldsymbol{w}_i$, say the jth element of $\boldsymbol{z}_i$, or z_i^j. Then observe that the jth element of $\boldsymbol{\beta}_{x,z}$, or $\beta_{x,z}^j$, is clearly identified from the reduced form linear regression in equation (3.2), and equation (3.6) reveals that the entire term $\beta_{x,z}^j\sigma_{\epsilon u}/\sigma_u^2$ can be identified as a parameter in the conditional mean $\mu_{y_i|x}$. It follows that the coefficient on the unique element of $\boldsymbol{z}$ that is not contained in $\boldsymbol{w}$ enables us to identify the term arising from unobserved confounding, $\sigma_{\epsilon u}/\sigma_u^2$. Once this ratio is identified, all model parameters, including the causal effect $\beta_{y,x}$, can be recovered.

3.2.1.1 Posterior Simulation As discussed in the introduction to this section, the most common and familiar approaches to estimating models like those in equations (3.1) and (3.2) employ IV and 2SLS methods. Given a description of the joint distribution as in equation (3.3), however, a full likelihood-based analysis or Bayesian analysis can also be conducted.

Below we therefore turn to discuss Bayesian estimation of the model parameters in equations (3.1), (3.2), and (3.3). We do so by employing the *Gibbs sampler* - an iterative simulation method that successively draws from the complete conditional posterior distributions of the model parameters. We do not review the details of Gibbs or MCMC algorithms here, as they are extensively covered elsewhere in this volume.

To begin, let us first stack the variables and parameters as follows:

$$\begin{bmatrix} y_i \\ x_i \end{bmatrix} = \begin{bmatrix} 1 & x_i & \boldsymbol{w}_i & 0 & \boldsymbol{0} \\ 0 & 0 & \boldsymbol{0} & 1 & \boldsymbol{z}_i \end{bmatrix} \begin{bmatrix} \beta_{y,0} \\ \beta_{y,x} \\ \boldsymbol{\beta}_{y,w} \\ \beta_{x,0} \\ \boldsymbol{\beta}_{x,z} \end{bmatrix} + \begin{bmatrix} \epsilon_i \\ u_i \end{bmatrix}, \tag{3.13}$$

or succinctly,

$$\tilde{\boldsymbol{y}}_i = \tilde{\boldsymbol{X}}_i \boldsymbol{\beta} + \tilde{\boldsymbol{\epsilon}}_i, \tag{3.14}$$

with $\tilde{\boldsymbol{y}}_i$, $\tilde{\boldsymbol{X}}_i$, $\boldsymbol{\beta}$, and $\tilde{\boldsymbol{\epsilon}}_i$ defined in the obvious ways. Once stacked in this form, the parameters of this model consist of the vector $\boldsymbol{\beta}$ and the elements of the covariance matrix $\boldsymbol{\Sigma}$. A Gibbs sampling algorithm, then, will require us to derive and sample from the posterior conditionals $\boldsymbol{\beta}|\boldsymbol{\Sigma}, \tilde{\boldsymbol{y}}$ and $\boldsymbol{\Sigma}|\boldsymbol{\beta}, \tilde{\boldsymbol{y}}$. To this end, suppose we are to employ priors of the form

$$\boldsymbol{\beta} \sim \mathcal{N}(\boldsymbol{\mu}_{\boldsymbol{\beta}}, \boldsymbol{V}_{\boldsymbol{\beta}}), \tag{3.15}$$

$$\boldsymbol{\Sigma}^{-1} \sim W\left([\kappa \boldsymbol{R}]^{-1}, \kappa\right), \tag{3.16}$$

with $\mathcal{N}(\cdot,\cdot)$ denoting a multivariate normal distribution with the given mean and variance and $W(\cdot,\cdot)$ denoting a Wishart distribution, parameterized as in (Koop et al., 2007, p. 339).

With this done, posterior simulation in our linear model with an endogeneity problem follows in a straightforward way. Specifically, a simple two-block Gibbs algorithm can be employed that iteratively samples from

$$\boldsymbol{\beta}|\boldsymbol{\Sigma}, \boldsymbol{y}, \boldsymbol{x} \sim \mathcal{N}(\boldsymbol{D}_{\boldsymbol{\beta}} \boldsymbol{d}_{\boldsymbol{\beta}}, \boldsymbol{D}_{\boldsymbol{\beta}}), \tag{3.17}$$

where

$$\boldsymbol{D}_{\boldsymbol{\beta}} = \left(\boldsymbol{V}_{\boldsymbol{\beta}}^{-1} + \sum_{i=1}^{n} \tilde{\boldsymbol{X}}_i{}' \boldsymbol{\Sigma}^{-1} \tilde{\boldsymbol{X}}_i\right)^{-1}, \quad \boldsymbol{d}_{\boldsymbol{\beta}} = \boldsymbol{V}_{\boldsymbol{\beta}}^{-1} \boldsymbol{\mu}_{\boldsymbol{\beta}} + \sum_{i=1}^{n} \left(\tilde{\boldsymbol{X}}_i{}' \boldsymbol{\Sigma}^{-1} \tilde{\boldsymbol{y}}_i\right), \tag{3.18}$$

and

$$\boldsymbol{\Sigma}^{-1}|\boldsymbol{\beta}, \boldsymbol{y}, \boldsymbol{x} \sim W\left(\left[\kappa \boldsymbol{R} + \sum_{i=1}^{n} \tilde{\boldsymbol{\epsilon}}_i \tilde{\boldsymbol{\epsilon}}_i{}'\right]^{-1}, \kappa + n\right). \tag{3.19}$$

For the interested reader, details of these derivations can be found in Lindley and Smith (1972) or Koop et al. (2007).

A posterior simulator for this model proceeds by iteratively sampling from equations (3.17) and (3.19), always conditioning on the most recent parameters produced from the sampling scheme. Once convergence to the posterior is determined to have been achieved, the subsequent samples can be used to obtain point estimates (e.g. posterior means), quantiles or entire posterior distributions for objects of interest. We provide an illustrative application of how this is done in Section 3.5. It is worth emphasizing, however, that estimation in this linear treatment-response model is nearly as simple as IV or 2SLS: all that is required is the generation of normal and Wishart variates, and the tiniest dose of patience as one waits a few seconds for the software to return a suitable post-convergence sample of parameters.

We close this section noting that the reader somewhat familiar with Bayes might observe, and perhaps be puzzled by, the connection of the simulator in equations (3.17) and (3.19) to the Gibbs algorithm one would obtain from a standard system of equations analysis such as a seemingly unrelated regressions (SUR) model. That is, one might find himself or herself asking: why does the simulator for this model with an endogeneity problem reduce to essentially the same simulator that would be used to estimate a bivariate SUR without any endogeneity concerns? The connection here critically relies on the Jacobian of transformation from (ϵ, u) to (x, y) being equal to one; such a result would not be obtained for a purely simultaneous equations model that is not triangular (e.g., the fully simultaneous income/health example mentioned previously).

3.3 Nonlinear Treatment Response Models

The previous section offered a model and estimation procedure for assessing the causal effect of treatment when both the outcome y and treatment variable x were continuous. In many (possibly most) cases, however, at least one of these variables will be discrete in nature. To provide just a few observational data examples, the treatment x might indicate participation in a job training program, or high school graduation. In each of these cases, some modification of the system in equations (3.1) and (3.2) is called for in order to properly account for the binary nature of the treatment.

If the outcome y in these cases (like a log wage) remains continuous, while the treatment x is binary, an extension of equations (3.2) and (3.1) would be to consider

$$y_i = \beta_{y,0} + x_i\beta_{y,x} + \boldsymbol{w}_i\boldsymbol{\beta}_{y,w} + \epsilon_i, \tag{3.20}$$

$$x_i^* = \beta_{x,0} + \boldsymbol{z}_i\boldsymbol{\beta}_{x,z} + u_i, \tag{3.21}$$

$$x_i = I(x_i^* > 0). \tag{3.22}$$

A key feature of the model in equations (3.20) and (3.21) is the addition of *latent data* (in this case x_i^*), and the adoption of a model that, like (3.1) and (3.2), is linear, but now is represented as linear in the latent-data x_i^* as opposed to the observed outcome x_i. In the above example one might interpret x_i^* as the (unobserved by the econometrician) net desire for receipt of treatment. The observed binary outcome x_i takes the value of one if this net desire is positive, and otherwise equals zero. This mapping between the latent construct x_i^* and the observed outcome x_i is formalized in equation (3.22) – the agent takes the treatment if her net desire for doing so is positive, and otherwise is left untreated.

The system in equations (3.20), (3.21), and (3.22) offers just one example of a nonlinear treatment-response system where y is continuous and x is binary. There are a variety of other cases, however, for us to consider - for example, the outcome y_i may be binary rather than continuous, leading us to consider a latent-variable version of equation (3.20) and adding another link between observed and latent outcomes, as in (3.22). Other cases that arise commonly in empirical work include *ordinal* outcomes. Below we consider a fairly generic nonlinear treatment-response system, applicable to a variety of data types.

3.3.1 A General Nonlinear Representation

We begin by defining the vectors: $\tilde{\boldsymbol{y}}_i^* = [y_i^* \; x_i^*]'$ and $\tilde{\boldsymbol{y}}_i = [y_i \; x_i]'$, which represent a 2×1 latent data vector and 2×1 observed data vector, respectively. With these definitions in place, a reasonably general nonlinear treatment-response system can be described as follows:

$$\tilde{\boldsymbol{y}}_i^* | \tilde{\boldsymbol{X}}, \boldsymbol{\beta} \overset{ind}{\sim} \mathcal{N}\left(\tilde{\boldsymbol{X}}_i \boldsymbol{\beta}, \boldsymbol{\Sigma}\right), \tag{3.23}$$

$$\tilde{\boldsymbol{y}}_i | \tilde{\boldsymbol{y}}_i^* = g(\tilde{\boldsymbol{y}}_i^*, \boldsymbol{\alpha}), \quad i = 1, 2, \ldots, n. \tag{3.24}$$

Equation (3.23) represents a latent-variable version of (3.14). Equation (3.24) then links the latent data to the observed outcome pair $\tilde{\boldsymbol{y}}_i$ through the function $g(\cdot)$ and, potentially, a vector of parameters $\boldsymbol{\alpha}$. For example, if both the outcome y_i and treatment x_i are binary, (3.24) would become:

$$y_i = I(y_i^* > 0), \quad x_i = I(x_i^* > 0).$$

If y_i was instead a positive variable censored at zero (such as expenditure or wages, for example) while x_i remained binary, we could write

$$y_i = \max\{0, y_i^*\}, \quad x_i = I(x_i^* > 0).$$

As a final example, if both x_i and y_i were ordinal responses, we could specify

$$\begin{aligned} y_i &= j \text{ if } \alpha_j^{(y)} < y_i^* \leq \alpha_{j+1}^{(y)}, \quad j = 1, 2, \ldots, J_y, \\ x_i &= l \text{ if } \alpha_l^{(x)} < x_i^* \leq \alpha_{l+1}^{(x)}, \quad l = 1, 2, \ldots, L_x. \end{aligned}$$

In this final case, unlike the binary and censored cases, the link function g in equation (3.24) depends on a vector of *cutpoint* parameters $\boldsymbol{\alpha}$. In specifications where these parameters are present, additional steps will be added to the posterior simulator in order to generate $\boldsymbol{\alpha}$ samples.

3.3.1.1 Gibbs Implementation Let $\boldsymbol{\theta} = [\boldsymbol{\beta}' \; vec(\boldsymbol{\Sigma})' \; \boldsymbol{\alpha}']'$ denote all the parameters in this specification. We will assume prior independence among these components, continue to use a multivariate normal prior for $\boldsymbol{\beta}$: $\boldsymbol{\beta} \sim \mathcal{N}(\boldsymbol{\mu_\beta}, \boldsymbol{V_\beta})$, as in (3.15) and a Wishart prior for $\boldsymbol{\Sigma}^{-1}$: $\boldsymbol{\Sigma}^{-1} \sim W([\kappa \boldsymbol{R}]^{-1}, \kappa)$ as in (3.16) and leave the prior for $\boldsymbol{\alpha}$ generically specified as $p(\boldsymbol{\alpha})$.

In order to implement a Gibbs algorithm for this model, we need to derive and sample from the complete conditional posterior distributions of the model parameters. If the $\tilde{\boldsymbol{y}}_i^*$ were "known," posterior inference could proceed very similarly to the continuous outcome model presented in equations (3.1) and (3.2). That is, $\boldsymbol{\beta}|\tilde{\boldsymbol{y}}^*, \boldsymbol{\Sigma}$ would be multivariate normal, and $\boldsymbol{\Sigma}^{-1}|\boldsymbol{\beta}, \tilde{\boldsymbol{y}}^*$ would remain a Wishart distribution, although possibly with some restrictions on its elements.

Given these appealing conveniences afforded by conditioning on $\tilde{\boldsymbol{y}}^*$, a possible approach is to *augment* the posterior distribution with the latent data $\tilde{\boldsymbol{y}}^*$ and sample the latent data vectors in the course of our posterior simulation (e.g., Tanner and Wong, 1987; Albert and Chib, 1993). We therefore consider the augmented posterior $p(\boldsymbol{\theta}, \tilde{\boldsymbol{y}}^*|\tilde{\boldsymbol{y}})$. A Gibbs implementation, then, requires us to derive and sample from the four posterior conditionals for $\boldsymbol{\beta}$, $\boldsymbol{\Sigma}^{-1}$, $\tilde{\boldsymbol{y}}^*$ and $\boldsymbol{\alpha}$ (and the last of these only when required).

The first two conditionals, as suggested earlier, are easily sampled conditional on the latent data:

$$\boldsymbol{\beta}|\boldsymbol{\Sigma}, \tilde{\boldsymbol{y}}^*, \tilde{\boldsymbol{y}} \sim \mathcal{N}\left(\boldsymbol{D}_{\boldsymbol{\beta}}\boldsymbol{d}_{\boldsymbol{\beta}}, \boldsymbol{D}_{\boldsymbol{\beta}}\right) \tag{3.25}$$

where

$$\boldsymbol{D}_{\boldsymbol{\beta}} \equiv \left[\left(\sum_{i=1}^{n} \tilde{\boldsymbol{X}}_i'\boldsymbol{\Sigma}^{-1}\tilde{\boldsymbol{X}}_i\right) + \boldsymbol{V}_{\boldsymbol{\beta}}^{-1}\right]^{-1}, \quad \boldsymbol{d}_{\boldsymbol{\beta}} \equiv \left(\sum_{i=1}^{n} \tilde{\boldsymbol{X}}_i'\boldsymbol{\Sigma}^{-1}\tilde{\boldsymbol{y}}_i^*\right) + \boldsymbol{V}_{\boldsymbol{\beta}}^{-1}\boldsymbol{\mu}_{\boldsymbol{\beta}}, \tag{3.26}$$

and

$$\boldsymbol{\Sigma}^{-1}|\boldsymbol{\beta}, \tilde{\boldsymbol{y}}^*, \tilde{\boldsymbol{y}} \sim W\left(\left[\kappa\boldsymbol{R} + \sum_{i=1}^{n}(\tilde{\boldsymbol{y}}_i^* - \tilde{\boldsymbol{X}}_i\boldsymbol{\beta})(\tilde{\boldsymbol{y}}_i^* - \tilde{\boldsymbol{X}}_i\boldsymbol{\beta})'\right]^{-1}, \kappa + n\right). \tag{3.27}$$

Note that, in terms of our coding, we act as if the latent data $\tilde{\boldsymbol{y}}^*$ are observed, and will simply update this vector at each step in our sampler. To complete a description of our posterior simulator, we note that the joint posterior distribution for the latent $\tilde{\boldsymbol{y}}^*$ and vector of parameters $\boldsymbol{\alpha}$ is given as

$$p(\tilde{\boldsymbol{y}}^*, \boldsymbol{\alpha}|\tilde{\boldsymbol{y}}, \boldsymbol{\beta}, \boldsymbol{\Sigma}) \propto p(\boldsymbol{\alpha}) \prod_{i=1}^{n} \phi(\tilde{\boldsymbol{y}}_i^*|\boldsymbol{X}_i\boldsymbol{\beta}, \boldsymbol{\Sigma}) I\left[\tilde{\boldsymbol{y}}_i = g(\tilde{\boldsymbol{y}}_i^*, \boldsymbol{\alpha})\right]. \tag{3.28}$$

To fix ideas and provide a little clarity to equation (3.28), consider the case discussed at the outset of this section where y is continuous and x is binary. In this instance, there are no parameters in $\boldsymbol{\alpha}$, and y_i^* is not needed, since that component of the model is fully observed. It remains, then, to sample just the latent data x_i^*. We do so by noting that equation (3.28) reduces to:

$$\begin{aligned} p(x_i^*|\boldsymbol{\beta}, \boldsymbol{\Sigma}, \boldsymbol{x}, \boldsymbol{y}) \quad &\propto \quad \phi(x_i^*|\mu_{x_i^*|y}, \sigma_{x|y}^2) \times \\ &\quad\; [I(x_i^* > 0)I(x_i = 1) + I(x_i^* \leq 0)I(x_i = 0)], \end{aligned} \tag{3.29}$$

for $i = 1, 2, \ldots, n$, where

$$\mu_{x_i^*|y} = \beta_{x,0} + \boldsymbol{z}_i\boldsymbol{\beta}_{x,z} + \frac{\sigma_{\epsilon u}}{\sigma_\epsilon^2}\left[y_i - \beta_{y,0} - x_i\beta_{y,x} - \boldsymbol{w}_i\boldsymbol{\beta}_{y,w}\right], \tag{3.30}$$

$$\sigma_{x|y}^2 = \sigma_u^2(1 - \rho_{u\epsilon}^2). \tag{3.31}$$

The structure of (3.29) clearly reveals that the sampling of each x_i^* can be conducted by generating draws from univariate truncated normal distributions. Simulations from the truncated normal are easily obtained, and can be produced directly via the method of inversion [see, e.g., (Koop et al., 2007, Exercise 11.20)].

When $\boldsymbol{\alpha}$ remains a component of the posterior simulator, owing to the consideration of, for example, ordinal data, an additional sampling step is required. While one could potentially proceed by sampling $\{\tilde{\boldsymbol{y}}_i^*|\boldsymbol{\alpha}, \boldsymbol{\beta}, \boldsymbol{\Sigma}, \tilde{\boldsymbol{y}}\}_{i=1}^n$ and $\boldsymbol{\alpha}|\tilde{\boldsymbol{y}}^*, \boldsymbol{\beta}, \boldsymbol{\Sigma}, \tilde{\boldsymbol{y}}$, this is usually not advisable given the high degree of correlation between the latent data $\tilde{\boldsymbol{y}}^*$ and the parameters in $\boldsymbol{\alpha}$. In this situation, $\tilde{\boldsymbol{y}}^*$ can often be integrated out of (3.28), $\boldsymbol{\alpha}$ can then be drawn from the resulting marginalized conditional, and finally the latent data can be drawn independently from its complete conditional:

$$p(\tilde{\boldsymbol{y}}_i^*|\boldsymbol{\beta}, \boldsymbol{\Sigma}, \boldsymbol{\alpha}, \tilde{\boldsymbol{y}}) \quad \propto \quad \phi(\tilde{\boldsymbol{y}}_i^*|\boldsymbol{X}_i\boldsymbol{\beta}, \boldsymbol{\Sigma})I[\tilde{\boldsymbol{y}}_i = g(\tilde{\boldsymbol{y}}_i^*, \boldsymbol{\alpha})], \quad i = 1, 2, \ldots, n,$$

which typically amounts, again, to truncated normal sampling. Jeliazkov et al. (2008) discuss a number of important issues related to identification of univariate and multivariate ordinal models and evaluate the performances of alternate schemes for sampling $(\boldsymbol{\alpha}, \tilde{\boldsymbol{y}}^*)$ in such models.

An important issue that, to this point, has only been briefly and casually referenced, concerns restrictions on elements of the covariance matrix $\boldsymbol{\Sigma}$. For example, in the case of binary observed outcomes, a common identification restriction is to impose that the diagonal elements of $\boldsymbol{\Sigma}$ are unity. In this case, the sampling of $\boldsymbol{\Sigma}^{-1}$ is not from (3.27), but rather, from (3.27) with restrictions on the main diagonal.

A few alternatives exist for dealing with this problem. Reparameterizations [e.g., Li (1998) and McCulloch et al. (2000)] are sometimes possible which allow the researcher to essentially sidestep the restrictions, sampling elements of $\boldsymbol{\Sigma}$ in blocks while preserving the positive definiteness of the covariance matrix. In the case of a single diagonal restriction, as would be the case if only one of the outcomes were binary, Nobile (2000) comments on the reparameterization scheme of McCulloch et al. (2000), and provides a way to directly sample from a Wishart, given a restriction on a diagonal element. Chan and Jeliazkov (2009) provide additional details regarding sampling from restricted covariance matrices. Finally, and consistent with our presentation in this chapter, one can choose to simply ignore the restrictions that are in place when implementing the simulator and post-process the simulations to focus on identified quantities. That is, one can simply sample $\boldsymbol{\Sigma}^{-1}$ as in (3.27) and appropriately adjust those simulations at each iteration of the sampler to calculate quantities that are identifiable. Rossi et al. (2005), particularly in Chapters 3 and 4, discuss applications of this approach and compare performances of samplers that navigate through identified and non-identified parameter spaces.

3.3.1.2 The Roy Model A popular specification in treatment-response modeling, and a slight generalization of the system in equations (3.20) and (3.21), is the Roy model (Roy, 1951) or model of *potential outcomes* (see also Heckman and Honoré, 1990). This specification again considers the case of a binary treatment, [and as such (3.21) and (3.22) remain unchanged], but adds to (3.20) by explicitly modeling outcome equations for both the treated and untreated regimes. Specifically, we might represent a slightly generalized version of the Roy model as

$$y_i^{(1)} = \beta_{y,0}^{(1)} + \boldsymbol{w}_i \boldsymbol{\beta}_{y,w}^{(1)} + \epsilon_i^{(1)}, \tag{3.32}$$

$$y_i^{(0)} = \beta_{y,0}^{(0)} + \boldsymbol{w}_i \boldsymbol{\beta}_{y,w}^{(0)} + \epsilon_i^{(0)}, \tag{3.33}$$

$$x_i^* = \beta_{x,0} + \boldsymbol{z}_i \boldsymbol{\beta}_{x,z} + u_i. \tag{3.34}$$

Variables and parameters associated with the treated state have been assigned a superscript of (1) in (3.32), while those associated with the untreated state are assigned a superscript of (0) in (3.33). Treatment effects are summarized by the outcome gain (or loss) resulting from receipt of treatment, denoted as $y^{(1)} - y^{(0)}$, and trivariate normality among $[\epsilon_i^{(1)} \ \epsilon_i^{(0)} \ u_i]'$ is assumed as a starting point, with the methods described in the following section used to relax this assumption, when needed.

In terms of posterior simulation, simple generalizations of the sampling steps used for the estimation of (3.20) and (3.21) can be employed to fit this model. Specifically, given the complete data, the posterior conditional for the stacked vector of parameters $\boldsymbol{\beta}$ will again remain normal with a structure identical to (3.25), while that for $\boldsymbol{\Sigma}^{-1}$ will remain Wishart, as in (3.27), with a diagonal restriction that sets $\sigma_u^2 = 1$. As discussed previously, this sampling step for $\boldsymbol{\Sigma}^{-1}$ can either be performed via a reparameterization, or a draw from a restricted Wishart can be directly obtained (e.g., Nobile, 2000), or the restriction can be ignored and the simulations post-processed to focus on identifiable quantities. As for the sampling of the latent data, the x_i^* are again sampled independently from their conditional truncated normal distributions, readily derived from the trivariate system of equations.

We also observe that for each observation i, either the treated or the untreated outcome is observed for each agent (but never both), and as such the observed outcome y_i can be expressed as

$$y_i = x_i y_i^{(1)} + (1 - x_i) y_i^{(0)}.$$

Thus, the treatment effect $\Delta \equiv y^{(1)} - y^{(0)}$ involves a *counterfactual* - the outcome the agent would have received had he or she made a different decision regarding the receipt of treatment. For every individual, then, exactly one of the $y^{(j)}$, $j = 0, 1$, is missing while the other is observed. In the spirit of data augmentation, we can thus add this missing outcome as an element of the joint posterior, and in the course of implementing the posterior simulator, sample the"missing" outcome for each i from its conditional normal distribution. Conditioned on this missing data - which is updated at every iteration of the sampler - the sampling of $\boldsymbol{\beta}$ and $\boldsymbol{\Sigma}$ proceeds as if we were faced with a fully observed system of linear equations (e.g., a seemingly unrelated regressions model). Full details regarding a Gibbs algorithm for this model are provided in Chib and Hamilton (1999) and (Exercise 14.11 Koop et al., 2007), and thus are not repeated here.

A final interesting issue arises in the Roy model setting, and is related to the potential outcomes structure it embraces. Since only $y^{(1)}$ *or* $y^{(0)}$ is observed for each individual, the correlation parameter between the errors of the treated and untreated equations, which we could denote as ρ_{10} to fix ideas, does not enter the likelihood function for the observed data and thus is not identified. Despite this, Vijverberg (1993) notes the possibility of learning about ρ_{10}, an idea formalized in the Bayesian setting by Koop and Poirier (1997). These authors clearly illustrate that the marginal priors and posteriors for ρ_{10} will not be the same, in general, even though this parameter is not identified by the likelihood function. The vehicle for this updating of marginal prior beliefs is prior dependence - information learned about identified correlation parameters, coupled with restrictions that $\mathbf{\Sigma}$ must be positive semidefinite, creates a knowledge spillover which restricts the conditional support of ρ_{10}. Chib and Hamilton (1999) address the non-identification of the cross-regime correlation ρ_{10} by setting it equal to zero and fitting the model subject to that restriction, while Poirier and Tobias (2003) and Li et al. (2004) allow for learning about this parameter and provide several applications. In a recent statement on this issue, Chib (2007) recommends working directly in the identified model space as opposed to the augmented potential outcomes space, as doing so improves the mixing properties of the posterior simulator and frees the researcher of the need to worry about the influence of the conditional prior for ρ_{10}- whose influence on posterior results does not vanish even asymptotically.

3.4 Other Issues and Extensions: Non-Normality, Model Selection, and Instrument Imperfection

3.4.1 Non-Normality

The reader may have observed that all the analysis to this point has been conducted based on the assumption of joint normality of the error terms. While the assumption of normality does serve as a useful starting point, and may suffice as an adequate description of the data in some cases, it is necessary nonetheless to have a modeling framework allowing for significantly greater flexibility.

When seeking out these more general treatments, it seems prudent to keep in mind a desire to keep things computationally tractable. What has been evident from our previous examples is that normal likelihoods combine nicely with conditionally conjugate normal/Wishart priors to yield conditional posterior distributions that are easily sampled, thus facilitating Gibbs estimation.

These computational conveniences of normality are something we would like to retain when considering more general specifications to take to the data.

One possibility, as noted by Carlin and Polson (1991) and Geweke (1993), and illustrated for the case of treatment-response modeling by Chib and Hamilton (1999), is to consider a class of Student-t sampling models by representing them as a scale mixture of multivariate Gaussian distributions. Specifically, one can generalize (3.23) by assuming

$$\tilde{\boldsymbol{y}}_i^* \stackrel{ind}{\sim} \mathcal{N}(\tilde{\boldsymbol{X}}_i\boldsymbol{\beta}, \lambda_i\boldsymbol{\Sigma}), \tag{3.35}$$

$$\lambda_i \stackrel{iid}{\sim} IG\left(\frac{\nu}{2}, \frac{2}{\nu}\right) \quad \Rightarrow p(\lambda_i) \propto \lambda_i^{-([\nu/2]+1)} \exp\left(-\frac{\nu}{2\lambda_i}\right), \tag{3.36}$$

where IG denotes an inverse gamma distribution, whose density function is provided up to proportionality.

When (3.35) is marginalized over the scale mixing variables λ_i, whose prior is given in (3.36), a multivariate-t sampling model (with ν degrees of freedom) for $\tilde{\boldsymbol{y}}_i^*$ is produced. This extension thus allows the researcher to address issues of fat tails beyond those allowed by the normal distribution, and the conditional normal representation of the sampling model leads to tractable conditional posteriors amenable to Gibbs sampling.

Gibbs implementation in the Student-t sampling model is fully described in Chib and Hamilton (1999) and we briefly overview those details here. Conditioned on $\boldsymbol{\lambda} = [\lambda_1, \lambda_2, \ldots, \lambda_n]'$, the sampling of $\tilde{\boldsymbol{y}}^*$, $\boldsymbol{\beta}$, $\boldsymbol{\Sigma}$, and $\boldsymbol{\alpha}$ follow similarly to what was described in the previous section. Specifically, the sampling of $\boldsymbol{\beta}$ and $\boldsymbol{\Sigma}$ will involve the simulation of normal and (possibly restricted) inverse Wishart variates, respectively, while the sampling of $\tilde{\boldsymbol{y}}^*$ in many cases will involve truncated normal sampling. The posterior conditional for each λ_i can be shown to be of the inverse gamma form, and draws from the inverse gamma can simply be obtained by inverting gamma random variates. Thus, extension to the Student-t case just adds an additional step to the Gibbs algorithm for the standard Gaussian model, and the conditional normal scale mixture representation of the Student-t in (3.35) and (3.36) retains the computational conveniences afforded by normality.

A more general representation that also allows for possible skewness or multimodality in the error distribution is to consider a finite mixture sampling model for the augmented data:

$$\tilde{\boldsymbol{y}}_i^*|c_i \sim \mathcal{N}\left(\boldsymbol{\beta}_{0c_i} + \tilde{\boldsymbol{X}}_i\tilde{\boldsymbol{\beta}}, \boldsymbol{\Sigma}_{c_i}\right), \tag{3.37}$$

$$\Pr(c_i = g|\boldsymbol{\pi}) = \pi_g, \quad g = 1, 2, \ldots, G, \quad \sum_{g=1}^{G} \pi_g = 1, \pi_g > 0 \ \forall g, \tag{3.38}$$

where $\boldsymbol{\beta}_{0c_i} = [\beta_{y,0c_i}, \beta_{x,0c_i}]'$ denotes component specific intercepts and $\tilde{\boldsymbol{\beta}}$ represents slope coefficients that are common to all components. The variable c_i is an integer-valued component labeling variable with support over

$g = 1, 2, \ldots, G$, representing which mixture component observation i is generated from. The vector $\boldsymbol{\pi} = [\pi_1, \pi_2, \ldots, \pi_G]'$ is a vector of component probabilities, and marginalized over c_i, the sampling model for $\tilde{\boldsymbol{y}}_i^*$ is easily seen as a mixture (or average) of multivariate normal models. To complete the model, a Dirichlet prior for $\boldsymbol{\pi}$ is commonly assumed.

In terms of posterior simulation, the sampling of $\boldsymbol{\beta} \equiv [\boldsymbol{\beta}_{01}', \boldsymbol{\beta}_{02}', \ldots, \boldsymbol{\beta}_{0G}', \tilde{\boldsymbol{\beta}}']'$ and $\boldsymbol{\Sigma}^{-1} \equiv [\boldsymbol{\Sigma}_1^{-1}, \boldsymbol{\Sigma}_2^{-1}, \ldots, \boldsymbol{\Sigma}_G^{-1}]$, follow essentially the same as in (3.25) and (3.27), although only the subset of observations currently assigned to the gth component are used in the respective inference for the component specific intercepts and covariance matrices. The component labeling variables themselves are drawn from a discrete distribution with support over $g = 1, 2, \ldots, G$, and the component probabilities are sampled from a Dirichlet. Details of these calculations are provided in Chib and Hamilton (1999) and Li et al. (2004), and are not repeated here for the sake of brevity.

Finite mixture models are quite flexible and can adapt to capture a variety of features of the data distribution, including skewness and multimodality. In some cases, prior knowledge of the problem at hand suggests that the population of interest consists of distinct groups, whose responses are likely to be homogeneous within each group but potentially different across groups. In this case it is natural to adopt a finite mixture model and to interpret its parameters. In other (perhaps most) cases, finite mixtures are simply used as a flexible modeling device to allow for departures from normality, and any ex-post interpretation of results as evidence for the existence of discrete "groups" in the population should be made with caution.

In practice, use of finite mixtures can produce concerns regarding parameter and component identification (i.e., the model parameters are not identified up to a relabeling of the components and the simulator can exhibit label-switching). Furthermore, the choice of the number of mixture components is a question that should be addressed, and we take up this issue in the following section.

Finally, an even more general approach, as described in Conley et al. (2008), is to consider an analysis based on Dirichlet process (DP) priors. In this approach, parameters are observation-specific, and assumed to be generated from a common distribution. A prior is placed over this common distribution, centered around an overall base measure. Although this approach avoids explicit selection of the number of components, one can think of this as being handled endogenously, as new parameter values (quite similar in spirit to components) are "born" when needed, "die" when unneeded, and are lumped together when appropriate to retain parsimony. Neal (2000) provides a number of useful details related to posterior simulation in DP models.

3.4.2 Model Comparison

As discussed in the previous section, when finite mixture models are adopted, there is a need to determine the appropriate number of mixture components.

Chib (1995) and Chib and Jeliazkov (2001) offer very useful, general-purpose methods for model comparison, model averaging and marginal likelihood calculation in a Bayesian setting. Below we discuss an alternative approach to this problem based on the bridge sampling technique.

A special case of the bridge sampling method is the reciprocal importance sampling approach utilized in Gelfand and Dey (1994):

$$p(\tilde{\boldsymbol{y}}) \approx \left[M^{-1} \sum_{m=1}^{M} \frac{q(\boldsymbol{\theta}^{(m)})}{p(\boldsymbol{\theta}^{(m)})p(\tilde{\boldsymbol{y}}|\boldsymbol{\theta}^{(m)})} \right]^{-1}, \tag{3.39}$$

where $\boldsymbol{\theta}^{(m)}$ is the mth draw of $\boldsymbol{\theta} = [\boldsymbol{\beta}', \text{vec}(\boldsymbol{\Sigma})', \boldsymbol{\alpha}', \boldsymbol{\pi}']'$ from a Gibbs algorithm for the model in (3.37) and (3.38). The denominator in the above, $p(\boldsymbol{\theta}^{(m)})p(\tilde{\boldsymbol{y}}|\boldsymbol{\theta}^{(m)})$, is an evaluation of the non-normalized posterior distribution of $\boldsymbol{\theta}$, which is the product of the prior distribution $p(\boldsymbol{\theta})$ and the likelihood function $p(\tilde{\boldsymbol{y}}|\boldsymbol{\theta})$ marginalized over the latent outcomes $\tilde{\boldsymbol{y}}^*$ and component labeling variables $\boldsymbol{c} = [c_1, c_2, \ldots, c_n]'$. The numerator in the sampling average $q(\boldsymbol{\theta}^{(m)})$ is an evaluation of the importance density $q(\boldsymbol{\theta})$. In the rare case where the posterior distribution $p(\boldsymbol{\theta}|\tilde{\boldsymbol{y}})$ has a known exact analytical form with an explicit expression of the normalizing constant, the posterior distribution $p(\boldsymbol{\theta}|\tilde{\boldsymbol{y}})$ serves as the best candidate for $q(\boldsymbol{\theta})$. In fact, in such a case, it is not even necessary to make the sampling average over the M draws of $\boldsymbol{\theta}^{(m)}$ and (3.39) holds exactly even for $M = 1$. Chib (1995) and Chib and Jeliazkov (2001) follow this strategy and make use of the Gibb algorithm to produce an accurate evaluation of the posterior distribution $p(\boldsymbol{\theta}|\tilde{\boldsymbol{y}})$ at a particular draw $\boldsymbol{\theta}^{(m)}$.

Often the importance density $q(\boldsymbol{\theta})$ chosen by a researcher has fatter tails than the posterior distribution $p(\boldsymbol{\theta}|\tilde{\boldsymbol{y}})$, making the reciprocal importance sampling method not very reliable. Various researchers, including Geweke (1999), Meng and Wong (1996), and Frühwirth-Schnatter (2004), have proposed important improvements over the reciprocal importance sampling method. For example, the bridge sampling technique examined and applied in Meng and Wong (1996) and Frühwirth-Schnatter (2004) generalize and replace (3.39) with

$$p(\tilde{\boldsymbol{y}}) \approx \frac{L^{-1} \sum_{l=1}^{L} \alpha(\tilde{\boldsymbol{\theta}}^{(l)}) p(\tilde{\boldsymbol{\theta}}^{(l)}) p(\tilde{\boldsymbol{y}}|\tilde{\boldsymbol{\theta}}^{(l)})}{M^{-1} \sum_{m=1}^{M} \alpha(\boldsymbol{\theta}^{(m)}) q(\boldsymbol{\theta}^{(m)})}, \tag{3.40}$$

where $\tilde{\boldsymbol{\theta}}^{(l)}$, for $l = 1, 2, \ldots, L$, are L independent draws from the importance density $q(\boldsymbol{\theta})$. Various choices of the function $\alpha(\boldsymbol{\theta})$ result in different versions of the bridge sampling method. For example, the reciprocal importance sampling approach in (3.39) is a special case of the bridge sampling method with $\alpha(\boldsymbol{\theta}) = [p(\boldsymbol{\theta})p(\tilde{\boldsymbol{y}}|\boldsymbol{\theta})]^{-1}$.

Meng and Wong (1996) propose an asymptotically optimal choice of $\alpha(\boldsymbol{\theta})$ under certain conditions:

$$\alpha(\boldsymbol{\theta}) \quad \propto \quad \frac{1}{Lq(\boldsymbol{\theta}) + Mp(\boldsymbol{\theta}|\tilde{\boldsymbol{y}})}. \tag{3.41}$$

Since the exact form of the posterior distribution $p(\boldsymbol{\theta}|\tilde{\boldsymbol{y}}) = [p(\boldsymbol{\theta})p(\tilde{\boldsymbol{y}}|\boldsymbol{\theta})]/p(\tilde{\boldsymbol{y}})$ in turn depends on the knowledge of the marginal likelihood $p(\tilde{\boldsymbol{y}})$, in practice, Meng and Wong (1996) suggest using some starting value of $p(\tilde{\boldsymbol{y}})$ such as the output from the reciprocal importance sampling approach in (3.39):

$$p_0(\tilde{\boldsymbol{y}}) \quad = \quad \left[M^{-1} \sum_{m=1}^{M} \frac{q(\boldsymbol{\theta}^{(m)})}{p(\boldsymbol{\theta}^{(m)})p(\tilde{\boldsymbol{y}}|\boldsymbol{\theta}^{(m)})} \right]^{-1},$$

and then, for $t = 1, 2, \ldots, T$, iteratively updating the values of $p_t(\boldsymbol{\theta}|\tilde{\boldsymbol{y}})$ and $p_t(\tilde{\boldsymbol{y}})$ using the following two steps which can converge rather quickly to the bridge sampling estimate of $p(\tilde{\boldsymbol{y}})$ with an asymptotically optimal choice of $\alpha(\boldsymbol{\theta})$:

$$p_t(\boldsymbol{\theta}|\tilde{\boldsymbol{y}}) \quad = \quad \frac{p(\boldsymbol{\theta})p(\tilde{\boldsymbol{y}}|\boldsymbol{\theta})}{p_{t-1}(\tilde{\boldsymbol{y}})}, \tag{3.42}$$

$$p_t(\tilde{\boldsymbol{y}}) \quad = \quad \left[L^{-1} \sum_{l=1}^{L} \frac{p(\tilde{\boldsymbol{\theta}}^{(l)})p(\tilde{\boldsymbol{y}}|\tilde{\boldsymbol{\theta}}^{(l)})}{Lq(\tilde{\boldsymbol{\theta}}^{(l)}) + Mp_t(\tilde{\boldsymbol{\theta}}^{(l)}|\tilde{\boldsymbol{y}})} \right] \times \left[M^{-1} \sum_{m=1}^{M} \frac{q(\boldsymbol{\theta}^{(m)})}{Lq(\boldsymbol{\theta}^{(m)}) + Mp_t(\boldsymbol{\theta}^{(m)}|\tilde{\boldsymbol{y}})} \right]^{-1}. \tag{3.43}$$

Meng and Wong (1996) and Frühwirth-Schnatter (2004) show many advantages of the bridge sampling method over the reciprocal importance sampling approach. However, for various finite mixture models, including the one in (3.37) and (3.38), the choice of the importance density $q(\boldsymbol{\theta})$ remains critical. It is well known that the finite mixture models suffer from an identification problem in the sense that both the prior $p(\boldsymbol{\theta})$ and the likelihood function $p(\tilde{\boldsymbol{y}}|\boldsymbol{\theta})$ in most cases are labeling-invariant. In such cases, the posterior distribution $p(\boldsymbol{\theta}|\tilde{\boldsymbol{y}})$ which is proportional to $p(\boldsymbol{\theta})p(\tilde{\boldsymbol{y}}|\boldsymbol{\theta})$ also will not change if we relabel all the G components completely. Unless there exist some restrictions, such as $\beta_{x,01} > \beta_{x,02} > \cdots > \beta_{x,0G}$, imposed on the prior $p(\boldsymbol{\theta})$, the likelihood function itself does not distinguish among all the $G!$ different ways of component labeling. Related to this identification problem, a typical Gibbs algorithm for finite mixture modeling also suffers from a slow mixing problem. In theory, the Gibbs sampler should converge to draws from the posterior distribution $p(\boldsymbol{\theta}|\tilde{\boldsymbol{y}})$ by visiting each of the $G!$ labeling subspaces. But in practice, as explained in Frühwirth-Schnatter (2001, 2004) and Geweke (2007), the Gibbs draws tend to stick in only one of the $G!$ labeling subspaces even with a large number of draws.

To combat the above mentioned identification problem in finite mixture modeling directly, Frühwirth-Schnatter (2001) proposes a random permutation method to force the Gibbs sampler to visit each of the $G!$ labeling subspaces. For the model in (3.37) and (3.38), this random permutation approach is achieved by, at the end of the mth draw of the Gibbs algorithm, reassigning the values of the parameters in $\boldsymbol{c}^{(m)}$, $\boldsymbol{\pi}^{(m)}$, $\boldsymbol{\beta}^{(m)}$ and $\boldsymbol{\Sigma}^{(m)}$ that are component-specific to one of the $G!$ labeling schemes that is randomly chosen. In this way, each of the $G!$ subspaces is visited with a probability of exactly $1/G!$. Correspondingly, to compute the marginal likelihood in ((3.43)), one can choose an importance density following the suggestions in Frühwirth-Schnatter (2004) and Kaufmann and Kaufmann and Frühwirth-Schnatter (2002):

$$
\begin{aligned}
q(\boldsymbol{\theta}) =& \frac{1}{S}\sum_{s=1}^{S} p(\boldsymbol{\pi}|\boldsymbol{c}^{(s)})p(\boldsymbol{\beta}|\boldsymbol{c}^{(s)}, \boldsymbol{\Sigma}^{(s-1)}, \tilde{\boldsymbol{y}}^{*(s-1)}) \\
& \times p(\boldsymbol{\Sigma}|\boldsymbol{c}^{(s)}, \boldsymbol{\beta}^{(s)}, \tilde{\boldsymbol{y}}^{*(s-1)})q(\boldsymbol{\alpha}|\boldsymbol{c}^{(s)}, \boldsymbol{\beta}^{(s)}, \boldsymbol{\Sigma}^{(s)}, \tilde{\boldsymbol{y}}^{*(s-1)}, \tilde{\boldsymbol{y}})
\end{aligned}
\tag{3.44}
$$

where $p(\boldsymbol{\pi}|\cdot)$, $p(\boldsymbol{\beta}|\cdot)$, $p(\boldsymbol{\Sigma}|\cdot)$, and $q(\boldsymbol{\alpha}|\cdot)$ correspond to the Gibbs steps of drawing $\boldsymbol{\pi}$, $\boldsymbol{\beta}$, $\boldsymbol{\Sigma}$, and $\boldsymbol{\alpha}$, respectively. Often Metropolis-Hastings steps are used in the sampling of parameters such as $\boldsymbol{\alpha}$. In such a case, following Kaufmann and Frühwirth-Schnatter (2002), $q(\boldsymbol{\alpha}|\cdot)$ now instead corresponds to the proposal density for drawing $\boldsymbol{\alpha}$ in a Metropolis-Hastings step. Assuming that we draw $\boldsymbol{c}$, $\boldsymbol{\pi}$, $\boldsymbol{\beta}$, $\boldsymbol{\Sigma}$, $\boldsymbol{\alpha}$, and $\tilde{\boldsymbol{y}}^{*}$ sequentially in the Gibbs algorithm, the values $\boldsymbol{c}^{(s)}$, $\boldsymbol{\beta}^{(s)}$, and $\boldsymbol{\Sigma}^{(s)}$ in (3.44) correspond to the sth draw and $\boldsymbol{\Sigma}^{(s-1)}$ and $\tilde{\boldsymbol{y}}^{*(s-1)}$ the $(s-1)^{th}$ draw of the Gibbs algorithm. Frühwirth-Schnatter (2001) demonstrates that the random permutation step forces the Gibbs algorithm to visit each of the $G!$ labeling subspaces. As a result, the importance density in ((3.44)) is a sampling average of densities with underlying parameters visiting all the $G!$ labeling subspaces. Without the random permutation step, the importance density in (3.44) may depend on parameters sticking in only one of the $G!$ labeling subspaces.

3.4.3 Instrument Imperfection

Procedures such as 2SLS and IV are commonly labeled as purely "classical" or "frequentist" and regarded as objective, free of the influences of priors. Despite this pervasive view, one can argue that such methods are, in fact, quite subjective and that beliefs surrounding the validity of such methods are rather personal.

To see this, consider a version of (3.1) and (3.2) where both $\boldsymbol{w}$ and $\boldsymbol{z}$ appear in the outcome and treatment equations:

$$
y_i = \beta_{y,0} + x_i\beta_{y,x} + \boldsymbol{w}_i\boldsymbol{\beta}_{y,w} + \boldsymbol{z}_i\boldsymbol{\beta}_{y,z} + \epsilon_i, \tag{3.45}
$$

$$
x_i = \beta_{x,0} + \boldsymbol{w}_i\boldsymbol{\beta}_{x,w} + \boldsymbol{z}_i\boldsymbol{\beta}_{x,z} + u_i. \tag{3.46}
$$

As argued in our introductory section, the model parameters are not fully identified without some form of additional structure. The approach behind instrumental variables is to determine a z or set of $\boldsymbol{z}$ for which $\boldsymbol{\beta}_{y,z} = 0$, and to achieve identification as a consequence of that restriction. In practice, we distinguish $\boldsymbol{w}$ from $\boldsymbol{z}$ via this belief.

What the IV-minded researcher does in practice, then, is to think carefully about the problem at hand and find a set of instruments which seem plausibly excludable. In papers of this type, an inordinate amount of time is spent in trying to convince the audience that it is reasonable to share his/her belief regarding this exclusion. But a *belief* is exactly what is being expressed here – an opinion of the investigator that an appropriate prior for the problem at hand is the dogmatic one in which $\boldsymbol{\beta}_{y,z} = 0$.

When thinking carefully about virtually all pursuits of this type based upon observational data, one can often come up with stories which undermine the perfect excludability of $\boldsymbol{z}$ from (3.45). For example, even quarter of birth, often used in past work to assess the causal impact of education on earnings, has recently been questioned as a legitimate instrument given its demonstrated correlation with maternal characteristics (Hungerman and Buckles, 2013). This leads researchers to instead consider a system like (3.45) and (3.46), and to see what we might learn about the causal effect $\boldsymbol{\beta}_{y,x}$ when $\boldsymbol{\beta}_{y,z}$ is allowed to be "small" but non-zero. Recent work by Kraay (2012) and Conley et al. (2012) discuss approaches based on this idea.

When allowing for instrument imperfection (that is, failing to impose $\beta_{y,z} = 0$), it is important to keep in mind the model becomes only partially identified. Given this partial identification, Chan and Tobias (2013) note that, when even moderately weak priors on $\boldsymbol{\beta}_{y,z}$ are employed, standard Gibbs algorithms for this model can mix extremely poorly - in some cases millions or even billions of simulations may be required in order to obtain a desired level of numerical accuracy. To circumvent this problem, they outline a semi-analytic approach to the calculation of marginal posterior moments that even offers an improvement over the gold standard of i.i.d. sampling from the posterior.

Although generalized treatment-response models like (3.45) and (3.46) have substantial appeal, given the reality that instruments chosen in practice are not likely to be perfectly excludable, the resulting partial identification implies that the priors employed can have considerable influence. In fact, asymptotically, the conditional posterior for the non-identified structural parameters - like the causal effect $\beta_{y,x}$ - will reduce to the conditional prior (e.g., Moon and Schorfheide, 2012), clearly revealing that the influence of the prior will not vanish as we acquire more data. As a result, it is incumbent upon the researcher to think very carefully about the priors selected, and perhaps to offer a menu of posterior results under a variety of different priors (one of which could be the common dogmatic choice $\boldsymbol{\beta}_{y,z} = 0$), in order to comprehensively document posterior sensitivity to different and reasonably maintained prior beliefs.

3.5 Illustrative Application

To illustrate how Bayesian methods can be used to estimate a treatment-response model, and how different models can be compared within that setting, we consider a version of the application conducted by Li et al. (2012).

This application considers analysis of data provided by an online payday loan lender, in operation across 38 different U.S. states. The variables to be modeled consist of the amount borrowed (typically around \$300, with payment in full due within two weeks), as well as whether or not the borrower ultimately defaults on the loan. Default is a common phenomenon in this industry, and happens in approximately 30% of the loans in our data. We consider a triangular system in which the agent chooses an amount to borrow (perhaps to cover an emergency or unanticipated expense), and subsequently, the amount borrowed may have a causal effect on the decision to default.

States regulate a number of the terms of such loans, including the maximum interest rate that can be charged (*Rate*, and in practice, our lender generally follows the strategy of charging a rate equal to the state maximum rate), the maximum amount that can be borrowed ($StateMAX$), penalties that can be assessed on borrowers defaulting on a \$300 loan ($StatePenalty$) as well as how long the lender (or collection agency) has to collect on the debt ($StatueLimit$). We use this variation in state policy to aid in identifying the parameters of our treatment-response model. Other variables used consist of the term of the loan (denoted $Term$, typically 14 days, as most borrowers in the sample are paid bi-weekly), and the monthly rent paid by the individual (*Rent*).

The specification we consider is given below:

$$\begin{aligned} LogAmt_i^* &= \beta_{A,0c_i} + \beta_{A,1}Rate_i + \beta_{A,2}Term_i \qquad (3.47)\\ &\quad + \beta_{A,3}LogRent_i + \beta_{A,4}LogStateMAX_i + \epsilon_i,\\ LogAmt_i &= \min\{LogAmt_i^*, LogStateMAX_i\}, \qquad (3.48)\\ Default_i^* &= \beta_{D,0c_i} + \beta_{D,1}LogAmt_i + \beta_{D,2}Rate_i + \beta_{D,3}Term_i\\ &\quad + \beta_{D,4}LogRent_i + \beta_{D,5}LogStatePenalty_i \qquad (3.49)\\ &\quad + \beta_{D,6}StatuteLimit_i + v_i,\\ Default_i &= I(Default_i^* > 0), \qquad (3.50)\\ \begin{pmatrix} \epsilon_i \\ v_i \end{pmatrix} \Bigg| \cdot &\overset{ind}{\sim} \mathcal{N}\left[\begin{pmatrix} 0 \\ 0 \end{pmatrix}, \Sigma_{c_i} = \begin{pmatrix} \sigma_{\epsilon\epsilon c_i} & \sigma_{\epsilon v c_i} \\ \sigma_{v\epsilon c_i} & \sigma_{vvc_i} = 1 \end{pmatrix} \right], \qquad (3.51)\\ \Pr(c_i = g) &= \pi_g, \text{ for } g = 1, 2, \ldots, G \text{ and } \sum_{g=1}^{G} \pi_g = 1. \qquad (3.52) \end{aligned}$$

Equation (3.47) can be interpreted as the amount that the agent would like to borrow, which is modeled as a function of the state-level interest rate, loan term, rent paid by the individual as well as the state-established maximum amount that can be borrowed. The desired amount is then linked to the observed (log) loan size through the restriction in (3.48) - actual amounts cannot exceed the state-established maximum, but otherwise are assumed to equal the desired loan size. (Provided the borrower meets a residency condition, a monthly income requirement, furnishes checking account information, and is not found to be present in a database of known defaulters, the requested loan amount is essentially approved by the lender).

Equation (3.49) can be interpreted as the net desire to default on the loan. The default decision is presumed to depend on loan size, the interest rate charged, the monthly rent paid by the individual and state-level penalties and collection terms on defaulting borrowers.

As discussed in our section on non-normality, we consider a finite mixture model, but allow only the intercepts and covariance matrices to vary across mixture components in order to avoid parameter proliferation. The notation above allows for G different mixture components; in practice we estimate specifications with up to 4 such components. Finally, note that the specification above represents a special case of the model described in (3.37) and (3.38), and as such, the Gibbs algorithm for that model can be readily applied here as well.

To determine the number of components to be included in the mixture modeling, we utilize the bridge sampling method explained in the previous section. In Table (3.1), we calculate the log marginal likelihood $\ln[p(\tilde{\boldsymbol{y}}|G\text{-component})]$ of a G-component mixture model, for $G = 1, 2, 3, 4$, and the Bayes factor $\frac{p(G\text{-component}|\tilde{\boldsymbol{y}})}{p(H\text{-component}|\tilde{\boldsymbol{y}})}$ of a G-component model as opposed to an H-component model, for $H = 1, 2, 3, 4$. The Bayes factors strongly suggest that the 2-component model is most favored by our payday loan data.

Table 3.1: Log marginal likelihoods and Bayes factors of G-component mixture models, for $G = 1, 2, 3, 4$.

G	$\ln[p(\tilde{\boldsymbol{y}}|G)]$	$\frac{p(\tilde{\boldsymbol{y}}|D)}{p(H=1|\tilde{\boldsymbol{y}})}$	$\frac{p(G|\tilde{\boldsymbol{y}})}{p(H=2|\tilde{\boldsymbol{y}})}$	$\frac{p(G|\tilde{\boldsymbol{y}})}{p(H=3|\tilde{\boldsymbol{y}})}$	$\frac{p(G|\tilde{\boldsymbol{y}})}{p(H=4|\tilde{\boldsymbol{y}})}$
1	-3338.8	1	8.741×10^{-11}	1.36×10^{-8}	5.41×10^{-6}
2	-3315.6	1.144×10^{10}	1	155.82	61907
3	-3320.7	7.342×10^{7}	0.0064176	1	397.29
4	-3326.7	1.848×10^{5}	1.62×10^{-5}	0.002517	1

Given this strong preference for a two-component specification, we focus our attention in the remainder of this discussion on parameter estimates obtained from this model. Posterior summary statistics for all model parameters are provided in Table (3.2).

Specifically, for each parameter, we report the posterior mean $\mathrm{E}(\theta_j|\tilde{\boldsymbol{y}})$, the posterior standard deviation $\mathrm{Std}(\theta_j|\tilde{\boldsymbol{y}})$, the posterior probability that the coefficient is positive $\Pr(\theta_j > 0|\tilde{\boldsymbol{y}})$, the numerical standard error (NSE) associated with the posterior mean and the marginal effect (ME) of the covariate associated with the parameter. These ME's are calculated for every observation in the sample and then averaged across those observations. For variables measured in dollars, the ME's represent changes resulting from a \$100 increase in the level of the given variable.

Table 3.2: Posterior summary statistics from the two-component model.

Parameter (θ_j)	$E(\theta_j\|\tilde{\boldsymbol{y}})$	$Std(\theta_j\|\tilde{\boldsymbol{y}})$	$\Pr(\theta_j > 0\|\tilde{\boldsymbol{y}})$	NSE	ME
$\beta_{A,Rate}$	−0.0051	0.298	0.492	0.00161	−0.0727
$\beta_{A,Term}$	0.0021	0.00223	0.83	1.23e-005	0.543
$\beta_{A,LogRent}$	0.0483	0.0283	0.957	0.000157	1.39
$\beta_{A,LogStateMAX}$	0.0712	0.0297	0.992	0.000162	11.1
$\beta_{D,LogAmt}$	0.382	0.266	0.929	0.00706	0.0296
$\beta_{D,Rate}$	1.91	0.905	0.982	0.00758	0.0484
$\beta_{D,Term}$	0.0189	0.00768	0.994	8.01e-005	0.00453
$\beta_{D,LogRent}$	−0.384	0.101	5e-005	0.00102	−0.0104
$\beta_{D,LogStatePenalty}$	−0.194	0.0493	0	0.000875	−0.0451
$\beta_{D,StatuteLimit}$	−0.0093	0.0124	0.226	9.04e-005	−0.0022
π_1	0.363	0.043	1	0.00156	
$\beta_{A,01}$	4.42	0.271	1	0.00147	
$\beta_{D,01}$	0.996	1.36	0.773	0.0273	
$\sigma_{\epsilon\epsilon 1}$	0.103	0.0103	1	0.000244	
$\rho_{\epsilon v1} = \sigma_{\epsilon v1}/\sqrt{\sigma_{\epsilon\epsilon 1}}$	0.0312	0.0865	0.644	0.00136	
π_2	0.637	0.043	1	0.00156	
$\beta_{A,02}$	5.02	0.276	1	0.00164	
$\beta_{D,02}$	−0.62	1.57	0.352	0.036	
$\sigma_{\epsilon\epsilon 2}$	0.285	0.0207	1	0.000518	
$\rho_{\epsilon v2} = \sigma_{\epsilon v2}/\sqrt{\sigma_{\epsilon\epsilon 2}}$	0.0934	0.116	0.79	0.00254	

These marginal impacts reveal interesting, and mostly expected results. For example, an increase in monthly rent of \$100 raises the expected loan amount by a modest \$1.39, while a similarly-sized \$100 increase in state borrowing limit leads to a much larger expected loan amount increase of \$11.10. The interest rate charged on the loan is found to have little effect on loan size, as the marginal posterior distribution of $\beta_{A,Rate}$ is nearly centered at zero with $\Pr(\beta_{A,Rate} > 0|\tilde{\boldsymbol{y}}) \approx .49$. This result is arguably consistent with the borrowers in our data seeking loans for emergency purposes, as that aspect of the loan seems to have little effect on loan size.

We also find that large loans are more likely to end in default, as a \$100 increase in the amount borrowed raises the default probability by approximately 3 percentage points. Loans with higher interest rates are also associated with increased incidences of default, as a 0.1 increase in the interest rate increases the default probability by 4.84 percentage points. A \$100 increase in monthly rent paid reduces the default probability by 1.04 percentage points, which is sensible given that individuals paying more in rent should be more likely to have the resources to pay off the loan when due. Finally, borrowers are sensitive to penalties associated with default; increasing the state-level penalty on default on a \$300 loan by \$100 decreases the default probability by 4.51 percentage points.

To focus on the "causal" effect of the amount borrowed on default probability, in Figure (3.1), we plot this marginal effect over the range of amount borrowed. While the solid line represents the point estimates of the marginal

Figure 3.1: Marginal effects of loan amount on default probability.

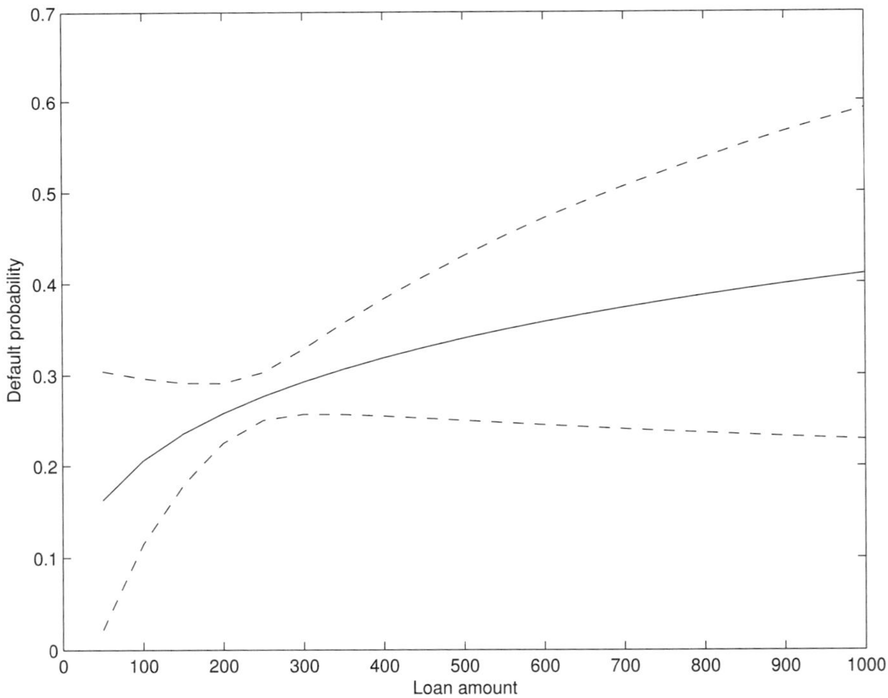

effects, the dashed lines correspond to the posterior means plus and minus two times the posterior standard deviations. Note that the standard deviations are easy to calculate given the post-convergence simulations from the Gibbs sampler: for every point in the loan amount space, we obtain a collection of marginal effect values, one for each post-convergence simulation. This collection represents a sample from the induced marginal effect posterior distribution, from which standard deviations are easily calculated.

3.6 Conclusion

This chapter has provided a brief overview of parametric Bayesian methods for treatment response modeling. We first considered the fully linear triangular model and discussed MCMC implementation in that context. We then extended this to allow for discrete possibilities among both the treatment and response, illustrating the value of data augmentation and the connection between posterior simulators for linear models and those for nonlinear models conditioned on the latent data. We reviewed standard procedures for extending models beyond the usual Gaussian case and discussed a procedure for model comparison and selection. Finally, we illustrated how Bayesian calculations are performed and interpreted, using data from an online payday loan lender.

Applications of these methods abound in the literature, including, for example, Li (1998), Li and Poirier (2003), Munkin and Trivedi (2003), Poirier and Tobias (2003), Li et al. (2004), Deb et al. (2006a,b), Hoogerheide et al. (2007), Chib and Jacobi (2007, 2008a,b), Conley et al. (2012), Kline and Tobias (2008), Hoogerheide et al. (2012) and Block et al. (2012), among many others. Furthermore, endogeneity and causality in Bayesian economic modeling is discussed in many recent Bayesian textbooks, including Koop (2003), Lancaster (2004), Geweke (2005), Rossi et al. (2005), Koop et al. (2007), and Greenberg (2013).

Bibliography

Albert, J. H. and S. Chib (1993). Bayesian analysis of binary and polychotomous response data. *Journal of the American Statistical Association 88*(422), 669–679.

Block, J. H., L. F. Hoogerheide, and A. R. Thurik (2012). Are education and entrepreneurial income endogenous? A Bayesian analysis. *Entrepreneurship Research Journal 2*(3).

Carlin, B. P. and N. G. Polson (1991). Inference for nonconjugate Bayesian models using the Gibbs sampler. *The Canadian Journal of Statistics 19*, 399–405.

Chan, J. and J. L. Tobias (2013). Priors and posterior computation in linear endogenous variables models with imperfect instruments. working paper.

Chan, J. C.-C. and I. Jeliazkov (2009). MCMC estimation of restricted covariance matrices. *Journal of Computational and Graphical Statistics 18*(2), 457–480.

Chib, S. (1995). Marginal likelihood from the Gibbs output. *Journal of the American Statistical Association 90*, 1313–1321.

Chib, S. (2007). Analysis of treatment response data without the joint distribution of potential outcomes. *Journal of Econometrics 140*, 401–412.

Chib, S. and B. Hamilton (1999). Bayesian analysis of cross-section and clustered data treatment models. *Journal of Econometrics 97*, 25–50.

Chib, S. and L. Jacobi (2007). Modeling and calculating the effect of treatment at baseline from panel outcomes. *Journal of Econometrics 140*, 781–801.

Chib, S. and L. Jacobi (2008a). Analysis of treatment response data from eligibility designs. *Journal of Econometrics 144*, 465–78.

Chib, S. and L. Jacobi (2008b). Causal effects from panel data in randomized experiments with partial compliance. *Advances in Econometrics, 23*, 183–215.

Chib, S. and I. Jeliazkov (2001). Marginal likelihood from the Metropolis-Hastings output. *Journal of the American Statistical Association 96*, 270–281.

Conley, T. G., C. B. Hansen, R. E. McCulloch, and P. E. Rossi (2008). A semi-parametric Bayesian approach to the instrumental variables problem. *Journal of Econometrics 144*(1), 276–305.

Conley, T. G., C. B. Hansen, and P. E. Rossi (2012). Plausibly exogenous. *Review of Economics and Statistics 94*(1), 260–272.

Deb, P., M. Munkin, and P. K. Trivedi (2006a). Bayesian analysis of the two-part model with endogeneity: Application to health care expenditure. *Journal of Applied Econometrics 21*, 1081–1099.

Deb, P., M. Munkin, and P. K. Trivedi (2006b). Private insurance, selection, and the health care use: A Bayesian analysis of a Roy-type model. *Journal of Business and Economic Statistics 24*, 403–415.

Frühwirth-Schnatter, S. (2001). Markov chain Monte Carlo estimation of classical and dynamic switching and mixture models. *Journal of the American Statistical Association 96*, 194–209.

Frühwirth-Schnatter, S. (2004). Estimating marginal likelihoods for mixture and Markov switching models using bridge sampling techniques. *Econometrics Journal 7*, 143–167.

Gelfand, A. E. and D. K. Dey (1994). Bayesian model choice: Asymptotics and exact calculations. *Journal of Royal Statistical Society B 56*, 501–514.

Geweke, J. (1993). Bayesian treatment of the independent student-*t* linear model. *Journal of Applied Econometrics 8*.

Geweke, J. (1999). Using simulation methods for Bayesian econometric models: Inference, development, and communication. *Econometric Reviews 18*, 1–73.

Geweke, J. (2005). *Contemporary Bayesian Econometrics and Statistics*. New York: John Wiley & Sons, Ltd.

Geweke, J. (2007). Interpretation and inference in mixture models: Simple MCMC works. *Computational Statistics and Data Analysis 51*, 3529–3550.

Greenberg, E. (2013). *Introduction to Bayesian Econometrics* (2nd ed.). Cambridge: Cambridge University Press.

Heckman, J. and B. Honoré (1990). The empirical content of the Roy model. *Econometrica 58*(5), 1121–1149.

Hoogerheide, L. F., J. H. Block, and A. R. Thurik (2012). Family background variables as instruments for education in income regressions: A Bayesian analysis. *Economics of Education Review 31*(5), 515–523.

Hoogerheide, L. F., J. F. Kaashoek, and H. K. van Dijk (2007). On the shape of posterior densities and credible sets in instrumental variable regression models with reduced rank: An application of flexible sampling methods using neural networks. *Journal of Econometrics 139*, 154–180.

Hungerman, D. and C. Buckles (2013). Season of birth and later outcomes: Old questions, new answers. *Review of Economics and Statistics*, forthcoming.

Jeliazkov, I., J. Graves, and M. Kutzbach (2008). Fitting and comparison of models for multivariate ordinal outcomes. *Advances in Econometrics, 23*, 115–156.

Kaufmann, S. and S. Frühwirth-Schnatter (2002). Bayesian analysis of switching ARCH models. *Journal of Time Series Analysis 23*, 425–458.

Kline, B. and J. L. Tobias (2008). The wages of BMI: Bayesian analysis of a skewed treatment response model with nonparametric endogeneity. *Journal of Applied Econometrics 23*, 767–793.

Koop, G. (2003). *Bayesian Econometrics*. New York: John Wiley & Sons, Ltd.

Koop, G. and D. J. Poirier (1997). Learning about the across-regime correlation in switching regression models. *Journal of Econometrics 78*, 217–227.

Koop, G., D. J. Poirier, and J. L. Tobias (2007). *Bayesian Econometric Methods*. Cambridge: Cambridge University Press.

Kraay, A. (2012). Instrumental variables regressions with uncertain exclusion restrictions: A Bayesian approach. *Journal of Applied Econometrics 27*, 108–128.

Lancaster, A. (2004). *An Introduction to Modern Bayesian Econometrics.* Blackwell Publishing.

Li, K. (1998). Bayesian inference in a simultaneous equations model with limited dependent variables. *Journal of Econometrics 85*, 387–400.

Li, K. and D. J. Poirier (2003). An econometric model of birth inputs and outputs for native americans. *Journal of Econometrics 113*(2), 337–361.

Li, M., K. Mumford, and J. L. Tobias (2012). A Bayesian analysis of payday loans and their regulation. *Journal of Econometrics 171*(2), 205–216.

Li, M., D. J. Poirier, and J. L. Tobias (2004). Do dropouts suffer from dropping out? Estimation and prediction of outcome gains in generalized selection models. *Journal of Applied Econometrics 19*, 203–25.

Lindley, D. V. and A. F. M. Smith (1972). Bayes estimates for the linear model. *Journal of the Royal Statistical Society, Series B 34*(1), 1–41.

McCulloch, R. E., N. G. Polson, and P. E. Rossi (2000). A Bayesian analysis of the multinomial probit model with fully identified parameters. *Journal of Econometrics 99*, 173–193.

Meng, X.-L. and W. H. Wong (1996). Simulating ratios of normalizing constants via a simple identity: A theoretical exploration. *Statistica Sinica 6*, 831–860.

Moon, H. and F. Schorfheide (2012). Bayesian and frequentist inference in partially identified models. *Econometrica 80*(2), 755–782.

Munkin, M. K. and P. K. Trivedi (2003). Bayesian analysis of self-selection model with multiple outcomes using simulation-based estimation: An application to the demand for healthcare. *Journal of Econometrics 114*, 197–220.

Neal, R. (2000). Markov chain sampling methods for Dirichlet process mixture models. *Journal of Computational and Graphical Statistics 9*(2), 249–265.

Nobile, A. (2000). Comment: Bayesian multinomial probit models with a normalization constraint. *Journal of Econometrics 99*, 335–345.

Poirier, D. J. and J. L. Tobias (2003). On the predictive distributions of outcome gains in the presence of an unidentified parameter. *Journal of Business and Economic Statistics 21*, 258–268.

Rossi, P. E., G. M. Allenby, and R. McCulloch (2005). *Bayesian Statistics and Marketing.* New York: John Wiley & Sons, Ltd.

Roy, A. D. (1951). Some thoughts on the distribution of earnings. *Oxford Economic Papers 3*(2), 135–146.

Tanner, M. and W. H. Wong (1987). The calculation of posterior distributions by data augmentation. *Journal of the American Statistical Association 82*, 528–49.

Vijverberg, W. (1993). Measuring the unidentified parameter of the extended Roy model of selectivity. *Journal of Econometrics 57*, 69–89.

CHAPTER 4

BAYESIAN ANALYSIS OF SAMPLE SELECTION MODELS

MARTIJN VAN HASSELT

RTI International, Research Triangle Park, USA.

In many empirical applications the outcome variable is not always observed, and the mechanism that determines whether an observation is "selected" into the sample or not also partially determines the outcome. The sample with observed outcomes is then no longer a representative sample from the population distribution of interest. Sample selection models are a useful tool in this context, because they adjust inferences to account for selection. This chapter provides an introduction to econometric selection models and Bayesian approaches to inference in these models.

4.1 Introduction

In many empirical situations the samples at hand are not randomly drawn from the population of interest. In this case we say that *sample selection* occurs. One of the most direct examples of sample selection involves sampling from a truncated distribution. A sample of low-income households clearly does not represent a random sample of all households in the U.S. Sample selection also occurs in stratified sampling designs, where some strata are under- or oversampled. For example, sampling for the National Survey on Drug Use and Health (NSDUH), a nationally representative survey of the non-institutionalized population in the U.S. ages 12 and older, is such that youth (ages 12-17 years old) and young adults (ages 18-25 years old) are oversampled Substance Abuse and Mental Health Services Administration (2013). This results in sample sizes for youth, young adults and adults (26 years and older) that are roughly equal, and hence the probability that a sampled individual falls into one of the three age categories is about $\frac{1}{3}$. However, for a randomly chosen individual from the U.S. population, the probability distribution over these age categories would be far from uniform.

A third situation in which sample selection arises is when the realizations of a (univariate or multivariate) process are only partially observed. That is, for some units in the sample the values of a subset of variables are not observed. In many cases a combination of observable individual features and unobserved heterogeneity determines whether a sampling unit is "fully observed" or not. It is this characterization of sample selection, which is closely related to missing data problems, that has become popular in the economics and econometrics literature since at least the 1970s. In an early, influential contribution, Gronau (1974) analyzed sample selection in the labor market, which arises when observed wages are used to make inferences about the distribution of wage offers. In this context an individual wage offer is observed only if the individual has been offered a job and decides to accept it. Observability of the wage offer therefore depends on decisions made by the individual and the firm, decisions that in turn can depend on observable and unobservable characteristics.

If the researcher's goal is to make inference about the underlying, partially latent process, then it is critical to understand the mechanism that generated the sample. Only using the observed sample and ignoring selection will typically lead to biased estimates and tests, a problem that is often broadly referred to as "sample selection bias." An early analysis of this type of bias in the context of a linear regression model was conducted in Heckman (1979), where the bias problem was framed in terms of model misspecification. This paper also showed how, under certain assumptions about the selection mechanism, the linear model should be modified to enable consistent estimation of the regression parameters by simple least squares. The resulting estimator has become known as Heckman's two-step estimator and is readily available in most statistical software packages (e.g., the command `heckman` in Stata and `proc qlim` in SAS).

Since Heckman's seminal contribution, sample selection models have been analyzed extensively in the frequentist literature, whereas Bayesian treatments of selection models have been much more limited in number. The Bayesian approaches that have been developed more recently, however, do span a wide range of both parametric and semiparametric models. This chapter aims to introduce the basic elements of selection models, as well as illustrate how Bayesian inference can be conducted in some of these models. The range of models discussed here is by no means exhaustive. A good (frequentist) survey of models that have been used in the literature is Vella (1998).

The remainder of this chapter is organized as follows. Section 4.2 introduces selection in univariate distributions. Section 4.3 discusses multivariate selection models. Some semiparametric model extensions are given in Section 4.4 and Section 4.5 concludes.

4.2 Univariate Selection Models

4.2.1 General Framework

In this section we present a general framework for characterizing selection in a univariate probability distribution. Throughout we assume that the random variable X is distributed according to a population density function $f(x|\theta)$ with support $\mathcal{X}$ and parameter vector θ. Let $S \in \{0, 1\}$ denote the selection indicator, equal to 1 if X is selected, and equal to 0 otherwise. We use $w(x)$ to denote the conditional probability of selection, $w(x) = \Pr(S = 1|X = x)$, so that

$$\begin{aligned}\Pr(S = 1) &= \int_{\mathcal{X}} w(x) f(x|\theta) dx \\ &= \mathrm{E}_\theta[w(X)].\end{aligned}$$

The density of the selected random variable is then a weighted version of the population density f:

$$\begin{aligned}g(x|\theta) &= f(x|S = 1, \theta) \\ &= \frac{w(x) f(x|\theta)}{\mathrm{E}_\theta[w(X)]}.\end{aligned} \tag{4.1}$$

Equation (4.1) shows that it is the dependence between S and X that gives rise to the weighted distribution. If the probability of selection was independent of X, for example if selection into the sample is determined by independent flips of a coin, then the distribution of the selected random variable is the same as the population distribution of interest.

EXAMPLE 4.1 Truncated Sampling

Suppose X is observed only if its realization falls in some set $A \subset \mathcal{X}$. The selection probability then takes a simple form:

$$w(x) = \begin{cases} 1 & \text{if } x \in A, \\ 0 & \text{if } x \notin A. \end{cases}$$

The density g in equation (4.1) is then

$$g(x|\theta) = \frac{f(x|\theta)}{\Pr(X \in A)}, \quad x \in A,$$

and 0 outside the set A. A special case of A is left-truncation at the point c, so that $A = \{x \in \mathbf{R}, x \geq c\}$. Letting F_θ be the cumulative distribution function (cdf) of X, the distribution of the selected random variable is

$$g(x|\theta) = \frac{f(x|\theta)}{1 - F_\theta(c)}, \quad x \geq c.$$

4.2.2 Likelihoods

Suppose we observe a selected sample $x_1, x_2, \ldots, x_n$ from $f(x|\theta)$. There are 2 underlying sampling mechanisms that are commonly considered: binomial and negative binomial (Lee and Berger, 2001). Under the binomial mechanism, there is some fixed $N(\geq n)$ and a random sample $X_1, X_2, \ldots, X_N$ from $f(x|\theta)$, such that each X_i is selected into the sample with probability $w(X_i)$. The likelihood function is then

$$\begin{aligned} f(x_1, \ldots, x_n|N, \theta) &= \binom{N}{n}(1 - \mathrm{E}_\theta[w(X)])^{N-n} \\ &\quad \prod_{i=1}^{n}(\mathrm{E}_\theta[w(X)]g(x_i|\theta)), \end{aligned} \tag{4.2}$$

where $g(x_i|\theta)$ is the weighted density in equation (4.1). On the other hand, suppose that n is considered fixed, N is unknown, and sampling from $f(x|\theta)$ continues until there are n selected observations. The likelihood of $x_1, \ldots, x_n$ and N is then the negative binomial likelihood

$$\begin{aligned} f(x_1, \ldots, x_n, N|n, \theta) &= \binom{N-1}{n-1}(1 - \mathrm{E}_\theta[w(X)])^{N-n} \\ &\quad \prod_{i=1}^{n}(\mathrm{E}_\theta[w(X)]g(x_i|\theta)). \end{aligned} \tag{4.3}$$

Integrating out N from (4.3) yields the likelihood of the selected sample alone:

$$f(x_1, \ldots, x_n|n, \theta) = \prod_{i=1}^{n} g(x_i|\theta), \tag{4.4}$$

which is simply the density of an i.i.d. sample from the weighted distribution in (4.1). In applications, if only an i.i.d. selected sample is available and N

is unknown (and not of interest), then the likelihood in (4.4) can be used. If, on the other hand, N and n are both observed, then for inference about θ the likelihood in either (4.2) or (4.3) can be used, because their kernels that depend on θ are identical.

4.2.3 Bayesian Inference

The basic components of a univariate selection model are the distribution f, its parameters θ, the weight function $w(x)$, or, in the case of truncation, the selection set A. Inference could therefore focus on any single component or a combination of components. For example, in the context of truncated sampling, Bayarri and DeGroot (1987) discuss situations where (i) θ is known but c is not, (ii) c is known but θ is not, and (iii) θ and c are both unknown but $\Pr(X \in A)$ is a known probability. Bayarri and Berger (1998) consider models with considerable uncertainty about $w(x)$ and derive bounds on posterior quantities (e.g., means, probabilities) as $w(x)$ varies over a given set of candidate weight functions. Lee and Berger (2001) construct a Gibbs sampling algorithm for θ and $w(x)$ in a model where $w(x)$ is unknown and has a Dirichlet process prior.

Below we present two examples in which the posterior is known or can easily be approximated via Gibbs sampling. Of course, in many applications the selection mechanism and $f(x|\theta)$ will be more complex, resulting in more elaborate Gibbs samplers or Metropolis-Hastings algorithms.

EXAMPLE 4.2 Truncated exponential sampling

Suppose X has an exponential distribution with parameter θ, $X \sim \exp(\theta)$, and that sample selection is based on left-truncation at some known point c. Then for $x > c$,

$$\begin{aligned} g(x|\theta) &= \frac{\theta \exp^{-\theta x}}{\Pr(X > c|\theta)} \\ &= \theta \exp^{-\theta(x-c)}, \end{aligned}$$

so that $X - c \sim \exp(\theta)$. Hence, the natural conjugate prior is a gamma distribution. Given a sample $x_1, \ldots, x_n$ and a $\mathcal{G}(a, b)$ prior for θ (the gamma distribution with mean $\frac{a}{b}$), it follows that

$$\theta | x_1, \ldots, x_n, a, b, c \sim \mathcal{G}\left(a + n, b + \sum_{i=1}^{n} [x_i - c]\right).$$

EXAMPLE 4.3 Binomial sampling with truncated selection

Consider a population distribution that is normal with mean μ, variance σ^2, and density $\phi(x|\mu,\sigma^2)$. Suppose that from an independent sample of size N a selected sample $x_1,\ldots,x_n$ of size $n < N$ is observed, where selection into the sample is determined by truncation from below at $c = 0$. Let $\Phi(x)$ be the cdf of the standard normal distribution. Assuming that N is known and the likelihood is binomial (see Section 4.2.2), it follows that

$$f(x_1,\ldots,x_n|N,\mu,\sigma^2) = \binom{N}{n}[1-\Phi(\mu/\sigma)]^{N-n}\prod_{i=1}^{n}\phi(x_i|\mu,\sigma^2).$$

While for a given prior for (μ,σ^2) the kernel of the posterior (and hence the posterior density) is known, calculating summary measures from the posterior is more complicated. In this case Gibbs sampling based on data augmentation is particularly useful (Tanner and Wong (1987)). The data augmentation step in the Gibbs sampler, or Markov chain Monte Carlo (MCMC) algorithm more generally, involves randomly drawing values for the missing data—the observations that were not selected—based on the likelihood and parameter values. Once the data are augmented in this way, and given convenient priors, the posterior distribution conditional on the observed and newly augmented data is often much more tractable.

To see how this approach works in this example, we use the following independent priors for μ and σ^2:

$$\mu|m_0,s_0^2 \sim \mathcal{N}(m_0,s_0^2), \qquad s_0^2 \sim \mathcal{IG}(a,b),$$

where $\mathcal{IG}$ is the inverse gamma distribution.[1] Let $X_{ns} = (X_{n+1},\ldots,X_N)$ be the vector of "non-selected" random variables. Conditional on (μ,σ^2), the elements of X_{ns} have a $\mathcal{N}(\mu,\sigma^2)$ distribution, truncated from above at 0. Also, conditional on the values $x = (x_1,\ldots,x_n,x_{n+1},\ldots,x_N)$, the conditional posteriors of μ and σ^2 are easy to derive. Below let $\mathrm{I}(A)$ be the indicator $(=0,1)$ of the event A. Given starting values (μ_0,σ_0^2) for the mean and variance, a Gibbs sampler with data augmentation involves the following steps.

The previous example provides a useful starting point for non-normal alternatives of the model. One such alternative is to use a Student-t population distribution with mean μ, scale parameter σ^2 and degrees of freedom ν. Following the steps outlined in Geweke (2005), a Gibbs sampler similar to the one in Example 4.3 can be constructed. Step 1 requires sampling from the truncated Student-t distribution, which can be done via acceptance-rejection sampling or the inverse cdf method. In the next 3 steps random draws from the conditional posteriors of μ, σ^2, and ν are generated. Geweke (2005) shows that an exponential prior distribution for ν leads to simple conditional posteriors of μ and σ^2. The conditional posterior of ν does not have a standard

[1] We use the convention that if $U \sim \mathcal{G}(a,b)$, then $1/U \sim \mathcal{IG}(a,b)$.

Algorithm 4.1: **Gibbs Sampler for Truncated Selection**

1. Given the current values of (μ, σ^2), generate values for the non-selected sample from the normal distribution truncated from above at 0, for $j = n + 1, \ldots, N$:

$$X_j | \mu, \sigma^2 \sim \frac{\phi(x|\mu, \sigma^2)}{1 - \Phi(\mu/\sigma)} \mathrm{I}(x \leq 0).$$

2. Given x and the sample average $\bar{x}$, update μ according to $\mu \sim \mathcal{N}(\bar{m}, s_p^2)$, where

$$\bar{m} = \left[\frac{1}{s_0^2} + \frac{n}{\sigma^2}\right]^{-1} \left[\frac{1}{s_0^2} m_0 + \frac{n}{\sigma^2} \bar{x}\right],$$

$$s_p^2 = \left[\frac{1}{s_0^2} + \frac{n}{\sigma^2}\right]^{-1}.$$

3. Given x and μ, update σ^2 according to

$$\sigma^2 | x, \mu \sim \mathcal{IG}\left(a + \frac{N}{2}, b + \frac{1}{2} \sum_{i=1}^{N} (x_i - \mu)^2\right).$$

---

form and a Metropolis step is needed to generate a random draw of ν. Hence, the resulting sampler is a hybrid Gibbs-Metropolis algorithm.

Selection with a known truncation point in either a normal or Student-t distribution is, of course, but one special case of a selection model in which both the distribution $f(x|\theta)$ and the weight function $w(x)$ are parametrically specified. As mentioned earlier, Lee and Berger (2001) have developed an MCMC algorithm for a model with a Dirichlet process prior for the unknown weight function. Lo (1993) analyzes a model with a known weight function but unknown distribution f.

4.3 Multivariate Selection Models

4.3.1 Motivation

Most econometric models in the social sciences, and economics in particular, are concerned with estimating the impact of one or more independent variables X on some outcome(s) of interest Y. Problems arise when Y and possibly X are not observed for some units in the sample and the mechanism that generates the missingness is not "entirely random." By this we mean that there is unobserved heterogeneity that affects both the outcome and the probability of observing it. The observed pairs (x, y) are then typically not a random sample from the population distribution. To conduct inference about population parameters it is then necessary to incorporate the selection mechanism into the statistical model.

Most empirical selection models are formulated as a system of two equations. One of these describes Y as a function of covariates X and unobserved heterogeneity. The second equation describes the mechanism, again as a function of X and unobserved heterogeneity, that determines whether $Y = y$ is observed or not. Dependence between the unobserved heterogeneity in both equations gives rise to sample selection.

In the following, we start with a discussion of Heckman's sample selection model (Heckman, 1979), arguably the best-known and most often used selection model in empirical applications. We also present a simple Gibbs sampler for the model parameters. We subsequently discuss a number of variations of this model, defined by different selection rules, in Sections 4.3.4 and 4.3.5. In Section 4.3.6 we briefly discuss incorporating endogeneity.

4.3.2 Heckman's Selection Model

Heckman's selection model can be written as follows:

$$\begin{aligned} S^* &= X_1'\beta_1 + U_1, \\ S &= \mathrm{I}(S^* > 0), \\ Y &= \begin{cases} X_2'\beta_2 + U_2 & \text{if } S = 1, \\ \text{missing} & \text{if } S = 0. \end{cases} \end{aligned} \tag{4.5}$$

Here X_1' and X_2' are row vectors of independent variables, β_1 and β_2 are column vectors of model coefficients, and U_1 and U_2 represent unobserved heterogeneity. Although not suggested by the notation, some or all of the independent variables contained in X_1 and X_2 could be the same. We could have, for example, $X_2 \subset X_1$ or $X_2 = X_1$. The binary variable S is the selection variable that indicates whether the latent variable S^* crosses the zero threshold or not. The outcome Y is observed and equal to $X_2'\beta_2 + U_2$ if and only if $S = 1$. In many treatments of this model, Y is set equal to 0 if there is no selection into the sample ($S = 0$). However, it should then be kept in mind that the zero is only a label and differs from a "corner solution" at 0 (e.g., Wooldridge, 2002). In equation (4.5) and throughout this chapter we instead prefer to note explicitly when Y is missing. Regarding observability of the other random variables in the model, X_1 and S are always observed, and X_2 is either always observed or only if Y is.

Let $X = (X_1', X_2')'$, $U = (U_1, U_2)'$. We assume that $\mathrm{E}(U|X) = 0$ and

$$\mathrm{V}(U|X) = \Sigma = \begin{bmatrix} 1 & \rho_{12}\sigma_2 \\ \rho_{12}\sigma_2 & \sigma_2^2 \end{bmatrix}.$$

The covariance matrix is parameterized in terms of the correlation (ρ_{12}) between U_1 and U_2 and the variance of U_2. Note that the variance of U_1 is normalized to 1, because this variance and β_1 are not separately identified. Given the model in (4.5) and the assumptions about U, it follows that the population regression function is $\mathrm{E}(Y|X) = X_2'\beta_2$. However, this is generally not the same as the regression function of the observed outcomes. This function can be explicitly calculated if we further assume that U has a bivariate normal distribution. Using the law of iterated expectations and the properties of the normal distribution, it can then be shown that (e.g., Wooldridge, 2002)

$$\begin{aligned}
\mathrm{E}(Y|X, S=1) &= \mathrm{E}(X_2'\beta_2 + U_2|X, U_1 > -X_1'\beta 1) \\
&= X_2'\beta_2 + \rho_{12}\sigma_2\mathrm{E}(U_1|U_1 > -X_1'\beta_1) \\
&= X_2'\beta_2 + \rho_{12}\sigma_2\lambda(X_1'\beta_1), \qquad (4.6)
\end{aligned}$$

where $\lambda(z) = \frac{\phi(z)}{\Phi(z)}$ is the inverse Mills ratio. Equation (4.6) shows in general regressing Y on X_2 does not yield unbiased estimates of β_2, because the inverse Mills ratio is incorrectly omitted. The one exception is the case where $\rho_{12} = 0$ and the regression function of the observed outcomes reduces to the population regression $\mathrm{E}(Y|X)$. In this case there is no sample selection bias associated with regressing the observed outcomes on X_2. In other words, it is the correlation between U_1 and U_2 that modifies the regression of the observed outcomes and creates the potential for sample selection bias. Equation (4.6) then forms the basis for the classical two-step estimator of β_2 that corrects for this bias: first, calculate the estimator $\hat{\beta}_1$ from a Probit regression and evaluate the Mills ratio $\lambda(X_1'\hat{\beta}_1)$; second, regress the observed Y on X_2 and $\lambda(X_1'\hat{\beta}_1)$.

4.3.3 Heckman's Selection Model: Bayesian Inference

As a first step toward inference, we calculate the likelihood function of the selection model under bivariate normality of U. As will become clear shortly, it is convenient to reparameterize the variance of U slightly:

$$\mathrm{V}(U|X) = \Sigma = \begin{bmatrix} 1 & \sigma_{12} \\ \sigma_{12} & \sigma_{12}^2 + \xi^2 \end{bmatrix}.$$

Here, σ_{12} is the covariance between U_1 and U_2 and ξ^2 is the conditional variance of U_2, given U_1. This parameterization, proposed in different contexts by Koop and Poirier (1997), Li (1998), and McCulloch et al. (2000), simplifies the construction of a Gibbs sampler greatly.

For simplicity of notation we collect all model parameters in the vector $\theta' = (\beta_1', \beta_2', \sigma_{12}, \xi^2)$. Suppose we observe a sample $\{x, s, y\} = \{x_{i1}', x_{i2}', s_i, y_i\}_{i=1}^n$.

The likelihood is then given by

$$\begin{aligned} f(y, s|x, \theta) &= \prod_{i:s_i=0} [1 - \Phi(x_{i1}'\beta_1)] \\ &\times \prod_{i:s_i=1} (\sigma_{12}^2 + \xi^2)^{-1/2} \phi\left(\frac{y_i - x_{i2}'\beta_2}{\sqrt{\sigma_{12}^2 + \xi^2}}\right) \\ &\times \prod_{i:s_i=1} \Phi\left(x_{i1}'\beta_1\sqrt{1 + \sigma_{12}^2/\xi^2} + \frac{\sigma_{12}(y_i - x_{i2}'\beta_2)}{\xi\sqrt{\sigma_{12}^2 + \xi^2}}\right). \end{aligned} \tag{4.7}$$

The first line in the expression of the likelihood is the contribution of the observations for which y_i is not observed. This is a product of probabilities that depend on x_{i1} and β_1. The second and third lines represent the contribution of observations for which y_i is observed. Their likelihood is factored as a normal density for y_i and the conditional probability of selection $\Pr(s_i = 1|y_i, x_i, \theta)$.

Clearly, the likelihood in (4.7) is non-standard, and there is no convenient choice of prior that allows us to sample directly from the posterior. However, van Hasselt (2011) shows that with a limited amount of data augmentation, there are convenient priors such that a Gibbs sampler is easy to construct. Data augmentation involves "completing" an incomplete data set by sampling values for missing observations from the posterior (Tanner and Wong, 1987). As such, the latent variables are treated in the same way as parameters, in the sense that they get updated in each iteration of the Gibbs sampler.

In the current model, we augment the data with the vector $s^* = (s_1^*, \ldots, s_n^*)$ of latent selection variables. The augmentation steps are similar to those in Albert and Chib (1993), except that when y_i is observed, sampling of s_i^* has to be done conditional on θ and y_i jointly. Thus, a value of s_i^* is drawn from the following conditional posterior:[2]

$$s_i^*|y_i, s_i, \theta \sim f(s_i^*|s_i, \theta)^{1-s_i} f(s_i^*|y_i, s_i, \theta)^{s_i}.$$

When $s_i = 0$ the outcome is not observed and we know that $s_i^* \leq 0$. The latent selection variable can then be sampled from $\mathcal{N}(x_{i1}'\beta_1, 1)\mathrm{I}(s_i^* \leq 0)$, a normal distribution with mean $x_{i1}'\beta_1$, variance 1 and truncated from above at 0. On the other hand, if $s_i = 1$, then y_i is observed and we know that $s_i^* > 0$. Conditional on (y_i, s_i, θ) the latent selection variable can be sampled from a conditional normal distribution, truncated from below at 0:

$$s_i^*|y_i, s_i = 1, \theta \sim \mathcal{N}\left(x_{i1}'\beta_1 + \frac{\sigma_{12}}{\sigma_{12}^2 + \xi^2}(y_i - x_{i2}'\beta_2), \frac{\xi^2}{\sigma_{12}^2 + \xi^2}\right) \mathrm{I}(s_i^* > 0).$$

For the parameters, suppose we can partition θ into B distinct "blocks" or subsets $\theta_1, \ldots, \theta_B$ and let θ_{-b} be the collection of all blocks except θ_b.

[2]For simplicity we omit conditioning on x_i.

Sampling θ_b for $b = 1, \ldots, B$ is then based on

$$\begin{aligned} f(\theta_b|y, s, s^*, \theta_{-b}) &= f(\theta_b|y, s^*, \theta_{-b}) \\ &\propto \left[\prod_{i:s_i=0} f(s_i^*|\theta) \prod_{i:s_i=1} f(y_i, s_i^*|\theta) \right] f(\theta), \end{aligned} \tag{4.8}$$

where $f(\theta)$ is the prior of θ and the equality of the first line follows because s is a function of s^* (hence, conditioning on s is redundant). Because $f(s_i^*|\theta)$ and $f(y_i, s_i^*|\theta)$ are simply univariate and bivariate normal likelihood functions, respectively, we use the following conditionally conjugate prior distributions: $\beta_j \sim \mathcal{N}(b_j, B_j), j = 1, 2$, $\xi \sim \mathcal{IG}(c_0, d_0)$ and $\sigma_{12}|\xi^2 \sim \mathcal{N}(0, \tau\xi^2)$, $\tau > 0$.[3]

Applying the expression in (4.8) to β_1, it follows from standard calculations that its conditional posterior is the $\mathcal{N}(\bar{b}_1, \bar{B}_1)$ distribution with

$$\begin{aligned} \bar{B}_1 &= \left(B_1^{-1} + \sum_{i:y_i=0} x_{i1}x_{i1}' + \left[\frac{\xi^2}{\xi^2 + \sigma_{12}^2} \right]^{-1} \sum_{i:y_i \neq 0} x_{i1}x_{i1}' \right)^{-1}, \\ \bar{b}_1 &= \bar{B}_1 \left(B_1^{-1}b_1 + \sum_{i:y_i=0} x_{i1}x_{i1}'\hat{\beta}_1(0) + \left[\frac{\xi^2}{\xi^2 + \sigma_{12}^2} \right]^{-1} \sum_{i:y_i \neq 0} x_{i1}x_{i1}'\hat{\beta}_1(1) \right), \end{aligned}$$

and $\hat{\beta}_1(0), \hat{\beta}_1(1)$ are two least squares estimators:

$$\begin{aligned} \hat{\beta}_1(0) &= \left[\sum_{i:s_i=0} x_{i1}x_{i1}' \right]^{-1} \sum_{i:s_i=0} x_{i1}s_i^*, \\ \hat{\beta}_1(1) &= \left[\sum_{i:s_i=1} x_{i1}x_{i1}' \right]^{-1} \sum_{i:s_i=1} x_{i1} \left(s_i^* - \frac{\sigma_{12}}{\xi^2 + \sigma_{12}^2}(y_i - x_{i2}'\beta_2) \right). \end{aligned}$$

To find the conditional posteriors of β_2, σ_{12}, and ξ^2, it is helpful to write the population model for Y as

$$Y = X_2'\beta_2 + \sigma_{12}(S^* - X_1'\beta_1) + \eta, \quad \eta|X \sim \mathcal{N}(0, \xi^2).$$

Thus, the remaining parameters are the coefficients and variance in a normal linear model. The conditional posteriors are then easy to find. Let

$$g_0 = \begin{bmatrix} b_2 \\ 0 \end{bmatrix}, \quad G_0 = \begin{bmatrix} B_2 & 0 \\ 0 & \tau\xi^2 \end{bmatrix},$$

[3]As noted in van Hasselt (2011), letting σ_{12} and ξ^2 be dependent in the prior is useful, because τ can be controled to allow for different (implicit) priors on the correlation ρ_{12}. In contrast, if σ_{12} and ξ^2 were *a priori* independent, the prior of ρ_{12} would have a concentration of mass near -1 and $+1$.

be the prior mean and variance of (β_2, σ_{12}). If W is the "data" matrix with rows $(x'_{i2}, s^*_i - x'_{i1}\beta_1)$ for all i with $s_i = 1$, then (β'_2, σ_{12}) has a conditional $\mathcal{N}(\bar{g}, \bar{G})$ posterior, where

$$\bar{G} = (G_0^{-1} + \xi^{-2}W'W)^{-1}, \quad \bar{g} = \bar{G}(G_0^{-1}g_0 + \xi^{-2}W'W\hat{g}),$$

and $\hat{g} = (W'W)^{-1}W'y^*$ is the least squares estimator.

Finally, applying equation (4.8) to ξ^2, we find that the conditional posterior of ξ^{-2} is the $\mathcal{IG}(\bar{c}, \bar{d})$ distribution, with

$$\begin{aligned} \bar{c} &= c_0 + \frac{1}{2}\left(1 + \sum_{i=1}^{n} s_i\right), \\ \bar{d} &= d_0 + \frac{\sigma_{12}^2}{2\tau} + \frac{1}{2}\sum_{i:s_i=1} (y_i - x'_{i2}\beta_2 - \sigma_{12}[s^*_i - x'_{i1}\beta_1])^2. \end{aligned}$$

We can now put all the pieces together and summarize the complete Gibbs sampler for this model.

Algorithm 4.2: **Gibbs Sampler for the Bivariate Normal Selection Model**

For a given starting value of θ, do the following:

1. For all i such $s_i = 0$, update s^*_i according to

$$\mathcal{N}(x'_{i1}\beta_1, 1)\mathrm{I}(s^*_i \leq 0).$$

For all i such $s_i > 0$, update s^*_i according to

$$\mathcal{N}\left(x'_{i1}\beta_1 + \frac{\sigma_{12}}{\sigma_{12}^2 + \xi^2}(y_i - x'_{i2}\beta_2), \frac{\xi^2}{\sigma_{12}^2 + \xi^2}\right)\mathrm{I}(s^*_i > 0) \text{ if } s_i = 1.$$

2. Update β_1 according to

$$\beta_1 | y, s^*, \beta_2, \sigma_{12}, \xi^2 \sim \mathcal{N}(\bar{b}_1, \bar{B}_1).$$

3. Update (β'_2, σ_{12}) according to

$$(\beta'_2, \sigma_{12}) | y, s^*, \beta_1, \xi^2 \mathcal{N}(\bar{g}, \bar{G}).$$

4. Update ξ^2 according to

$$\xi^2 | y, s^*, \beta_1, \beta_2, \sigma_{12} \sim \mathcal{IG}(\bar{c}, \bar{d}).$$

5. Return to #1 and repeat.

Sample selection models naturally lend themselves to data augmentation. We note however, that the "amount" of data augmentation is limited because we do not simulate all unobserved random variables. In particular, random draws for the missing values of y_i are not generated. This is what makes Algorithm 4.2 different from the one in Li (1998) and similar to that given in Chib et al. (2009). As argued in the latter paper, the advantage of not simulating the missing outcomes is twofold. First, Algoritm 2does not require x_{i2} to be observed when y_i is missing. Second, reducing the number of data augmentation steps reduces computational time and improves the mixing properties of the Markov chain.

4.3.4 A Model with Tobit Selection

In the model of Section 4.3.2 selection was based on a probit equation. In some cases we may observe more of S^* than just a binary indicator. This is the case when selection is based on the Tobit model. For example, consider a case where S^* is the number of hours worked, only observed if it is positive, and Y is wage income. We now modify the model in (4.5) as follows.

$$\begin{aligned} S^* &= X_1'\beta_1 + U_1, \\ S &= S^*\mathrm{I}(S^* > 0), \\ Y &= \begin{cases} X_2'\beta_2 + U_2 & \text{if } S > 0, \\ \text{missing} & \text{if } S = 0. \end{cases} \end{aligned} \tag{4.9}$$

Thus, when S^* is positive, $S = S^*$ and both S^* and Y are observed. On the other hand, when $S^* < 0$, S^* and Y are not observed and S is set equal to 0. Assume as before that $U|X \sim \mathcal{N}(0, \Sigma)$, where it will now be convenient to parameterize Σ as

$$\Sigma = \begin{bmatrix} \sigma_1^2 & h\sigma_1^2 \\ h\sigma_1^2 & \xi^2 + h^2\sigma_1^2 \end{bmatrix}.$$

Here, ξ^2 is again the conditional variance of U_2 given U_1 and $h = \sigma_{12}/\sigma_1^2$ is the coefficient in the regression of U_2 on U_1. Note also that σ_1^2 is no longer normalized at 1: the variance is identified and can be estimated because at least a subset of values s_i^* are actally observed in the data. Given a sample $\{x_i, s_i, y_i\}_{i=1}^n$ from the model in (4.9), the likelihood function can be written as

$$\begin{aligned} f(y, s|x, \theta) &= \prod_{i:s_i=0} [1 - \Phi(x_{i1}'\beta_1/\sigma_1)] \\ &\quad \times \prod_{i:s_i>0} \phi\left(\frac{s_i - x_{i1}'\beta_1}{\sigma_1}\right) \\ &\quad \times \prod_{i:s_i>0} \phi\left(\frac{y_i - x_{i2}'\beta_2 - h[s_i - x_{i1}'\beta_1]}{\xi}\right). \end{aligned} \tag{4.10}$$

We can construct a Gibbs sampler for this model, similar to Algorithm 4.2, by augmenting the data with values for s_i^* when $s_i = 0$:

$$s_i^*|y_i, s_i = 0, \theta \sim \mathcal{N}(x_{i1}'\beta_1, \sigma_1^2).$$

We partition θ into blocks θ_b and update each block according to

$$f(\theta_b|y, s, s^*, \theta_{-b}) \propto \left[\prod_{i:s_i=0} f(s_i^*|\theta) \prod_{i:s_i>0} f(y_i, s_i^*|\theta)\right] f(\theta). \tag{4.11}$$

Similar to before, convenient priors are $\beta_j \sim \mathcal{N}(\bar{b}_j, \bar{B}_j)$ for $j = 1, 2$, $\xi^2 \sim \mathcal{IG}(c_0, d_0)$. In addition, we take $h|\xi^2 \sim \mathcal{N}(0, \tau\xi^2)$ and $\sigma_1^2 \sim \mathcal{IG}(a_0, a_1)$. To find the conditional posterior of β_1, we use (4.11) and decompose $f(y_i, s_i^*|\theta) = f(y_i|\theta)f(s_i^*|y_i, \theta)$ because the marginal likelihood of y_i does not depend on β_1. It can then be shown that the conditional posterior is a $\mathcal{N}(\bar{b}_T, \bar{B}_T)$ distribution (the subscript T is used to indicate "Tobit" and differentiate with the posteriors in Algorithm 4.2).

$$\begin{aligned}
\bar{B}_T &= \left(B_1^{-1} + \sigma_1^{-2} \sum_{i:s_i=0} x_{i1}x_{i1}' + \left[\frac{\xi^2}{\xi^2/\sigma_1^2 + h^2}\right]^{-1} \sum_{i:s_i>0} x_{i1}x_{i1}'\right)^{-1}, \\
\bar{b}_T &= \bar{B}_T\left(B_1^{-1}b_1 + \sigma_1^{-2} \sum_{i:s_i=0} x_{i1}x_{i1}'\tilde{\beta}_1(0) + \left[\frac{\xi^2}{\xi^2/\sigma_1^2 + h^2}\right]^{-1} \sum_{i:s_i>0} x_{i1}x_{i1}'\tilde{\beta}_1(1)\right),
\end{aligned}$$

and $\tilde{\beta}_1(0), \tilde{\beta}_1(1)$ are two least squares estimators:

$$\begin{aligned}
\tilde{\beta}_1(0) &= \left[\sum_{i:s_i=0} x_{i1}x_{i1}'\right]^{-1} \sum_{i:s_i=0} x_{i1}s_i^*, \\
\tilde{\beta}_1(1) &= \left[\sum_{i:s_i>0} x_{i1}x_{i1}'\right]^{-1} \sum_{i:s_i>0} x_{i1}\left(s_i^* - \frac{h}{\xi^2/\sigma_1^2 + h^2\sigma_1^2}(y_i - x_{i2}'\beta_2)\right).
\end{aligned}$$

Using (4.11), we see that (β_2', h) can be updated jointly from their conditional posterior $\mathcal{N}(\bar{g}, \bar{G})$, where $\bar{g}$ and $\bar{G}$ are the same as in Algorithm 4.2. The conditional posterior of ξ^2 can be found from (4.11) as $\mathcal{IG}(\tilde{c}, \tilde{d})$, where

$$\begin{aligned}
\tilde{c} &= c_0 + \frac{1}{2} + \frac{1}{2}\sum_{i=1}^{n} \mathrm{I}(s_i > 0), \\
\tilde{d} &= d_0 + \frac{h^2}{2\tau} + \frac{1}{2} \sum_{i:s_i>0} (y_i - x_{i2}'\beta_2 - h[s_i^* - x_{i1}'\beta_1])^2.
\end{aligned}$$

It remains to find the conditional posterior of σ_1^2. Because $f(y_i, s_i^*|\theta) = f(s_i^*|\theta)f(y_i|s_i^*, \theta)$ and the second term in the product is a function of (h, ξ^2) but not σ_1^2, the conditional posterior from (4.11) becomes

$$f(\sigma_1^2|y, s^*, \beta_1, \beta_2, h, \xi^2) \propto f(\sigma_1^2) \prod_{i=1}^{n} f(s_i^*|\beta_1, \sigma_1^2).$$

It then follows this posterior is the $\mathcal{IG}(\bar{a}_0, \bar{a}_1)$ distribution with

$$\bar{a}_0 = a_0 + \frac{n}{2}, \quad \bar{a}_1 = a_1 + \frac{1}{2}\sum_{i=1}^{n}(s_i^* - x_{i1}'\beta_1)^2.$$

To summarize, Gibbs sampling in the bivariate normal model with Tobit selection involves the following steps.

Algorithm 4.3: **Gibbs Sampler for the Bivariate Normal Model with Tobit Selection**
For a given starting value of θ, do the following:
1. For i such that $s_i = 0$, update s_i^* according to

$$s_i^*|s_i = 0, \beta_1, \sigma_1^2 \sim \mathcal{N}(x_{i1}'\beta_1, \sigma_1^2)\mathrm{I}(s_i^* \leq 0).$$

2. Update β_1 according to

$$\beta_1|y, s^*, \beta_2, h, \sigma_1^2, \xi^2 \sim \mathcal{N}(\bar{b}_T, \bar{B}_T)$$

3. Update (β_2', h) according to

$$(\beta_2', h)|y, s^*, \beta_1, \sigma_1^2, \xi^2 \sim \mathcal{N}(\bar{g}, \bar{G}).$$

4. Update ξ^2 according to

$$\xi^2|y, s^*, \beta_1, \beta_2, \sigma_{12} \sim \mathcal{IG}(\tilde{c}, \tilde{d}).$$

5. Update σ_1^2 according to

$$\sigma_1^2|y, s^*, \beta_1, \beta_2, h, \xi^2 \sim \mathcal{IG}(\bar{a}_0, \bar{a}_1).$$

6. Return to #1 and repeat.

4.3.5 Alternative Specifications

The models in Sections 4.3.2 and 4.3.4 consisted of a single outcome equation for Y and a single equation for S describing the selection mechanism. There are of course many empirical cases in which more elaborate models with more than one outcome and/or selection mechanism are needed. In a labor market analysis referred to earlier, the researcher may have information on labor force participation status, in addition to hours worked and (the logarithm of) wages earned. A joint model for these three variables can be the following:

$$\begin{aligned}
P^* &= X_1'\beta_1 + U_1, \\
P &= \mathrm{I}(P^* > 0), \\
S^* &= X_2'\beta_2 + U_2, \\
S &= P \cdot \mathrm{I}(S^* > 0)S^*, \\
Y &= \begin{cases} X_3'\beta_3 + U_3 & \text{if } (P = 1, S > 0), \\ \text{missing} & \text{otherwise.} \end{cases}
\end{aligned}$$

In this model, an individual participates in the labor force if the latent variable P^* is positive, so that $P = 1$. Hours worked S^* are observed if and only if there is labor force participation ($P = 1$) and $S^* > 0$. Finally, wages Y are observed if and only if $P = 1$ and $S^* > 0$. This model would apply if we have a sample with 3 types of individuals: (1) those not participating in the labor force ($P = S = Y = 0$); (2) those who are part of the labor force but are not employed ($P = 1$, $S = Y = 0$); and (3), those who are in the labor force and are employed ($P = 1$, $S = S^* > 0$, $Y \neq 0$).

Letting $U = (U_1, U_2, U_3)'$ and $X' = (X_1', X_2', X_3')$, a parametric normal model is based on a trivariate normal distribution $U|X \sim \mathcal{N}(0, \Sigma)$. Data augmentation in this model involves generating sample values P^* from an appropriate truncated normal distribution. Using convenient normal and inverse gamma priors and the general arguments used in Sections 4.3.2 and 4.3.4, a Gibbs sampler based on data augmentation can be constructed.

A recent example of a Bayesian analysis in a more elaborate selection model is Li (2011). The empirical question that the paper investigates is to determine the relation between residential density, the number of cars and trucks owned by a household, and the number of miles driven annually in cars and trucks. The model for cars alone can be written as follows

$$\begin{aligned} S^* &= X_1'\beta_1 + U_1, \\ S &= \sum_{t=0}^{T} t \cdot \mathrm{I}(\alpha_t < S^* \leq \alpha_{t+1}), \\ Y &= \begin{cases} X_2'\beta_2 + U_2 & \text{if } S > 0, \\ \text{missing} & \text{if } S = 0. \end{cases} \end{aligned}$$

The equations for S^* and S represent an ordered multinomial choice model: a household has $t(= 0, 1, \ldots, T)$ cars if the latent variable S^* falls between the two cutoff points (α_t, α_{t+1}). The variable Y is the number of car miles driven annually, which is observed only if the household has at least 1 car.

The model described above can be expanded with 3 additional equations for the number of trucks owned and truck miles driven. These would involve the random variables (U_3, U_4) that represent unobserved heterogeneity. Li (2011) presents a hybrid Gibbs-Metropolis algorithm based on the assumption that the $U = (U_1, U_2, U_3, U_4)|X \sim \mathcal{N}(0, \Sigma)$. In his algorithm, updating the β's and Σ and the data augmentation scheme are fairly similar to the updating in Algorithms 4.2 and 4.3. Updating the cutoff parameters α_t presents an additional complication because due to the ordering constraints $(\alpha_t < \alpha_{t+1})$ a simple Gibbs iteration is not feasible. To remove the ordering constraints, Li (2011) reparameterizes the α's and samples the new parameters in a Metropolis-Hastings step.

4.3.6 Endogeneity

The models discussed to this point all assumed that the covariates in the outcome and selection equations were not correlated with the unobserved heterogeneity U.[4] The assumption that covariates are exogeneous, while convenient, is of course not always appropriate. Surprisingly, despite the large numbers of papers that have examined either sample selection or endogeneity in isolation, few papers have developed models that incorporate endogenous covariates into a sample selection framework. An exception is Li (1998) who analyzed Heckman's selection model with endogeneity. His model is obtained by adding the selection indicator S as a covariate in the outcome equation in (4.5). Since S depends on U_1, and U_1 and U_2 are dependent, S is endogenous in the equation for Y. More recently, Chib et al. (2009) also incorporate endogeneity into a sample selection framework. In their model there are possibly multiple endogenous variables in the outcome equation, and like the outcome, some of these variables may be only partially observed.

In general, endogeneous variables can appear in the selection equation, the outcome equation, or both, and can be binary, (ordered) categorical, or continuous. As such, the selection models with endogeneity will be as varied as the MCMC algorithms needed to conduct inference. Overcoming the endogeneity problem requires finding a variable, the "instrument," that partly explains the endogenous variable, is not correlated with (or independent of) the unobserved heterogeneity, and that is properly excluded from the equation (selection and/or outcome) in which the endogeneous variable appears (Wooldridge, 2002). The selection model is then expanded by adding an equation for each endogenous variable, a "first-stage" equation, that relates the endogenous variable to the instrument and the remaining exogenous variables.[5]

4.4 Semiparametric Models

The models discussed so far are based on the assumptions of a linear functional form and multivariate normal likelihood functions. These assumptions are widely used in applied Bayesian work and very convenient in terms of constructing Gibbs samplers and Metropolis-Hastings algorithms from scratch. Such models are, of course, subject to the criticism that in many cases there are no good reasons to impose such strong assumptions. Moreover, severe model misspecification may lead to posteriors that yield misleading inferences. Consequently, much research, Bayesian and frequentist, has been devoted to relaxing these assumptions. In the frequentist literature on sample selection models, most papers from the early 1990s onwards have used a semipara-

[4] In fact, to construct the Gibbs sampler the stronger assumption of independence was used.

[5] See equations (2.1)-(2.4) in Chib et al. (2009). Note that in the model in Li (1998), the selection equation also serves as the first-stage equation for the endogenous variable.

metric modeling framework in which distributions and functional forms were no longer fully specified. Examples include Ahn and Powell (1993), Cosslett (1991), and Lee (1994).

The first Bayesian paper on semiparametric sample selection modeling was Chib et al. (2009). In this paper the linearity assumption is relaxed but the conditional normality of the errors is retained. Specifically, the model that Chib et al. (2009) analyze in detail can be written as

$$\begin{aligned}
S^* &= X_0'\beta_0 + g_0(Z_0) + U_0, \\
S &= S^* \mathrm{I}(S^* > 0), \\
Y_1 &= X_1'\beta_1 + g_1(Z_1) + U_1, \\
Y_2 &= X_2'\beta_2 + g_2(Z_2) + U_2, \\
Y_3 &= X_3'\beta_3 + g_3(Y_1, Y_2, Z_3) + U_3.
\end{aligned}$$

In this model X_j,Z_j, $j = 0, 1, 2, 3$ are exogenous variables. As before S^* is the latent selection variable and S is a Tobit selection rule: if $S > 0$, (Y_1, Y_2, Y_3) is observed; if $S = 0$, (Y_1, Y_2, Y_3) is missing. The random vector $U = (U_0, U_1, U_2, U_3)$ has a multivariate normal distribution, conditional on the exogenous variables. Consequently, Y_1 and Y_2 are endogenous variables in the equation for Y_3. The functions g_j are unknown, so that the model is only partially linear and the effect, for example, of Y_1 on Y_3 is nonparametric. If g_j has a q_j-dimensional argument u, this function is modeled additively as

$$g_j(u) = \sum_{k=1}^{q_j} g_{jk}(u_k).$$

Chib et al. (2009) model the unique function evaluations of $g_j(u)$—for different values of u—as a realization of a second-order Markov process prior. In this prior, the arguments u are first ordered. The functional value at the first (smallest) ordered value of the argument is normalized at 0, whereas the value at the second ordered value has a normal prior. The functional values at the third and following ordered values of the argument then have a prior that depends on the previous two functional values. Intuitively, this prior induces a certain amount of smoothness in the function g_j. Chib et al. (2009) derive and present the complete MCMC algorithm based on this prior for the nonparametric g_j functions. They also present evidence through a series of simulation experiments that their approach is very effective in estimating a range of highly non-linear covariate effects.

A different approach to relaxing strong parametric assumptions in Bayesian modeling is to use a class of likelihood functions that is flexible enough to approximate a wide range of distributions. Finite mixture distributions form such a class (Geweke, 2005; Frühwirth-Schnatter, 2006). A mixture distribution is often constructed as a convex combination of, say, k distributions that belong to the same parametric family but have different parameters. Mixtures of normal distributions are analytically tractable and have therefore been a popular tool for approximating continuous distributions. As illustrated in Geweke (2005), even mixtures with only 3 or 4 components can already be quite good in approximating severely skewed and heavy-tailed distributions.

Finite mixture modeling can be incorporated into a sample selection framework through the distribution of U. Suppose, for example, that

$$U|X \sim p\mathcal{N}(\mu_1, \Sigma_1) + (1-p)\mathcal{N}(\mu_2, \Sigma_2),$$

where $p \in (0, 1)$ is a mixture probability, μ_1, μ_2 are 2-dimensional vectors of means and Σ_2, Σ_2 are distinct covariance matrices. The vector U has this 2-component mixture distribution if with probability p it has a $\mathcal{N}(\mu_1, \Sigma_1)$ distribution, and with probability $(1-p)$ a $\mathcal{N}(\mu_2, \Sigma_2)$ distribution. In addition to the larger number of mean and variance parameters, the component probability p is now also part of the parameter space. A convenient prior for p is the Beta distribution or, when there are more than 2 mixture components, the Dirichlet distribution (Frühwirth-Schnatter, 2006).

The paper by van Hasselt (2011) applies mixture modeling to the sample selection model discussed in Section 4.3.2. In particular, it uses a Dirichlet process (DP) prior (Ferguson, 1973; Escobar, 1994; Escobar and West, 1995) which results in a likelihood function that is a Dirichlet mixture of normals distribution. Unlike the fixed, finite mixture models discussed above, Dirichlet mixtures have a random number of mixture components. Hence, the number of mixture components is an additional unknown parameter that can be estimated. The Dirichlet process prior used in van Hasselt (2011) can be briefly described as follows. Let ω be the vector containing the means, variances and covariance of U_1 and U_2. The random variables for observation i satisfy

$$\begin{aligned} U_i|X_i, \omega_i &\sim \mathcal{N}(\mu_i, \Sigma_i), \\ \omega_i &\sim F, \\ F|\alpha &\sim \mathcal{DP}(\alpha, F_0). \end{aligned}$$

The first line indicates that the error for observation i are normally distributed with means and variances that can differ between observations. In other words, the (conditional) distributions of U_i and U_j are not necessarily the same when $i \neq j$. The second line states that the parameters ω_i are drawn from a distribution F. Finally, F itself is not a fixed distribution but rather has a DP prior (α, F_0) with "base measure" F_0 and scale α. A sample from the DP prior can be generated sequentially. First, sample ω_1 from F_0. Then, for $i = 2, \ldots, n$:

$$\omega_i|(\omega_1, \ldots, \omega_{i-1}) \begin{cases} = \omega_j & \text{with probability } \frac{1}{\alpha+i-1}, j = 1, \ldots, i-1, \\ \sim F_0 & \text{with probability } \frac{\alpha}{\alpha+i-1}. \end{cases} \tag{4.12}$$

Thus, ω_i is either equal to an existing parameter value or a new random draw from F_0. The number of unique values among $\omega_1, \ldots, \omega_n$ is then the number of mixture components and therefore also random.

van Hasselt (2011) derives the Gibbs sampler for Heckman's selection model based on (4.12). As is evident from Algorithm 2 in that paper, updating β_1, β_2, and s_i^* is similar to the steps of Algorithm 4.2 in this chapter, except that the various conditional distributions involved are now conditional on a much larger number of mean and variance parameters ω_i. The posterior distribution of $\omega_1, \ldots, \omega_n$ is of a similar nature as equation (4.12). It can be shown that with some probability ω_i is equal to a "pre-existing" value, and with some probability ω_i is a new, unique draw from the posterior associated with prior base distribution F_0 (e.g., MacEachern, 1998).

4.5 Conclusion

This chapter has introduced the basic elements of selection models, both in a univariate and multivariate setting. We have provided Gibbs sampling algorithms for two selection models that have been extensively applied in frequentist empirical work. Of course, the range of selection models that is relevant for empirical work is much wider than could be discussed here. In particular, in many cases extensions to accommodate endogenous variables are important. However, the methods presented in this chapter, together with some recent semiparametric Bayesian analyses of selection models, can serve as a starting point for developing new MCMC algorithms for alternative selection models.

Bibliography

Ahn, H. and J. L. Powell (1993). Semiparametric estimation of censored selection models with a nonparametric selection mechanism. *Journal of Econometrics 58*, 3–29.

Albert, J. H. and S. Chib (1993, June). Bayesian analysis of binary and polychotomous response data. *Journal of the American Statistical Association 88*(422), 669–679.

Bayarri, M. and J. Berger (1998). Robust Bayesian analysis of selection models. *Annals of Statistics 26*, 645–659.

Bayarri, M. and M. DeGroot (1987). Bayesian analysis of selection models. *The Statistician 36*, 137–146.

Chib, S., E. Greenberg, and I. Jeliazkov (2009). Estimation of semiparametric models in the presence of endogeneity and sample selection. *Journal of Computational and Graphical Statistics 18*(2), 321–348.

Cosslett, S. R. (1991). Semiparametric estimation of a regression model with sample selectivity. In W. A. Barnett, J. Powell, and G. Tauchen (Eds.), *Nonparametric and Semiparametric Methods in Econometrics and Statistics*, pp. 175–197. UK: Cambridge.

Escobar, M. D. (1994, March). Estimating normal means with a Dirichlet process prior. *Journal of the American Statistical Association 89*(425), 268–277.

Escobar, M. D. and M. West (1995, June). Bayesian density estimation and inference using mixtures. *Journal of the American Statistical Association 90*(430), 577–588.

Ferguson, T. S. (1973, March). A Bayesian analysis of some nonparametric problems. *Annals of Statistics 1*(2), 209–230.

Frühwirth-Schnatter, S. (2006). *Finite Mixture and Markov Switching Models.* Springer.

Geweke, J. (2005). *Contemporary Bayesian Econometrics and Statistics.* New York: Wiley.

Gronau, R. (1974). Wage comparisons – a selectivity bias. *The Journal of Political Economy 82*(6), 1119–1143.

Heckman, J. J. (1979, January). Sample selection as a specification error. *Econometrica 47*(1), 153–162.

Koop, G. and D. J. Poirier (1997). Learning about the across-regime correlation in switching regression models. *Journal of Econometrics 78*, 217–227.

Lee, J. and J. Berger (2001). Semiparametric Bayesian analysis of selection models. *Journal of the American Statistical Association 96*, 1269–1276.

Lee, L. F. (1994). Semiparametric two-stage estimation of sample selection models subject to tobit-type selection rules. *Journal of Econometrics 61*, 305–344.

Li, K. (1998). Bayesian inference in a simultaneous equation model with limited dependent variables. *Journal of Econometrics 85*, 387–400.

Li, P. (2011). Estimation of sample selection models with two selection mechanisms. *Computational Statistics and Data Analysis 55*, 1099–1108.

Lo, A. (1993). A Bayesian method for weighted sampling. *The Annals of Statistics 21*(4), 2138–2148.

MacEachern, S. N. (1998). Computational methods for mixture of Dirichlet process models. In D. Dey, P. Müller, and D. Sinha (Eds.), *Practical Nonparametric and Semiparametric Bayesian Statistics*, pp. 23–44. New York: Springer.

McCulloch, R. E., N. G. Polson, and P. E. Rossi (2000). A Bayesian analysis of the multinomial probit model with fully identified parameters. *Journal of Econometrics 99*, 173–193.

Substance Abuse and Mental Health Services Administration (2013). Results from the 2012 national survey on drug use and health: Summary of national findings. NSDUH Series H-46 HHS Publication No. (SMA) 13-4795, Substance Abuse and Mental Health Services Administration, Rockville, MD.

Tanner, M. A. and W. H. Wong (1987). The calculation of posterior distributions by data augmentation. *Journal of the American Statistical Association 82*, 528–550.

van Hasselt, M. (2011). Bayesian inference in a sample selection model. *Journal of Econometrics 165*(2), 221–232.

Vella, F. (1998). Estimating models with sample selection bias: a survey. *Journal of Human Resources 33*, 127–169.

Wooldridge, J. M. (2002). *Econometric Analysis of Cross Section and Panel Data.* MIT Press.

CHAPTER 5

MODERN BAYESIAN FACTOR ANALYSIS

HEDIBERT FREITAS LOPES

George Washington University, D.C., USA

5.1 Introduction

The origin of factor analysis can be traced back to Spearman's (1904) seminal paper on general intelligence. At the time, psychologists were trying to define intelligence by a single, all-encompassing unobservable entity, the g factor. Spearman studied the influence of the g factor on examinees' test scores on several domains: pitch, light, weight, classics, French, English, and mathematics. At the end of the day, the g factor would provide a mechanism to detect common correlations among such imperfect measurements. More precisely, Spearman's (1904) one-factor model based on p test domains (measurements) and n examinees (individuals) can be written as

$$y_{ij} = \mu_j + \beta_j g_i + \varepsilon_{ij}, \tag{5.1}$$

for $i = 1, \ldots, n$, $j = 1, \ldots, p$, where y_{ij} is the score of examinee i on test domain j, μ_j is the mean of test domain j, g_i is the value of the intelligence factor for person i, β_j is the loading of test domain j onto the intelligence factor g, and ϵ_{ij} is the random error term for person i and test domain j.

Spearman spends part of his 90-page paper defending his one-factor model of general intelligence, arguably his main, seminal contribution to the fields of psychometrics as well as statistical modeling at this time. Nonetheless, Bartholomew's (1995) review paper starts by stating that "*Spearman invented factor analysis but his almost exclusive concern with the notion of a general factor prevented him from realizing its full potential.*"

For subsequent developments, mainly in psychology studies, see Burt (1940), Holzinger and Harman (1941), and Thomson (1953), amongst others, where the factors had an *a priori* known structure. The extension to multiple factors as well as its formal statistical framework came many decades later. Multiple factor analysis was first introduced by Thurstone (1935,1947) and Lawley (1940,1953), along with estimation via the centroid method and maximum likelihood, respectively. Hotelling (1955) proposed a more robust method of estimation, the method of principal components, while Anderson and Rubin (1956) formalized and elevated factor analysis to the realm of statistically and probabilistically sound modeling schemes.

Computationally speaking, maximum likelihood estimation became practical in the late 1960s through the work of Jöreskog (1967,1969). A further improvement was achieved in the early 1980s through the EM algorithms of Rubin and Thayer (1983,1983); see also Bentler and Tanaka (1983). In the late 1980s, Anderson and Amemiya (1988) and Amemiya and Anderson (1990) studied the asymptotic behavior of estimation and hypothesis testing for a large class of factor analysis under general conditions, while Akaike (1987) proposed an information criterion to selecting the proper number of common factors.

Table 5.1 illustrates the fast increase in the use of factor analysis in a few areas of science and industry over the last century. This chapter discusses the Bayesian contribution to this table, most of which appearing after the mid-1990s due to both the increasing access to faster, smaller, cheaper and sharable computers and processors and the revival and/or introduction of efficient Monte Carlo schemes for posterior inference in highly structured stochastic systems (Gamerman and Lopes, 2006), of which factor analysis is an increasingly popular member.

Factor Analysis at 100: To celebrate the centennial of Spearman (1904), The L. L. Thurstone Psychometric Laboratory, University of North Carolina at Chapel Hill, hosted in May 2004 a workshop entitled *Factor Analysis at 100: Historical Developments and Future Directions*, with David Bartholomew, Michael Browne, Robert Jennrich, Karl Jöreskog, Yasuo Amemiya and many other distinguished invited speakers. A thought-provoking historical account is the paper *Three Faces of Factor Analysis* presented at the meeting by

Table 5.1: Distribution of papers on factor analysis on the Internet. Reproduced from Kaplunovsky (2004), who presents more areas of science and industry and more disaggregated time intervals.

	1904-1985 80 years	1986-1995 10 years	1996-2004 10 years
Psychology	179	378	723
Psychiatry	29	100	236
Medicine	62	131	225
Spectroscopy	38	90	198
Chemistry	26	89	165
Biology	35	43	88
Physiology	46	77	80
Geriatry	14	19	56
Economics	26	11	46
Chromatography	11	38	39

Bartholomew. Another historical account was neatly organized by a group of students and faculty at the Thurstone Psychometric Lab entitled *Factor Analysis Genealogy* and *Factor Analysis Timeline*. See the conference webpage for these and additional information at `http://www.fa100.info`. The papers presented at the meeting appeared in Cudeck and MacCallum (2007).

The rest of the chapter is organized as follows. Section 5.2 reviews and explores the basic normal linear factor model, identification issues, estimation of parameters and factors as well as estimation/selection of the number of common factors. Section 5.3 deals with factor variance models, commonly present in financial econometrics contexts. Factor models for spatial and space-time problems are introduced in Section 5.4. Section 5.5 presents recent developments in factor analysis, such as prior and posterior robustness, mixture of factor analyzers, factor analysis in time series and macroeconometric modeling and sparse factor structures. Some of the recent contributions to the literature on non-Bayesian (large dimensional and/or dynamic) factor analysis are presented in Section 5.6. The chapter concludes with Section 5.7.

5.2 Normal Linear Factor Analysis

Let $y_i = (y_{i1}, \ldots, y_{ip})'$, for $i = 1, \ldots, n$, be a p-dimensional vector with the measurements on p related variables (Spearman's tests, attributes, macroeconomic or financial time series, census sectors, monitoring stations, to name a few examples). The basic normal linear factor model assumes that ys are independent and identically distributed $N(0, \Omega)$, i.e., a zero-mean multivariate normal with a $p \times p$ non-singular variance matrix Ω. Loosely speaking, a factor model usually rewrites Ω, which depends of $q = p(p+1)/2$ variance and covariance components, as a function of d parameters, where d is potentially many orders of magnitude smaller than q. Figure 5.1 illustrate a 3-factor model for a $p = 9$-dimensional vector of attributes.

More specifically, for any positive integer $k \leq p$, a standard normal linear k-factor model for y_i is written as

$$y_i|f_i, \beta, \Sigma \sim N(\beta f_i, \Sigma), \quad (5.2)$$

$$f_i|H \sim N(0, H), \quad (5.3)$$

where f_i is the k-dimensional vector of common factors, β is the $p \times k$ matrix of factor loadings, $\Sigma = \text{diag}(\sigma_1^2, \ldots, \sigma_p^2)$ is the covariance of the specific factors and $H = \text{diag}(h_1, \ldots, h_k)$ is the covariance matrix of the common factors. The uniquenesses σ_i^2s, also known as idiosyncratic or specific variances, measure the residual variability in each of the data variables once that contributed by the factors is accounted for. Conditionally on the common factors, f_k, the measurements in y_i are independent. In other words, the common factors explain all the dependence structure among the p variables and, based on equations (5.2) and (5.3), the unconditional, constrained covariance matrix of y_i becomes

$$\Omega = \beta H \beta' + \Sigma. \quad (5.4)$$

The matrix Ω depends on $d = (p + 1)(k + 1) - 1$, the number of elements of β, H and Σ, a number usually considerably smaller than $q = p(p + 1)/2$, the number of elements of the unconstrained Ω.

5.2.1 Parsimony

In practical problems, especially with larger values of p, the number of factors k will often be small relative to p, so most of the variance-covariance structure is explained by a small number of common factors. For example, when $p = 100$ and $k = 10$, a configuration commonly found in modern applications of factor analysis, $q = 5050$ and $d = 1110$, or roughly $q = 5d$. Similarly, when $p = 1000$ and $k = 50$, if follows that $q = 500500$ and $d = 51050$, or roughly $q = 10d$. Such a drastic reduction in the number of unrestricted parameters renders factor modeling inherently a parsimony-inducing technique.

5.2.2 Identifiability

The k-factor model of equations (5.2) and (5.3) is invariant under transformations of the form $\beta^* = \beta P'$ and $f_i^* = P f_i$, where P is any orthogonal $k \times k$ matrix. Likelihood identification is achieved by assuming that β is a block lower triangular matrix of rank k and with ones in the main diagonal. It is worth noting that the ones in the main diagonal of β can be replaced

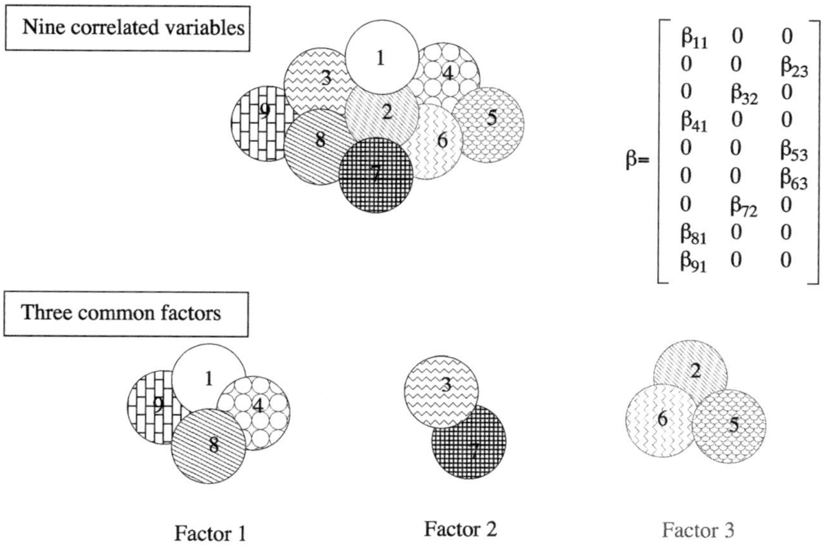

Figure 5.1: Illustration of a 3-factor model structure for $p = 9$ correlated variables. Notice that each variable is explained by only one of the $k = 3$ common factors. Therefore, $\Omega = \beta H \beta' + \Sigma$ is block-diagonal, with independent blocks defined by variables (y_1, y_4, y_8, y_9), (y_3, y_7) and (y_2, y_5, y_6).

by strictly positive values, as long as H, the common factors' covariance matrix, is replaced by I_k. This form is used, for example, in Geweke and Zhou (1996) and Lopes and West (2004), as well as the majority of the Bayesian factor modelers after them, and provides both identification and, often, useful interpretation of the factor model.

With Ω of full rank p, the resulting factor form of Ω has $d = p(k+1) - k(k-1)/2$ parameters, compared with the total $q = p(p+1)/2$ in an unconstrained (or $p = k$) model, leading to the constraint that

$$p(p+1)/2 - p(k+1) + k(k-1)/2 \geq 0, \tag{5.5}$$

which provides an upper bound on k. Even for small p, the bound will often not matter as relevant k values will not be so large. In realistic problems, with p in double digits or more, the resulting bound will rarely matter. Finally, note that the number of factors can be increased beyond such bounds by reducing the rank of Σ. See also Ihara and Kano (1995), who establish and discuss conditions for full, marginal and conditional model identification are discussed.

5.2.3 Invariance

Three important results for the class of standard normal linear factor models are shown below: *(i)* non-diagonal H is irrelevant, *(ii)* they are invariant to the order of the variables, and *(iii)* reduced rank factor loadings and non-identifiability.

Result (i): Let us assume that H is non-diagonal and that $H = LL'$. Then, equations (5.2) could be rewritten as $y \sim N(\beta^* f^*, \Sigma^*)$, where $\beta^* = \beta L$, $f^* = L^{-1} f$ and $\Sigma^* = \Sigma$ unchanged. The rotated common factors f^* are still zero-mean multivariate normal with covariance matrix $I_q = L^{-1} H (L^{-1})'$. The loading matrix β^* can be transformed into a lower block triangular matrix by letting $\beta^* = ((\beta_1^*)', (\beta_2^*)')'$, β_1^* a $(k \times k)$ matrix, β_2^* a $(p - k \times k)$ matrix. So,

$$\tilde{\beta} = \beta^* (Z')^{-1} = \begin{pmatrix} Z \\ \beta_2^* (\beta_1^*)' (Z')^{-1} \end{pmatrix} \quad \text{and} \quad \tilde{f} = Z' (\beta_1^*)^{-1} f^*,$$

where $\beta_1^* (\beta_1^*)' = ZZ'$, for lower triangular Z. More compactly, $\tilde{\beta} = \beta P_1$ and $\tilde{f} = P_2 f$ for $P_1 = L(Z')^{-1}$ and $P_2 = Z'(\beta_1^*)^{-1} L^{-1}$. Now, it is straightforward to see that

$$\begin{aligned} P_2 H P_2' &= Z'(\beta_1^*)^{-1} L^{-1} H (L^{-1})' ((\beta_1^*)^{-1})' Z \\ &= Z'(\beta_1^*)^{-1} ((\beta_1^*)^{-1})' Z = Z' (\beta_1^* (\beta_1^*)')^{-1} Z \\ &= Z'(ZZ')^{-1} Z = Z'(Z')^{-1} Z^{-1} Z = I_k, \end{aligned}$$

so equations (5.2) and (5.3) can be recovered by letting β be $\tilde{\beta}$ with column i normalized by $\tilde{\beta}_{ii}$, for $i = 1, \ldots, q$, and $H = \mathrm{diag}(\tilde{\beta}_{11}^2, \ldots, \tilde{\beta}_{kk}^2)$.

Result (ii): It follows directly and similarly since reordering the components of y is equivalent to pre-multiplying it by a permutation matrix Q, accordingly pre-multiplying the factor loadings matrix by Q and post-multiplying it by L, i.e., $\beta^* = Q\beta L$, while $f^* = L^{-1} f$ and $\Sigma^* = Q\Sigma Q'$ are still diagonal. β^* is transformed into a lower block triangular matrix and f rotated to produce orthogonal factors by repeating the steps of result *(i)*.

Result (iii): It is important to emphasize that the above two results apply only when k is the right number of factors, i.e., when the rank of the loading matrix β is k. More precisely, when $H = I_k$, Geweke and Singleton (1980) showed that, if β has rank $r < k$, then there exists a matrix Q such that $\beta Q = 0$ and $Q'Q = I_k$ and, for any $(p \times k)$ orthogonal matrix M, it follows that

$$\beta\beta' + \Sigma = (\beta + MQ')'(\beta + MQ') + (\Sigma - MM'). \qquad (5.6)$$

This translation invariance of Ω under the factor model implies a lack of identification and, in application, induces symmetries and potential multimodalities in the resulting likelihood functions. This issue relates directly to the question of uncertainty of the number of factors. Figure 5.3 of Example 5.1 illustrates this phenomenon. See also Lopes and West (2004) for further discussion and additional empirical evidence. In Frühwirth-Schnatter and Lopes (2009), we take advantage of the result *(iii)* and propose a new stochastic search scheme that navigates the joint space of the sparse factor loadings and the number of common factors.

5.2.4 Posterior Inference

Early MCMC-based posterior inference in standard factor analysis appears in, among others, Geweke and Zhou (1996) and Lopes and West (2004). They basically propose and implement a standard Gibbs sampler that cycles through the full conditional distributions of $p(\beta|f, \Sigma, y)$, $p(\Sigma|f, \beta, y)$ and $p(f|\beta, \Sigma, y)$, which following well known distributions when conditionally conjugate priors are used. Here we assume that $H = I_k$ and that β is block lower triangular, for identification, with diagonal strictly positive diagonal elements.

Prior Specification. The unconstrained components of β are independent and identically distributed (i.i.d.) $N(m_0, C_0)$, the diagonal components β_{ii}s are i.i.d. truncated normal from below at zero, denoted here by $N_{(0,\infty)}(m_0, C_0)$, and the idiosyncratic variances, σ_i^2 are i.i.d. $IG(\nu/2, \nu s^2/2)$. The hyperparameters m_0, C_0, ν and s^2 are known. It is worth mentioning that the above prior specification has been extended and modified many times over to accommodate specific characteristics of the scientific modeling under consideration. Lopes et al. (2008), for example, utilize spatial proximity to parameterize the columns of β when modeling pollutants across Eastern US monitoring stations. See the following sections for more structured factor loading matrices, as well as common and specific factor dynamics.

Full Conditional Distributions. The factor model can be seen as a standard multivariate regression model when deriving the full conditionals $(f|\beta, \sigma, y)$ and $(\beta|\Sigma, f, y)$ (Box and Tiao, 1973; Press, 1982; Broemeling, 1985; Zellner, 1971; Gamerman and Lopes, 2006). More precisely, let us start by denoting $y = (y_{(1)}, \ldots, y_{(p)})$, $\beta = (\beta_1', \ldots, \beta_p')'$ and $F_i = (f_{(1)}, \ldots, f_{(i)})$, with $\beta_i = (\tilde{\beta}_i', 0_{k-i})'$ and $F_i = (f_{(1)}, \ldots, f_{(i)})$, for $i = 1, \ldots, k-1$. Therefore, for $i = 1, \ldots, n$, $(f_i|\beta, \Sigma, y) \sim N((I_k + \beta'\Sigma^{-1}\beta)^{-1}\beta'\Sigma^{-1}y_i, (I_k + \beta'\Sigma^{-1}\beta)^{-1})$. For $i = 1, \ldots, k$, $(\tilde{\beta}_i'|\Sigma, f, y) \sim N(m_i, C_i)\delta_{\beta_{ii}>0}$, where $m_i = C_i(C_0^{-1}\mu_0 1_i + \sigma_i^{-2}F_i' y_{(i)})$, $C_i^{-1} = C_0^{-1}I_i + \sigma_i^{-2}F_i'F_i$, 1_i is an i-dimensional vector of ones and δ_x is the indicator function at x. For $i = k+1, \ldots, p$, $(\beta_i|\Sigma, f, y) \sim N(m_i, C_i)$, where $m_i = C_i\left(C_0^{-1}\mu_0 1_k + \sigma_i^{-2}F_k' y_{(i)}\right)$ and $C_i^{-1} = C_0^{-1}I_k + \sigma_i^{-2}F_k'F_k$. Finally, $(\sigma_i^2|\beta, f, y) \sim IG((\nu+n)/2, (\nu s^2+d_i)/2)$, where $d_i = (y_{(i)} - F_k\beta_i')'(y_{(i)} - F_k\beta_i')$.

Chib et al. (2006), in the context of multivariate stochastic volatility modeling, propose an MCMC scheme that jointly samples the factor loadings matrix β and the common factor scores f_is. They basically integrate the factors out of the full conditional distribution of β and use a Metropolis-Hastings step to sample β. Then they sample f given β, as described in our scheme above. See also Ghosh and Dunson (2009) for a slight modification of our scheme, and Song and Lee (2001) for a factor analysis combining continuous and polytomous data. Paisley and Carin (ICML) propose a nonparametric extension to the factor analysis problem using a beta process prior.

EXAMPLE 5.1

Lopes and West (2004) fit one-, two- and three-factor models to monthly international exchange rate data spanning from January of 1975 to December 1986 (a total of $n = 144$ observations). The time series are the exchange rates in British pounds of the US dollar (US), Canadian dollar (CAN), Japanese yen (JAP), French franc (FRA), Italian lira (ITA) and (West) German (Deutsch)mark (GER). Here we replicate the setup used in Lopes and West (2004), with prior hyperparameters $m_0 = 0$, $C_0 = 1$, $\nu = 2.2$ and $s^2 = 0.0455$. For the Gibbs sampler, we burn-in the algorithms for 10,000 iterations, and then save equally spaced samples of 5,000 draws from a longer run of 100,000. It takes only about one minute to run a two-factor model on my MacBook Pro with a 2.6GHz Intel Core i7 processor, 8 GB 1600 MHz DDR3 Memory running a Mac OS X Lion 10.7.5. The posterior means of Σ and β in a two-factor model are $E(\Sigma|y) = \text{diag}(0.05, 0.13, 0.62, 0.04, 0.25, 0.26)$ and

$$E(\beta'|y) = \begin{pmatrix} \mathbf{1.00} & \mathbf{0.96} & 0.46 & 0.39 & 0.42 & 0.41 \\ 0.00 & 0.05 & 0.43 & \mathbf{0.92} & \mathbf{0.78} & \mathbf{0.78} \end{pmatrix}.$$

Apart from the third row of $E(\beta|y)$, corresponding to the Japanese yen, which is equally explained by both common factors, one can argue that the first common factor groups North American currencies (US and Canadian dollars) and the second common factor groups European currencies. In factor analysis, it is fairly standard to summarize the importance of a common factor by its percentage contribution to the variability of a given attribute. Figure 5.2 presents the variance decomposition for the example and enhances the above statements regarding the interpretation of the two latent factors. Finally, Figure 5.3 illustrates the multimodality implied by overfitting the number of factors, regardless of using more or less informative prior specifications.

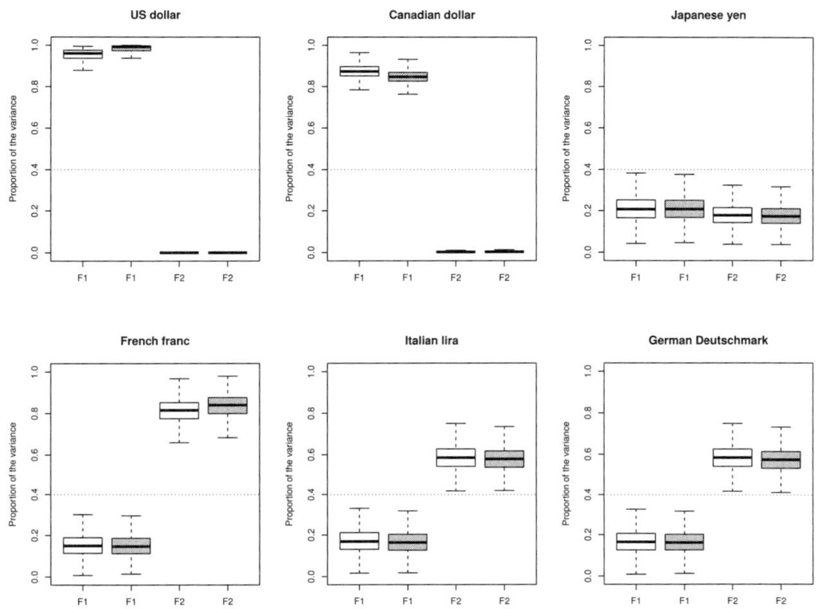

Figure 5.2: Example 1 – variance decomposition. The proportion of the variance of currency i, for $i = 1, \ldots, 6$, attributed to common factor j, for $j = 1, 2$ is given by $\nu_{ij} = \beta_{ij}^2/(\beta_{i1}^2 + \beta_{i2}^2 + \sigma_i^2)$. The boxplots summarize the posterior distributions of the ν_{ij}s. Informative prior (white boxplots). Noninformative prior (gray boxplots).

5.2.5 Number of Factors

Bayesian and non-Bayesian references that tackle the estimation/choice of the number of common factors are, amongst many others, Lawley and Maxwell (1963), Jöreskog (1967), Martin and McDonald (1975), Bartholomew (1981), Press (1982) (Chapter 10), Lee (1981), Akaike (1987), Bartholomew (1987), Press and Shigemasu (1989), Press and Shigemasu (1994). The book by Bartholomew (1987) is an excellent overview of the field right before MCMC tools became available.

Polasek (1997) uses Chib's methods (Chib, 1995) to approximate marginal likelihoods, and therefore posterior model probabilities, by running MCMC methods for factor models with a different number of common factors. Lopes and West (2004) introduced, developed and explored MCMC methods for factor models that treat the number of factors as unknown. Building on prior work on MCMC methods for a given number of factors, we introduced a reversible jump Markov chain Monte Carlo (RJMCMC, see Green, 1995)

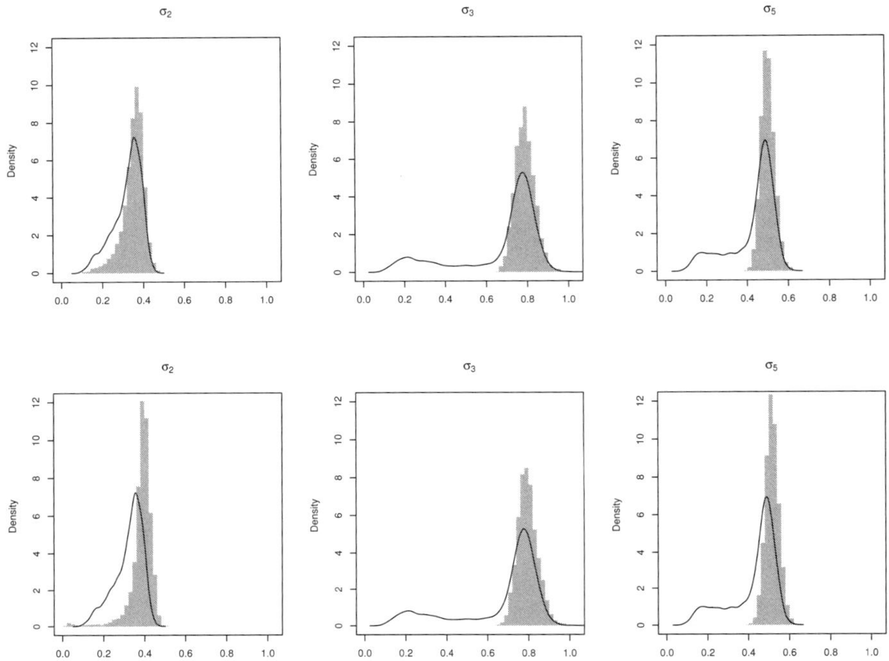

Figure 5.3: Example 1 – posterior inference. Posterior distributions (MCMC approximation) of $p(\sigma_i|y)$, $i = 2, \ldots, 5$ of Example 5.1, when estimating a two-factor model (histograms) and a (overfitted) three-factor model (solid lines). *Informative prior (top row):* The prior hyperparameters are $(m_0, C_0, \nu_0, s^2) = (0, 1, 2.2, 0.0455)$. Prior mode and median of σ_i are 0.154 and 0.252, respectively. *Noninformative prior (bottom row):* The prior hyperparameters are $(m_0, C_0, \nu_0, s^2) = (0, \infty, 0.001, 1)$. Prior mode and median of σ_i^2 are 0.032 and $1.4e^{149}$, respectively.

algorithm for moving between models with different numbers of factors, which avoids the computation of marginal likelihoods by treating the number of factors as a parameter.

Lopes and West (2004) served as the motivation to a handful of significant contributions to the discussion regarding the estimation of the number of common factors. West (2003) and Carvalho et al. (2008), for example, introduced high-dimensional factor analysis for modeling gene expression data with highly sparse factor loadings. The sparse representation induces a probability distribution over the number of factors. They model the factor scores via a Dirichlet. See also Lee and Song (2002) and Chow et al. (2011). Lopes et al.

(2008) extend Lopes and West's (2004) RJMCMC to the context of normal dynamic factor analysis, particularly when the loadings are spatially modeled. Bhattacharya and Dunson (2011), also focusing on high-dimensional problems, propose a gamma process shrinkage prior on the factor loadings matrix that allows the introduction of infinitely many factors of decaying importance.

We proposed, in Frühwirth-Schnatter and Lopes (2009), a new stochastic search strategy for estimating the number of factors and show that our approach encompasses and improves upon Carvalho et al. (2008), Bhattacharya and Dunson (2011) and most of the existing strategies. In Conti et al. (2011), we use and extend the new the strategy to examine the effect of early-life conditions and education on health but incorporating discrete attributes as well as limited dependent variable structures, commonly present when dealing with endogeneity in microeconometric studies.

5.3 Factor Stochastic Volatility

The basic, and certainly the most used and cited, *stochastic volatility* (SV) model can be described by the following non-linear dynamic model (West and Harrison, 1997):

$$y_t = \exp\{x_t/2\}\varepsilon_t, \tag{5.7}$$

$$x_t = \beta_0 + \beta_1 x_{t-1} + \tau\eta_t, \tag{5.8}$$

where y_t are log-returns and log-variances $x_t = \log v_t$, ε_t and η_t i.i.d. standard normal errors. We take $\mu = 0$ for simplicity, $\beta_0 = \kappa\gamma$, $\beta_1 = 1 - \kappa$. The initial log-volatility state $x_0 \sim N(m_0, C_0)$, for known prior moments m_0 and C_0. An alternative specification assumes that $(x_0|\beta_0, \beta_1, \tau^2) \sim N(\beta_0/(1 - \beta_1), \tau^2/(1-\beta_1^2))$ with $|\beta_1| < 1$; see Kalayloglu and Ghosh (2009) for Bayesian unit root tests regarding β. The centering parameterization moves β_0 to the observation equation and centers log-variances. This parameterization only marginally affects posterior inference in most cases while creating an unnecessary computational burden. We will then keep the simpler, less restrictive, more general specification with m_0 and C_0.

The SV model is completed with a conjugate prior distribution for $\theta = (\beta, \tau^2)$, i.e., $p(\theta) = p(\beta|\tau^2)p(\tau^2)$, where $(\beta|\tau^2) \sim N(b_0, \tau^2 B_0)$ and $\tau^2 \sim IG(c_0, d_0)$, for known hyperparameters b_0, B_0, c_0 and d_0. An alternative specification where β and τ^2 are independent *a priori* can be easily implemented with negligible additional computational cost.

Given a set of observed asset returns $y^n = (y_1, \ldots, y_n)$ and equations (5.7) and (5.8), the posterior distribution of the hidden volatility states and parameters (x^n, θ) is given by Bayes rule

$$p(x^n, \theta|y^n) \propto p(\theta)\prod_{t=1}^{n} p(y_t|x_t, \theta)p(x_t|x_{t-1}, \theta), \tag{5.9}$$

which is analytically intractable because of the nonlinearity of equation (5.7). Jacquier, Polson and Rossi (JPR, 1994) performed fully Bayesian inference through an MCMC scheme, with Kim et al. (1998) improving upon JPR's scheme based on the well known forward filtering-backward sampling (FFBS) algorithm of Carter and Kohn (1994) and Frühwirth-Schnatter (1994). Also, Jensen (2004) develops semiparametric inference for long-memory SV processes, while So et al. (1998) and Carvalho and Lopes (2004) accommodate Markov jumps in the log-volatilities.

5.3.1 Factor Stochastic Volatility

The literature on multivariate stochastic volatility models is now abundant, with Harvey et al. (1994), Pitt and Shephard (1999), Aguilar and West (2000), Lopes and Migon (2002), Chib et al. (2006) and Lopes and Carvalho (2007) representing only a few. Roughly speaking, they model the levels (or first differences) of a set of (financial) time-series by a standard normal factor model (Lopes and West, 2004) in which both the commonfactor variances and the specific (or idiosyncratic) time-series variances are modeled as univariate or multivariate (of low dimension) SV processes. The main practical and computational advantage of the factor stochastic volatility (FSV) model is its parsimony, where all the variances and covariances of a vector of time-series are modeled by a low dimensional stochastic volatility structure dictated by common factors. It is fairly common to find that, for large vectors of time series, the number of common factors is usually one or two orders of magnitude smaller, which speeds up computation and estimation considerably.

In this more general context, the simple normal linear factor model of equations (5.2) and (5.3) are replaced by

$$(y_t|f_t, \beta_t, \Sigma_t) \sim N(\beta_t f_t; \Sigma_t), \tag{5.10}$$

$$(f_t|H_t) \sim N(0; H_t), \tag{5.11}$$

where $H_t = \text{diag}(h_{1t}, \ldots, h_{kt})$ contain the variances of the factors and $\Sigma_t = \text{diag}(\sigma_{1t}^2, \ldots, \sigma_{pt}^2)$. The main, nontrivial departure from the standard normal linear factor model lies in the time-varying structure of β_t, Σ_t and H_t. Log idiosyncratic variances, $\eta_{it} = \log \sigma_{it}^2$, are modeled by first-order autoregressions, AR(1):

$$(\eta_{it}|\eta_{i,t-1}, \tilde{\alpha}_i, \rho_i, \tau_i^2) \sim N(\tilde{\alpha}_i + \rho_i \eta_{i,t-1}, \tau_i^2), \tag{5.12}$$

for $i = 1, \ldots, p$. This is one of the simplest but certainly the most used specification in the literature (Jacquier et al., 1994). Similarly, the k-dimensional vector of factors' log variances, $\lambda_t = (\lambda_{1t}, \ldots, \lambda_{kt})'$, where $\lambda_{it} = \log h_{it}^2$, is modeled by a first-order vector autoregression, VAR(1), as

$$(\lambda_t|\lambda_{t-1}, \alpha, \phi, U) \sim N(\alpha + \phi\lambda_{t-1}; U), \tag{5.13}$$

with correlated innovations characterized by the non-diagonal matrix U (see Aguilar and West, 2000). When U is a diagonal matrix, the above multivariate model is reduced to k univariate conditionally independent autoregressive models (Pitt and Shephard, 1999). Both Pitt and Shephard (1999) and Aguilar and West (2000) consider $\beta_t = \beta$ for all t time periods. Lopes and Migon (2002) and Lopes and Carvalho (2007) extend the previous works by modeling the evolution of the unconstrained loadings, $\tilde{\beta}_t$, as

$$(\tilde{\beta}_t|\tilde{\beta}_{t-1}, \zeta, \Theta, W) \sim N(\zeta + \Theta\tilde{\beta}_{t-1}; W), \tag{5.14}$$

therefore allowing changes in covariances that are not exclusively associated to changes in the individual factor variances.

Philipov and Glickman (2006a,b) extend the above FSV model (with $\Sigma_t = \Sigma$) and model H_t as a full covariance matrix via their Wishart random process. They implement their model on return series 324 monthly observations of 88 individual companies from the S&P500 and use $k = 2$ common factors. Han (2006) implements a similar FSV model to form a portfolio based on 36 stocks, 1200 observations collected from the Center for Research in Security Prices (CRSP). Chib et al. (2006) introduce fat-tailed errors and jumps in the FSV model as well as efficient and fast MCMC algorithm. They implement their extension to simulated data ($p = 50$) and real data on international weekly stock index returns where $p = 10$ (see also Nardari and Scruggs, 2007). Lopes and Carvalho (2007) extend the FSV model to allow for Markovian regime shifts in the dynamic of the variance of the common factors and apply their model to study Latin America's main markets ($p = 5$). More recently, Nakajima and West (2013) extend the basic factor stochastic volatility model by allowing time-varying patterns of occurrence of zero elements in factor loadings matrices, which potentially leads to more interpretable, dynamic sparsity.

5.3.2 Financial Index Models

Carvalho et al. (2011) consider *financial index models* (FIM) appropriate choices for the purpose of covariance estimation and asset allocation. They develop a dynamic factor model encompass both the BARRA[1] and Fama-French[2] strategies in a simple yet flexible modeling setup. The fact that size, book-to-price and momentum are relevant to explain covariation among stocks is exploited in two common ways: (*i*) as individual regressors in a multivariate linear model, and (*ii*) as ranking variables used to construct portfolios that are used as indices. A very large body of literature is dedicated to selecting and testing the indices (Cochrane, 2001; Tsay, 2005).

[1]The BARRA strategy, after the company BARRA, Inc., founded by Barr Rosenberg, whose ideas can be found in Rosenberg and McKibben (1973).

[2]Fama and French, in a series of papers, identified a significant effect of market capitalization and book-to-price ratio into expected returns. This has lead to the now famous *Fama-French* three-factor model where, besides the market, two indices are built as portfolios selected on the basis of firms' size and book-to-price ratio.

FIM leads to tractable and parsimonious estimates of the covariances and it is economically interpretable and theoretically justified. From a methodological viewpoint, our models can be seen as a "structured" extension of current factor model ideas as developed in Aguilar and West (2000), West (2003), Lopes and West (2004), Lopes et al. (2008) and Carvalho et al. (2008). On the applied side our goal is to propose a model-based strategy that creates better FIM, helps deliver better estimates of time-varying covariances, and leads to more effective portfolios.

The general form of an index model assumes that stock returns follows $r_t = \alpha_t + \beta_t f_t + \varepsilon_t$, where, as in our general factor model of equation (5.2), f_t is a vector of common factors at time t, β_t is a matrix of factor loadings (or exposures) and ε_t is a vector of idiosyncratic residuals. Therefore, if $\text{Var}(f_t) = H_t$ and $\text{Var}(\varepsilon_t) = \Sigma_t$, then $\text{Var}(r_t) = \beta_t H_t \beta_t' + \Sigma_t$, from equation (5.4).

Carvalho et al. (2011) take a dynamic, model-based perspective and assume that at time t one observes the vector (r_t, x_t, Z_t), where r_t is a p-dimensional vector of stock returns; Z_t is a $p \times k$ matrix of firm specific information; and x_t is the market return (or some equivalent measure). The index model is then defined by a dynamic factor model as

$$r_t = \alpha_t + \gamma_t x_t + \beta_t f_t + \varepsilon_t, \tag{5.15}$$

where γ_t is a p-dimensional vector of market loadings, ε_t is the vector of idiosyncratic residuals, and f_t is a k-dimensional vector of common factors. Each element of both α_t and γ_t follows a standard first-order dynamic linear model (West and Harrison, 1997) and that ε_t is defined by a set of independent stochastic volatility models (Jacquier et al., 1994; Kim et al., 1998). They also assume that f_t is zero-mean multivariate normal with diagonal covariance matrix H_t driven by univariate stochastic volatility models. Finally, through β_t, company specific information will be used to help uncover relevant latent structures representing the risk factors. Taking $\gamma_t = 0$ and fixing the loading through time gets us to the factor stochastic volatility models of Section 5.3.

They consider five models, three of which we briefly list here for illustration. In all cases α_t and β_t follow standard first order dynamic linear models and ϵ_t follows standard stochastic volatility AR(1) models.

Dynamic CAPM: $\beta_t = 0$ and $r_t = \alpha_t + \gamma_t x_t + \varepsilon_t$.

Dynamic BARRA: $\beta_t =$(market size, book-to-price ratio, momentum) with $r_t = \alpha_t + \gamma_t x_t + \beta_t f_t + \varepsilon_t$ and $f_t \sim N(0, H_t)$.

Sparse Dynamic BARRA: We extend sparse factor analysis[3] to the dynamic BARRA by setting $\beta_{tj} = 0$ with probability $1 - \pi_{tj}$, where π_{tj}s are the inclusion probabilities.

[3]See West (2003), Carvalho et al. (2008) and Frühwirth-Schnatter and Lopes (2009).

5.4 Spatial Factor Analysis

In this section we discuss the extension of the basic normal linear factor model from equations (5.2) and (5.3) to accommodate spatially oriented data. Wang and Wall (2003), for instance, fit a spatial one-factor model to the mortality rates for three major diseases in nearly one hundred counties of Minnesota. In their case, y_i is the vector of p observed variables at each location s_i in region $\mathcal{D}$ and f_i is, as usual, the underlying common spatial factor at location s_i. The spatial structure comes in through the joint distribution of $f = (f_1, \ldots, f_n)'$ for the n locations:

$$f|\gamma \sim N(0, H(\gamma)), \tag{5.16}$$

where $H(\gamma)$ is the covariance matrix representing the spatial structure, and γ is the vector of parameters in the covariance structure. Two common covariance structures are the exponential model and the conditional autoregressive (CAR) model. In the exponential model, $H(\gamma)$ has components h_{ij} given by $h_{ij} = \alpha \exp\{-\phi d_{ij}\}$, where $d_{ij} = |s_i - s_j|$ is the distance between location s_i and s_j, α is the correlation-free variance and ϕ is a range parameter, which represents the decrease in correlation between two locations as the distance increases. Here, therefore, $\gamma = (\alpha, \phi)$. In the CAR model, f is discretely indexed over a partitioned area (areal data) and the correlations depend on the neighboring structure. In this case, $H(\gamma) = \tau^2(I_n - \rho W)^{-1}$, with spatial association parameter ρ, while τ^2 is the conditional variance of $f_i|f_{-i}$. W is a neighborhood matrix of the lattice with w_{ij} an indicator for whether areas i and j share a boundary. Here, therefore, $\gamma = (\rho, \tau^2)$. They also generalize their model to the Poisson common spatial factor analysis, which is applied to model the number of deaths due to lung, pancreas, esophagus and stomach cancers at the county level of Minnesota between the years of 1991 and 1998.

Christensen and Amemiya (2002, 2003) proposed what they called the shift-factor analysis method to model multivariate spatial data with temporal behavior modeled by autoregressive components, while Hogan and Tchernis (2004) and Lopes et al. (2012) fit a one-factor spatial model and entertained several forms of spatial dependence through the single common factor. See also Mezzetti and Billari (2005) and Mezzetti (2012) for other applications of Bayesian factor analysis to spatially correlated data in the contexts of socio-demographic and cancer incidence data.

The next two sections give further details on the spatially hierarchical factor model of Lopes et al. (2012), where we propose a model-based vulnerability index of the population from Uruguay to vector-borne diseases that combines different sources of information via a set of microenvironmental indicators and geographical location in the country. Lopes et al. (2008) propose a new class of nonseparable and nonstationary space-time models that resembles a standard dynamic factor model (Peña and Poncela, 2004, for instance), where the temporal dependence is modeled by latent factors while the spatial dependence is modeled by the factor loadings. Our model is applied to model the space-time structure of pollutants in the northern US.

5.4.1 Spatially Hierarchical Factor Analysis

We propose, in Lopes et al. (2012), a model-based vulnerability index of the population from Uruguay to vector-borne diseases. More specifically, the index is derived form a spatially oriented hierarchical factor model structure that combines both (Departamental) capital level data and census tract level data (see Figure 5.4). Apart form Bella Unión and Montevideo, with 11 and 1031 census tracts, respectively, the number of census tracts per capital varies roughly between 20 and 40. The $p = 11$ indicators used to construct the one-factor (vulnerability) index are standardized to represent percentages, averages, and densities observed at the census tract level: illiteracy rate, percentage of the population with access to public health care care, unemployed males, owed houses, households headed by a woman, households without sewage system, average number of persons per household, households with more than two persons per room, households without access to treated, drinkable water, households with air conditioner, and households poorly built.

For each one of the n_i census tracts of capital i, a p-dimensional vector of variables (social-economical, environmental, demographical, etc.) is observed, namely $y_{ij} = (y_{ij1}, \ldots, y_{ijp})'$, for $i = 1, \ldots, I$ and $j = 1, \ldots, n_i$. The spatially hierarchical factor model (SHFM) we proposed can be described according to the following hierarchy levels:

Observational level: Observations y_{ink} are linked to the vulnerability factor f_{ij},

$$y_{ijk} = \mu_k + \beta_k f_{ij} + \sigma_k \varepsilon_{ijk} \quad k = 1, \ldots, p, \tag{5.17}$$

where μ_k represents the overall grand mean vector for measurement k. The factor loadings vector $\beta = (1, \beta_2, \ldots, \beta_p)'$ plays an important role in understanding the role and the composition of the common factor. Its first element is set to one in order to ensure likelihood identifiability (see Lopes and West, 2004). The specific factors ε_{ijk} are standard normally distributed and independent across capitals, census tracts and measurements.

Vulnerability index level: The vector of factors $f_i = (f_{i1}, \ldots, f_{in_i})'$ within capital i is decomposed as the sum of two spatially structured components: one that captures the overall mean of the capital, and the other one captures the local structure of the index, in the census tracts' level, and also accounts for possible effects of neighboring census tracts. More precisely, we assume

$$f_{ij} = \theta_i + \tilde{f}_{ij} + \sqrt{\omega_i} u_{ij}, \tag{5.18}$$

Figure 5.4: Spatially hierarchical factor analysis. The state level data is summarized by the 19 departmental capitals. Census tract data is exemplified here by the city of Melo, which is the capital of Cerro Lago. This figure reproduces Figure 1 from Lopes et al. (2012).

where θ_i is the common factor for capital i, $\tilde{f}_{ij}$ is the specific factor for census tract j and capital i, and u_{ij} are independent standard normals. The error term u_{ij} accounts for unanticipated, location specific idiosyncrasies.

Within capital variation: As the capitals are divided into census tracts defining irregular subregions, we model the within capital factors $\tilde{f}_i = (\tilde{f}_{i1}, \ldots, \tilde{f}_{in_i})'$, for $i = 1, \ldots, I$, by a proper conditionally autoregressive (CAR) specification (Sun, Tsutakawa, and Speckman, 1999):

$$\tilde{f}_i \sim N(0, \tau_i^2 P_i), \tag{5.19}$$

where $P_i = P_i(\phi) = (I_{n_i} + \phi M_i)^{-1}$, $M_i = D_i - W_i$, with w_{ijl}, the (j,l) component of W_i, given by $w_{ijl} = 1/d_{jl}$ if j and l are neighbors (denoted here by $j \sim l$) and zero otherwise, $d_{jl} = ||s_j - s_l||$ is the Euclidean distance between centroids of capitals j and l, $D_i = \text{diag}(w_{i1+}, \ldots, w_{in_i+})$ and $w_{ij+} = \sum_{l \sim j} w_{ijl}$. The inverse matrix P_i^{-1} is diagonally dominant and positive definite (Harville, 1997). The parameter ϕ controls the strength of the association between the components of $\tilde{f}_i$, with $\phi = 0$ implying independence. Equation (5.19) approaches the intrinsic autoregressive model when ϕ approaches infinity (Besag, York, and Mollié, 1991; Besag and Kooperberg, 1995).

Between capitals variation: The θ_is are conditionally independent and Gaussian with common baseline vulnerability factor θ_0 and covariance structure driven by the Euclidean distances between the centroids of the capitals, i.e.,

$$\theta \sim N\left(1_I\theta_0, \delta^2 H(\lambda)\right), \tag{5.20}$$

where $\theta = (\theta_1, \cdots, \theta_I)$. Although each capital i has its own vulnerability factor, the above model allows borrowing-strength across neighboring regions. The correlation matrix H is fully specified by a Matérn correlation function, i.e., $H_{ij} = \rho(\lambda, d_{ij}) = 2^{1-\lambda_2}\Gamma(\lambda_2)^{-1}(d_{ij}/\lambda_1)^{\lambda_2}\mathcal{K}_{\lambda_2}(d_{ij}/\lambda_1)$ where $\mathcal{K}_{\lambda_2}$ is the modified Bessel function of the second kind and of order λ_2, $\lambda = (\lambda_1, \lambda_2)$ and $d_{ij} = ||s_i - s_j||$ is the Euclidean distance between the centroids s_i and s_j of capitals i and j, for $i, j = 1, \ldots, I$.

Figure 5.5 illustrates the performance of the SHFM in producing the vulnerability indexes to all 1031 census tracts of Montevideo. The figure helps discriminating very opposing situations within the city and captures the local effects of the factor/index. In other words, the richest Southeast area of the city has fewer census tracts with low values of the index, representing low vulnerability. Land in these regions is irregularly owned. Overall, the levels of the index in Montevideo are in accordance to what is anticipated by experts, showing high values (more vulnerability) towards the North-West region which comprise more rural areas.

The next model uses factor analysis to reduce the dimension of multiple time-series representing the dynamics of dozens of locations. In this case, the attributes are, in fact, univariate measurements at the locations. We will see that the structure of the matrix of factor loadings plays an important role in capturing conditional spatial variation.

5.4.2 Spatial Dynamic Factor Analysis

Lopes et al. (2008) propose a new class of nonseparable and nonstationary space-time models that resembles a standard dynamic factor model (Peña and Poncela, 2004, for instance),

$$
\begin{aligned}
y_t &= \mu_t^{y*} + \beta f_t + \epsilon_t, & \epsilon_t &\sim N(0, \Sigma), & (5.21)\\
f_t &= \Gamma f_{t-1} + \omega_t, & \omega_t &\sim N(0, \Lambda), & (5.22)\\
\beta_{(j)} &\sim GRF(\mu_j^{\beta^*}, \tau_j^2 \rho_{\phi_j}(\cdot)) \equiv N(\mu_j^{\beta^*}, \tau_j^2 R_{\phi_j}), & & & (5.23)
\end{aligned}
$$

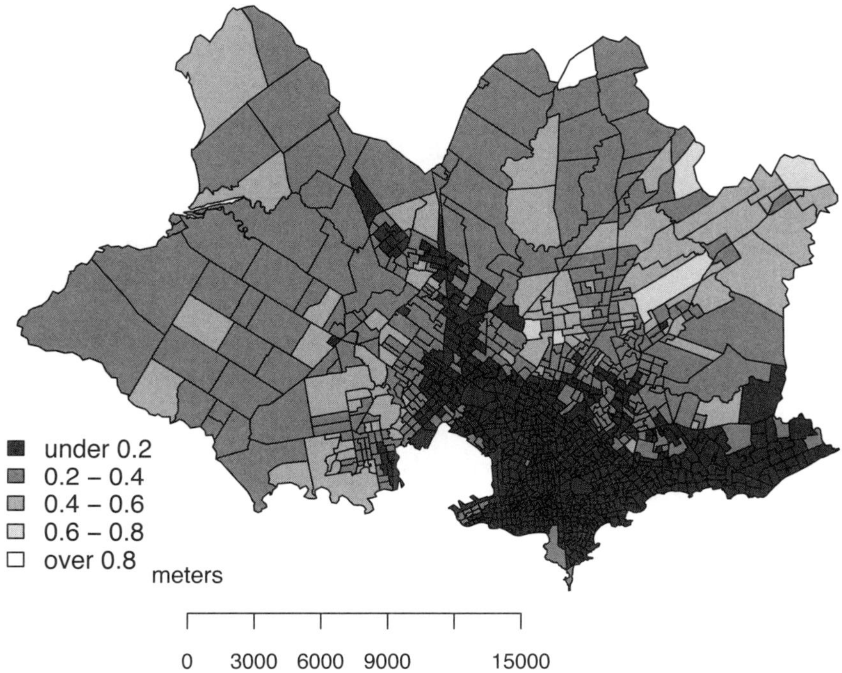

Figure 5.5: Spatially hierarchical factor analysis. Within-city posterior (standardized) vulnerability index per census tract of Montevideo. This is Figure 5 from Lopes et al. (2012).

where $y_t = (y_{1t}, \ldots, y_{Nt})'$ is the N-dimensional vector of observations (locations $s_1, \ldots, s_N$ and times $t = 1, \ldots, T$), μ_t^{y*} is the mean level of the space-time process, f_t is an m-dimensional vector of common factors, for $m < N$ (m is potentially several orders of magnitude smaller than N) and the jth column of the factor loadings matrix,

$$\beta_{(j)} = (\beta_{(j)}(s_1), \ldots, \beta_{(j)}(s_N))',$$

for $j = 1, \ldots, m$, is modeled as a conditionally independent, distance-based Gaussian process or a Gaussian random field (GRF). The matrix Γ characterizes the evolution dynamics of the common factors, while Σ and Λ are observational and evolutional variances. $\mu_j^{\beta*}$ is a N-dimensional mean vector. The (l, k)-element of R_{ϕ_j} is given by $r_{lk} = \rho_{\phi_j}(|s_l - s_k|)$, $l, k = 1, \ldots, N$, for suitably defined correlation functions $\rho_{\phi_j}(\cdot)$, $j = 1, \ldots, m$, for instance, exponential, power exponential or Matérn. The parameters ϕ_js are typically scalars or low dimensional vectors (for details, see Cressie, 1993 and Stein, 1999).

The univariate measurements from all observed locations, either areal or point-referenced, at any given time, are grouped together in what otherwise would be the vector of attributes in standard factor analysis and the spatial dependence is introduced by the columns of the factor loadings matrix. Therefore, common dynamic factors can be thought of as describing temporal similarities amongst the time series, such as common annual cycles or (stationary or nonstationary) trends, while the importance of common factors in describing the measurements in a given location is captured by the components of the factor loadings matrix. More general time series models can be entertained, through the common factors, without imposing additional constraints to the current spatial characterization of the model, and vice-versa (Lopes et al., 2008).

Lopes et al. (2008) model the spatial and temporal variations in the concentration levels of sulfur dioxide, SO_2, across 24 monitoring stations. Weekly measurements in $\mu g/m^3$ are collected by the Clean Air Status and Trends Network (CASTNet), which is part of the Environmental Protection Agency (EPA) of the United States. Measurements span from the first week of 1998 to the 30th week of 2004, a total of 342 observations. Figure 5.6 shows the posterior mean surface interpolation corresponding to the seasonal factor when fitting one of their SDFM. The loadings for the seasonal factor are smaller in the highly industrialized state of Ohio.

Calder (2007) also used SO_2 along with three other pollutants in a related dynamic factor model where the columns of the factor loadings matrix is model via deterministic smoothed kernels (see also Sansó et al., 2004). Several other models are particular cases of the our SDFM: principal kriging (Sahu and Mardia, 2005, and Lasinio, Sahu and Mardia, 2005), kriged Kalman filter (Mardia et al., 1998), and orthonormal basis loadings (Wikle and Cressie, 1999).

5.5 Additional Developments

5.5.1 Prior and Posterior Robustness

Lee and Press (1998) studies posterior robustness of the loadings, common factors and idiosyncratic covariance, while Lopes (2003) studies prior specification and sensitivity in factor models via the expected posterior prior setup of Pérez and Berger (2002).

5.5.2 Mixture of Factor Analyzers

Mixture of factor analyzers (MFA) is a nonlinear, more flexible extension of the linear factor analysis of Section 5.2. The basic structure of a MFA model is given by

$$(y|\beta, f, \Sigma) \sim \sum_{j=1}^{m} \pi_j N(y; \beta_j f; \Sigma), \tag{5.24}$$

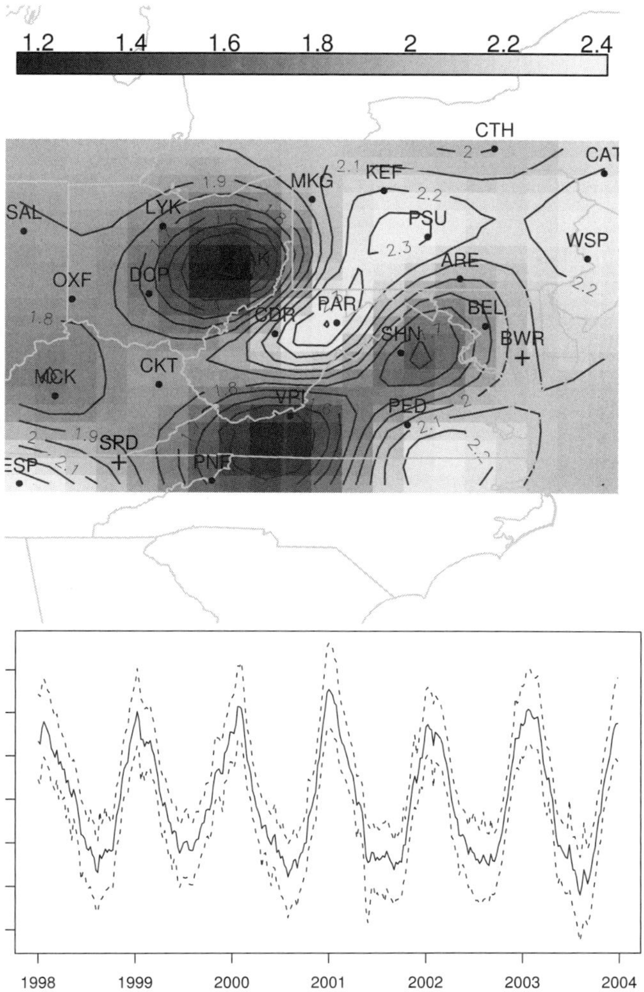

Figure 5.6: Spatial dynamic factor model. Posterior Bayesian interpolation for loadings factors or the northeast part of the U.S. (top frame). Posterior means (and 95% credibility intervals) of the seasonal common factors (bottom frame). These are taken from Figures 5 and 6 from Lopes et al. (2008).

where m is the number of *analyzers*. Conditional on j, a standard normal linear factor model arises. Ghahramani and Beal (2000) introduce an algorithm that fits mixture of factor analyzers models via variational approximation to full Bayesian integration over model parameters. Utsugi and Kumagai (2001), Fokoué and Titterington (2003) and Fokoué (2004) propose MCMC-based posterior inference for MFA models. McLachlan et al. (2007) and Andrews and

McNicholas (2011) extend the MFA to the multivariate t family and uses the Expectation-Maximation (EM) algorithm for parameter estimation.

5.5.3 Factor Analysis in Time Series Modeling

Peña and Box (1987) introduce factor models with common, independent or dependent, factors following ARMA processes. Similarly, Engle (1987) proposes a multivariate ARCH with factor structures. Diebold and Nerlove (1989) model multivariate GARCH structures through a one-factor model to study the dynamics of exchange rate volatilities. Engle et al. (1990), who use Factor-ARCH to model a conditional covariance matrix of asset returns, while Ng and Rothschild (1992) relates dynamic and static factors to portfolio allocation in financial markets. Lin (1992) compares four frequentist estimators for factor GARCH models: two-stage univariate GARCH, two-stage quasi-maximum likelihood, quasi-maximum likelihood with known factor weights and quasi-maximum likelihood with unknown factor weights. Molenaar et al. (1992) employ nonstationary multivariate time series dynamic factor analysis with lagged common factors to account for the persistence in time series trends.

Bollerslev and Engle (1993) introduce a k-factor GARCH model and study co-persistence in multivariate integrated GARCH models. Harvey et al. (1994) introduce one of the first factor stochastic volatility models (see Section 5.3.1). Escribano and Peña (1994) establish the relationship between cointegrated vectors and common factors via Peña and Box's (1987) dynamic factor models. Demos and Sentana (1998) introduces an EM algorithm for conditionally heterescedastic factor models, while Sentana (1998) investigates the similarities and the differences of Engle's (1987) factor GARCH model and Diebold and Nerlove's (1989) latent factor ARCH model. See also Fiorentini et al. (2004). Vrontros et al. (2003) proposes a full-factor multivariate GARCH model (2003) and treats the order of variable via Bayesian model averaging. More recently, Sentana et al. (2008) derive indirect estimators of conditionally heteroskedastic factor models in which the volatilities of common and idiosyncratic factors depend on their past unobserved values. See also the recent paper by Zhou et al. (2012) that introduces correlated dynamic latent factor structures into a new class of latent threshold dynamic factor models for multivariate volatility analysis and forecasting of financial time series.

5.5.4 Factor Analysis in Macroeconometrics

Stock and Watson (2002b,2002a) implement principal components analysis project a large number of predictors (about 215 in their case) on a few principal components, or *diffusion indexes*. Then, the diffusion indexes are used as explanatory variables when forecasting a macroeconomic time series variable. More precisely, let z_{t+1} donate the macroeconomic time series and y_t a p-dimensional vector with (possibly many highly correlated) predictors. They assume that (z_{t+h}, y_t) admit a factor model representation with k common factors f_t, whose simplest version is

$$y_t = \beta f_t + \varepsilon_t, \quad (5.25)$$

$$z_{t+h} = \theta' f_t + u_{t+h}. \quad (5.26)$$

Roughly speaking, they propose a two-step estimation procedure. First, the diffusion indexes are estimated from Equation (5.25) by principal component analysis. Second, the estimated diffusion indexes are plugged in the forecasting Equation (5.26). Their empirical applications aim at forecasting major monthly macroeconomic variables for the United States (1959:1 to 1998:12), such as total industrial production, real personal income less transfers, real manufacturing and trade sales, and number of employees on nonagricultural payrolls. The predictors represent main categories of macroeconomic time series, including real output and income, employment and hours, real retail, manufacturing, and trade sales and consumption, amongst many others. They showed empirically that only $k = 6$ factors account for much of the variance of our $p = 215$ time series. Bernanke et al. (2005) and Stock and Watson (2005) combined factor models with vector autoregressive models, commonly known as *factor-augmented VAR* (FAVAR) models, while Negro and Otrok (2008) and Korobilis (2013) extend these models, from a Bayesian viewpoint, to include time-varying coefficients. The review paper by Stock and Watson (2006) provides an important review to the area of forecasting with many predictors.

The Bayesian approach to factor analysis applied to macroeconomic problems has grown considerably over the last decade. A few important contributions are the following. Koop and Potter (2004) revisits Stock and Watson (2002b) and implement Bayesian model averaging on the above dynamic factor structure. Otrok and Whiteman (1998) designs and implements a Bayesian dynamic latent factor and produce coincident and leading indicators for a local US economy based on the posterior mean of the latent factors. Kose et al. (2003) proposes a dynamic factor model for international business cycles whose common factors are divided into world, region and country specific ones (see also Koop and Korobilis, 2009). Uhlig and Ahmadi (2012) proposes a Bayesian factor-augmented VAR, or BFAVAR, to study the effects of monetary policy shocks. A recent overview of Bayesian macroeconometrics is provided by Del Negro and Schorfheide (2011).

5.5.5 Term Structure Models

The yield curve is defined as the relationship between τ and $\tau^{-1} \log p_t(\tau)$, where $p_t(\tau)$ is the price, at time t, of a zero-coupon bond with payoff 1 at maturity $t + \tau$. See, for instance, Diebold and Li (2006) and Diebold et al. (2008) for further details. Diebold and Li (2006), for example, argue that two popular approaches to the term structure modeling are *(i)* no-arbitrage models and *(ii)* equilibrium models, with significant contributions to the former by Hull and White (1990) and Heath et al. (1992), and to the latter by Vasicek (1977), Cox et al. (1985) and Duffie and Kan (1996), amongst others.

Diebold and Li (2006), however, use variations of the Nelson and Siegel (1987) exponential components framework to model the yield curve as a three-factor model (level, slope and curvature) that evolves over time dynamically. Nelson-Siegel framework imposes a parsimonious structure on the factor loadings matrix:

$$y_t(\tau) = \beta_{1t} + \beta_{2t}\left(\frac{1 - e^{-\lambda_t \tau}}{\lambda_t \tau}\right) + \beta_{3t}\left(\frac{1 - e^{-\lambda_t \tau}}{\lambda_t \tau} - e^{-\lambda_t \tau}\right), \tag{5.27}$$

with λ_t characterizing the decaying rate; small values of λ_t produce slow decay and can better fit the curve at long maturities, while large values of λ_t produce fast decay and can better fit the curve at short maturities.

Chib and Ergashev (2009) presented a Bayesian approach for the fitting of affine yield curve models with macroeconomic factors that emphasizes the use of a prior on the parameters of the model which implies an upward-sloping yield curve. Chib and Kang (2012) extend the model to accommodate change-points in affine term structure models.

5.5.6 Sparse Factor Structures

We finish the chapter by summarizing the current literature on Bayesian sparse factor analysis. We have already listed a few of these contributions, such as West (2003), Carvalho et al. (2008) and Frühwirth-Schnatter and Lopes (2009). Mayrink and Lucas (2013), for instance, study gene expression data by extending Carvalho et al. (2008) to entertain interactions between the common factors in their sparse factor model. Similarly, Conti et al. (2011) implement Frühwirth-Schnatter and Lopes' (2009) parsimonious factor analysis strategy to examine the effect of early-life conditions and education on health but incorporating discrete attributes as well as limited dependent variable structures, commonly present when dealing with endogeneity in microeconometric studies.

Additional references are Hahn et al. (2012), Pati et al. (2012), Cron and West (2012) and Hahn et al. (2013), who propose sparse factor probit modeling, sparse factor analysis for massive covariance matrices, random sparse orthogonal matrices, and predictor-dependent shrinkage partial factor analysis, respectively. See also Yoshida and West (2013) for sparse graphical factor models, and Knowles and Ghahramani (2011) for nonparametric Bayesian sparse factor models applied, once again, to gene expression modeling. Sparse dynamic factor models are proposed, for instance, by Kaufmann and Schumacher (2013) who introduce a new, order-independent identification strategy based on semi-orthogonal loadings. See also Zhang and Nesselroade (2007), Kaufmann and Schumacher (2012) and Cui and Dunson (2012).

5.6 Modern non-Bayesian factor analysis

The literature on modern factor analysis is growing, as Table 5.1 suggested, as expected outside the realm of Bayes. For those more eager to explore more non-Bayesian factor analysis alternatives, the following paragraphs bring some of the recent papers by Bai and Ng's group and by Forni, Hallin, Lippi and Reichlin's groups. The list is narrow and limited, which reflects the author's own limitations. These few papers, as well as their reference lists, we believe, will provide the reader with a start-up for her own search.

Bai (2003), Bai and Li (2012) and Li (2013) consider ML estimation of high-dimensional (static and dynamic) factor models when p is at least as large as n. Bai and Ng (2002) propose tools to select the number of common factor in the above large n and large p scenario. Bai and Ng (2008) survey the main theoretical results, including how to determine the number of factors, how to conduct inference factor-regression models (see also Bai and Ng, 2006, Amengual and Watson, 2007). Bai and Ng (2010) propose a class of factor instrumental variables in the context of data rich environment, where a large number of endogenous variables are driven by a small number of unobservable exogenous common factors. Moench et al. (2011) propose a (inherently Bayesian) large dimensional hierarchical factor model that takes into account within- and between-block variability. See also Fan et al. (2008), Fan et al. (2011), Onatski (2009) and Onatski (2012).

Forni et al. (2000), Forni and Lippi (2001), Forni et al. (2005), Forni and Lippi (2011) introduce generalized (dynamic) factor models and discuss extensively identifiability, estimation and forecasting, while Forni et al. (2009) talk structural factor model with large cross-sections. Hallin and Liska (2007) tackle the estimation of the number of common factors in the above generalized dynamic factor model, while Hallin and Liska (2011) decompose large sets of macroeconomic variables into smaller, but still quite large, blocks. Additional works tackling high dimensional (dynamic) factor models in econometrics are Kapetanios and Marcellino (2009), Doz et al. (2011, 2012) and Hallin et al. (2011).

5.7 Final Remarks

We started reviewing the basic normal linear factor model, which is considered the spinal cord of all factor models presented later on in the chapter. The one hundred plus references listed at the end of the chapter represent a biased fraction of the existing literature. Factor models appear now in virtually all areas of sciences, despite its start in psychometric studies back in the 1900s with the seminal work of Spearman. Amongst its various extensions, we discussed and referenced several factor stochastic volatility models, dynamic factor models, spatial factor models and sparse factor models.

We believe the next several years will witness the multiplication of more complex and massively large factor structures and applications to study related objects other than neatly organized rectangle matrices of observations (rows) and attributes (columns). The interactions between statisticians, econometricians, machine learners and data miners will lead to highly sophisticated and computationally efficient and fast factor-based modeling.

Bibliography

Aguilar, O. and M. West (2000). Bayesian dynamic factor models and variance matrix discounting for portfolio allocation. *Journal of Business and Economic Statistics 18*, 338–357.

Akaike, H. (1987). Factor analysis and AIC. *Psychometrika 52*, 317–332.

Amemiya, Y. and T. W. Anderson (1990). Asymptotic chi-square tests for a large class of factor analysis models. *The Annals of Statistics 18*, 1453–1463.

Amengual, D. and M. W. Watson (2007). Consistent estimation of the number of dynamic factors in a large n and t panel. *Journal of Business and Economic Statistics 25*, 91–96.

Anderson, T. W. and Y. Amemiya (1988). The asymptotic normal distribution of estimators in factor analysis under general conditions. *The Annals of Statistics 16*, 759–771.

Anderson, T. W. and H. Rubin (1956). Statistical inference in factor analysis. In J. Neyman (Ed.), *Proceedings of the Third Berkeley Symposium of Mathematical Statistics and Probability*, Volume 5, pp. 111–150. University of California Press, Berkeley.

Andrews, J. L. and P. D. McNicholas (2011). Extending mixtures of multivariate t-factor analyzers. *Statistics and Computing 21*, 361–373.

Bai, J. (2003). Inferential theory for factor models of large dimensions. *Econometrica 71*, 135–71.

Bai, J. and K. Li (2012). Statistical analysis of factor models of high dimension. *The Annals of Statistics 40*, 436–465.

Bai, J. and S. Ng (2002). Inferential theory for factor models of large dimensions. *Econometrica 71*, 135–171.

Bai, J. and S. Ng (2006). Confidence intervals for diffusion index forecasts and inference for factor-augmented regressions. *Econometrica 74*, 1133–1150.

Bai, J. and S. Ng (2008). Large dimensional factor analysis. *Foundations and Trends in Econometrics 3*, 89–163.

Bai, J. and S. Ng (2010). Instrumental variable estimation in a data rich environment. *Econometric Theory 26*, 1577–1606.

Bartholomew (1995). Spearman and the origin and development of factor analysis. *Journal of Mathematical and Statistical Psychology 48*, 211–220.

Bartholomew, D. (1981). Posterior analysis of the factor model. *The British Journal of Mathematical and Statistical Psychology 34*, 93–99.

Bartholomew, D. (1987). *Latent Variable Models and Factor Analysis*. London: Charles Griffin.

Bentler, P. M. and J. S. Tanaka (1983). Problems with EM algorithms for ML factor analysis. *Psychometrika 48*, 247–251.

Bernanke, B. S., J. Boivin, and P. Eliasz (2005). Measuring the effects of monetary policy: A factor-augmented vector autoregressive (FAVAR) approach. *Quarterly Journal of Economics 7*, 387–422.

Besag, J. and C. Kooperberg (1995). On conditional and intrinsic autoregressions. *Biometrika 82*, 733–746.

Besag, J., J. C. York, and A. Mollié (1991). Bayesian image restoration, with two applications in spatial statistics (with discussion). *Annals of the Institute of Statistical Mathematics 43*, 1–59.

Bhattacharya, A. and D. B. Dunson (2011). Sparse Bayesian infinite factor models. *Biometrika 98*, 291–306.

Bollerslev, T. and R. F. Engle (1993). Common persistence in conditional variances. *Econometrica 61*, 167–186.

Box, G. E. P. and G. C. Tiao (1973). *Bayesian Inference in Statistical Analysis.* Massachusetts: Addison-Wesley.

Broemeling, L. (1985). *Bayesian Analysis of Linear Models.* New York: Marcel Dekker.

Burt, C. (1940). *The Factors of the Mind: An Introduction to Factor Analysis in Psychology.* University of London Press, London.

Calder, C. (2007). Dynamic factor process convolution models for multivariate space-time data with application to air quality assessment. *Environmental and Ecological Statistics 14*, 229–47.

Carter, C. and R. Kohn (1994). On Gibbs sampling for state space models. *Biometrika 81*, 541–553.

Carvalho, C. and H. Lopes (2004). Simulation-based sequential analysis of Markov switching stochastic volatility models. Technical report, Graduate School of Business, University of Chicago.

Carvalho, C. M., J. Chang, J. Lucas, Q. Wang, J. Nevins, and M. West (2008). High-dimensional sparse factor modelling: Applications in gene expression genomics. *Journal of the American Statistical Association 103*, 1438–1456.

Carvalho, C. M., H. F. Lopes, and O. Aguilar (2011). Dynamic stock selection strategies: A structured factor model framework. In J. M. Bernardo, M. J. Bayarri, J. O. Berger, A. P. Dawid, D. Heckerman, A. F. M. Smith, and M. West (Eds.), *Bayesian Statistics 9*, Volume 9, pp. 69–90. Oxford University Press, Oxford.

Chib, S. (1995). Marginal likelihood from the Gibbs output. *Journal of the American Statistical Association 90*, 1313–1321.

Chib, S. and B. Ergashev (2009). Analysis of multifactor affine yield curve models. *Journal of the American Statistical Association 104*, 1324–1337.

Chib, S. and K. H. Kang (2012). Change-points in affine arbitrage-free term structure models. *Journal of Financial Econometrics (available online).*

Chib, S., F. Nardari, and N. Shephard (2006). Analysis of high dimensional multivariate stochastic volatility models. *Journal of Econometrics 134*, 341–371.

Chow, S.-M., N. Tang, Y. Yuan, X. Song, and H. Zhu (2011). Bayesian estimation of semiparametric nonlinear dynamic factor analysis models using the Dirichlet process prior. *British Journal of Mathematical and Statistical Psychology 64*, 69–106.

Christensen, W. and Y. Amemiya (2002). Latent variable analysis of multivariate spatial data. *Journal of the American Statistical Association 97*(457), 302–317.

Christensen, W. and Y. Amemiya (2003). Modeling and prediction for multivariate spatial factor analysis. *Journal of Statistical Planning and Inference 115*, 543–564.

Cochrane, J. (2001). *Asset Pricing*. Princeton University Press.

Conti, G., J. J. Heckman, H. F. Lopes, and R. Piatek (2011). Constructing economically justified aggregates: an application of the early origins of health. Discussion paper, University of Chicago Booth School of Business.

Cox, J. C., J. E. Ingersoll, and S. A. Ross (1985). A theory of the term structure of interest rates. *Econometrica 53*, 385–407.

Cressie, N. (1993). *Statistics for Spatial Data*. New York: Wiley.

Cron, A. and M. West (2012). Models of random sparse eigenmatrices with application to Bayesian factor analysis. Discussion Paper 2012-01, Department of Statistical Science, Duke University.

Cudeck, R. and R. C. MacCallum (2007). *Factor Analysis at 100: Historical Developments and Future Directions*. Routledge.

Cui, K. and D. B. Dunson (2012). Generalized dynamic factor models for mixed-measurement time series. *Journal of Computational and Graphical Statistics*.

Del Negro, M. and F. Schorfheide (2011). Bayesian macroeconometrics. In J. Geweke, G. Koop, and H. V. Dijk (Eds.), *Handbook of Bayesian Econometrics*. Oxford University Press.

Demos, A. and E. Sentana (1998). An EM algorithm for conditionall heterescedastic factor models. *Journal of Business and Economic Statistics 16*, 357–361.

Diebold, F. X. and C. Li (2006). Forecasting the term structure of government bond yields. *Journal of Econometrics 130*, 337–364.

Diebold, F. X., C. Li, and V. Z. Yue (2008). Global yield curve dynamics and interactions: A dynamic Nelson-Siegel approach. *Journal of Econometrics 146*, 3511–363.

Diebold, F. X. and M. Nerlove (1989). The dynamics of exchange rate volatility: a multivariate latent ARCH model. *Journal of Applied Econometrics 4*, 1–21.

Doz, C., D. Giannone, and L. Reichlin (2011). A two-step estimator for large approximate dynamic factor models based on Kalman filtering. *Journal of Econometrics 164*, 188–205.

Doz, C., D. Giannone, and L. Reichlin (2012). A quasi-maximum likelihood approach for large, approximate dynamic factor models. *The Review of Economics and Statistics 94*, 1014–1024.

Duffie, D. and R. Kan (1996). A yield-factor model of interest rates. *Mathematical Finance 6*, 379–406.

Engle, R. (1987). Multivariate ARCH with factor structures - cointegration in variance. Discussion paper, University of California, San Diego, Department of Economics.

Engle, R. F., V. K. Ng, and M. Rothschild (1990). Asset pricing with a factor ARCH covariance structure: empirical estimates for treasury bills. *Journal of Econometrics 45*, 213–238.

Escribano, A. and D. Peña (1994). Cointegration and common factors. *Journal of Time Series Analysis 15*, 577–586.

Fan, J., Y. Fan, and J. Lv (2008). High dimensional covariance matrix estimation using a factor model. *Journal of Econometrics 147*, 186–197.

Fan, J., Y. Liao, and M. Mincheva (2011). High-dimensional covariance matrix estimation in approximate factor models. *Annals of Statistics 39*, 3320–3356.

Fiorentini, G., E. Sentana, and N. Shephard (2004). Likelihood-based estimation of latent generalized ARCH structures. *Econometrica 72*, 1481–1517.

Fokoué (2004). Mixtures of factor analysers: An extension with covariates. *Journal of Multivariate Analysis 95*, 370–384.

Fokoué, E. and D. M. Titterington (2003). Mixtures of factor analysers: Bayesian estimation and inference by stochastic simulation. *Machine Learning 50*, 73–94.

Forni, M., D. Giannone, M. Lippi, and L. Reichlin (2009). Opening the black box: Structural factor models with large cross-sections. *Econometric Theory 25*, 1319–1347.

Forni, M., M. Hallin, M. Lippi, and L. Reichlin (2000). The generalized factor model: Identification and estimation. *The Review of Economics and Statistics 82*, 540–554.

Forni, M., M. Hallin, M. Lippi, and L. Reichlin (2005). The generalized dynamic factor model one-sided estimation and forecasting. *Journal of the American Statistical Association 100*, 830–840.

Forni, M. and M. Lippi (2001). The generalized factor model: Representation theory. *Econometric Theory 17*, 1113–1141.

Forni, M. and M. Lippi (2011). The general dynamic factor model: One-sided representation results. *Journal of Econometrics 163*, 23–28.

Frühwirth-Schnatter, S. (1994). Data augmentation and dynamic linear models. *Journal of Time Series Analysis 15*, 183–202.

Frühwirth-Schnatter, S. and H. F. Lopes (2009). Parsimonious Bayesian factor analysis when number of factors is unknown. Discussion paper, University of Chicago Booth School of Business.

Gamerman, D. and H. F. Lopes (2006). *Markov Chain Monte Carlo - Stochastic Simulation for Bayesian Inference* (2nd ed.). Chapman & Hall/CRC.

Geweke, J. and G. Zhou (1996). Measuring the pricing error of the arbitrage pricing theory. *The Review of Financial Studies 9*, 557–587.

Ghahramani, Z. and M. Beal (2000). Variational inference for Bayesian mixtures of factor analysers. In S. A. Solla, T. K. Leen, and K. Müller (Eds.), *Advances in Neural Information Processing Systems*, Volume 12, pp. 449–455. MIT Press.

Ghosh, J. and D. Dunson (2009). Default prior distributions and efficient posterior computation in Bayesian factor analysis. *Journal of Computational and Graphical Statistics 18*, 306–320.

Green, P. (1995). Reversible jump Markov chain Monte Carlo computation and Bayesian model determination. *Biometrika 82*, 711–732.

Hahn, P. R., C. M. Carvalho, and S. Mukherjee (2013). Partial factor modeling: Predictor-dependent shrinkage for linear regression. *Journal of the American Statistical Association 108*, 999–1008.

Hahn, P. R., C. M. Carvalho, and J. Scott (2012). A sparse factor analytic probit model for congressional voting patterns. *Applied Statistics 61*, 619–635.

Hallin, M. and R. Liska (2007). Determining the number of factors in the general dynamic factor model. *Journal of the American Statistical Association 102*, 603–617.

Hallin, M. and R. Liska (2011). Dynamic factors in the presence of blocks. *Journal of Econometrics 163*, 29–41.

Hallin, M., C. Mathias, H. Pirotte, and D. Veredas (2011). Market liquidity as dynamic factors. *Journal of Econometrics 163*, 42–50.

Han, Y. (2006). Asset allocation with a high dimensional latent factor stochastic volatility model. *The Review of Financial Studies 19*, 237–271.

Harvey, A., E. Ruiz, and N. Shephard (1994). Multivariate stochastic variance models. *Review of Economic Studies 61*, 247–264.

Harville, D. A. (1997). *Matrix Algebra from a Statistician's Perspective.* Springer-Verlag, New York.

Heath, D., R. Jarrow, and A. Morton (1992). Bond pricing and the term structure of interest rates: a new methodology for contingent claims valuation. *Econometrica 60*, 77–105.

Hogan, J. W. and R. Tchernis (2004). Bayesian factor analysis for spatially correlated data, with application to summarizing area-level material deprivation from census data. *Journal of the American Statistical Association 99*, 314–324.

Holzinger, K. J. and H. H. Harman (1941). *Factor analysis: A Synthesis of Factorial Methods.* University of Chicago Press, Chicago.

Hotelling, H. (1955). Analysis of a complex of statistical variables into principal components. *Journal of Educational Psychology 24*, 417–441.

Hull, J. and A. White (1990). Pricing interest-rate-derivative securities. *Review of Financial Studies 3*, 573–592.

Ihara, M. and Y. Kano (1995). Identifiability of full, marginal, and conditional factor analysis model. *Statistics and Probability Letters 23*, 343–350.

Jacquier, E., N. Polson, and P. Rossi (1994). Bayesian analysis of stochastic volatility models. *Journal of Business and Economic Statistics 12*, 371–389.

Jensen, M. (2004). Semiparametric Bayesian inference of long-memory stochastic volatility models. *Journal of Time Series Analysis 25*, 895–922.

Jöreskog, K. G. (1967). Some contributions to maximum likelihood factor analysis. *Psychometrika 32*, 443–382.

Jöreskog, K. G. (1969). A general approach to confirmatory maximum likelihood factor analysis. *Psychometrika 34*, 183–220.

Kapetanios, G. and M. Marcellino (2009). A parametric estimation method for dynamic factor models of large dimensions. *Journal of Time Series Analysis 30*, 208–238.

Kaufmann, S. and C. Schumacher (2012). Finding relevant variables in sparse Bayesian factor models: Economic applications and simulation results. Discussion Papers 29/2012, Deutsche Bundesbank, Research Centre.

Kaufmann, S. and C. Schumacher (2013). Bayesian estimation of sparse dynamic factor models with order-independent identification. Working Papers 13.04, Swiss National Bank, Study Center Gerzensee.

Kim, S., N. Shephard, and S. Chib (1998). Stochastic volatility: likelihood inference and comparison with ARCH models. *Review of Economic Studies 65*, 361–393.

Knowles, D. and Z. Ghahramani (2011). Nonparametric Bayesian sparse factor models with application to gene expression modeling. *Annals of Applied Statistics 5*, 1534–1552.

Koop, G. and D. Korobilis (2009). Bayesian multivariate time series methods for empirical macroeconomics. *Foundations and Trends in Econometrics 3*, 267–358.

Koop, G. and S. Potter (2004). Forecasting in dynamic factor models using Bayesian model averaging. *Econometrics Journal 7*, 550–565.

Korobilis, D. (2013). Assessing the transmission of monetary policy shocks using time-varying parameter dynamic factor models. *Oxford Bulletin of Economics and Statistics 75*, 157–179.

Kose, A., C. Otrok, and C. H. Whiteman (2003). International business cycles: World, region, and country specific factors. *American Economic Review 93*, 1216–1239.

Lasinio, G. J., S. K. Sahu, and K. V. Mardia (2005). Modeling rainfall data using a Bayesian kriged-Kalman model. Technical report, School of Mathematics, University of Southampton.

Lawley (1940). The estimation of factor loadings by the method of maximum likelihood. *Proceedings of the Royal Society of Edinburgh 60*, 64–82.

Lawley (1953). Further investigations in factor estimation. *Proceedings of the Royal Society of Edinburgh 60*, 64–82.

Lawley, D. and A. Maxwell (1963). *Factor Analysis as a Statistical Method.* London: Butterworth.

Lee, S. (1981). A Bayesian approach to confirmatory factor analysis. *Psychometrika 46*, 153–160.

Lee, S. E. and S. J. Press (1998). Robustness of Bayesian factor analysis estimates. *Communications in Statistics, Theory and Methods 27*, 1871–1893.

Lee, S. Y. and X. Y. Song (2002). Bayesian selection on the number of factors in a factor analysis model. *Behaviormetrika 29*, 23–40.

Li, K. (2013). Fixed-effects dynamic panel models, a factor analytical method. *Econometrica 81*, 285–314.

Lin, W.-L. (1992). Alternative estimators for factor GARCH models - a Monte Carlo comparison. *Journal of Applied Econometrics 7*, 259–279.

Lopes, H. F. (2003). Expected posterior priors in factor analysis. *Brazilian Journal of Probability and Statistics 17*, 91–105.

Lopes, H. F. and C. M. Carvalho (2007). Factor stochastic volatility with time varying loadings and Markov switching regimes. *Journal of Statistical Planning and Inference 137*, 3082–3091.

Lopes, H. F. and H. S. Migon (2002). Comovements and contagion in emergent markets: Stock indexes volatilities. In C. Gatsonis, R. Kass, A. Carriquiry, A. Gelman, Verdinelli, D. Pauler, and D. Higdon (Eds.), *Case Studies in Bayesian Statistics, Volume 6*, pp. 287–302.

Lopes, H. F., E. Salazar, and D. Gamerman (2008). Spatial dynamic factor models. *Bayesian Analysis 5*, 1–30.

Lopes, H. F., A. Schmidt, E. Salazar, M. Gomez, and M. Achkar (2012). Measuring vulnerability via spatially hierarchical factor models. *Annals of Applied Statistics 6*, 284–404.

Lopes, H. F. and M. West (2004). Bayesian model assessment in factor analysis. *Statistica Sinica 14*, 41–67.

Mardia, K. V., C. R. Goodall, E. Redfern, and F. J. Alonso (1998). The kriged Kalman filter (with discussion). *Test 7*, 217–85.

Martin, J. and R. McDonald (1975). Bayesian estimation in unrestricted factor analysis: A treatment for Heywood cases. *Psychometrika 40*, 505–517.

Mayrink, V. D. and J. E. Lucas (2013). Sparse latent factor models with interactions: Analysis of gene expression data. *The Annals of Applied Statistics 7*, 799–822.

McLachlan, G., R. Bean, and L. Jones (2007). Extension of the mixture of factor analyzers model to incorporate the multivariate t-distribution. *Computational Statistics and Data Analysis 11*, 5327–5338.

Mezzetti, M. (2012). Bayesian factor analysis for spatially correlated data: application to cancer incidence data in Scotland. *Statistical Methods & Applications 21*, 49–74.

Mezzetti, M. and F. C. Billari (2005). Bayesian correlated factor analysis of socio-demographic indicators. *Statistical Methods & Applications 14*, 223–241.

Moench, E., S. Ng, and S. Potter (2011). Dynamic hierarchical factor models. *Review of Economics and Statistics (to appear).*

Molenaar, P. C. M., J. G. D. Gooijer, and B. Schmitz (1992). Dynamic factor analysis of nonstationary multivariate time series. *Psychometrika 57*, 333–349.

Nakajima, J. and M. West (2013). Dynamic factor volatility modeling: A Bayesian latent threshold approach. *Journal of Financial Econometrics 11*, 116–153.

Nardari, F. and J. T. Scruggs (2007). Bayesian analysis of linear factor models with latent factors, multivariate stochastic volatility, and APT pricing restrictions. *Journal of Financial and Quantitative Analysis 42*, 857–892.

Negro, M. D. and C. Otrok (2008). Dynamic factor models with time-varying parameters: Measuring changes in international business cycles. Staff reports, no. 326, Federal Reserve Bank of New York.

Nelson, C. R. and A. F. Siegel (1987). Parsimonious modeling of yield curve. *Journal of Business 60*, 473–489.

Ng, V. and Engle, R. F. and M. Rothschild (1992). A multi-dynamic factor model for stock returns. *Journal of Econometrics 52*, 245–266.

Onatski, A. (2009). Testing hypotheses about the number of factors in large factor models. *Econometrica 77*, 1447–1479.

Onatski, A. (2012). Asymptotics of the principal components estimator of large factor models with weak factors and i.i.d. Gaussian noise. *Journal of Econometrics 168*, 244–258.

Otrok, C. and C. H. Whiteman (1998). Bayesian leading indicators: Measuring and predicting economic conditions in Iowa. *International Economic Review 39*, 997–1014.

Paisley, J. and L. Carin (ICML). Nonparametric factor analysis with beta process priors. *Annals of Applied Statistics*, 777–784.

Pati, D., A. Bhattacharya, N. S. Pillai, and D. B. Dunson (2012). Posterior contraction in sparse Bayesian factor models for massive covariance matrices. Discussion Paper 2012-05, Department of Statistical Science, Duke University.

Peña, D. and G. E. P. Box (1987). Identifying a simplifying structure in time series. *Journal of the American Statistical Association 82*, 836–843.

Peña, D. and P. Poncela (2004). Forecasting with nonstationary dynamic factor models. *Journal of Econometrics 119*, 291–321.

Pérez, J. M. and J. M. Berger (2002). Expected posterior prior distributions for model selection. *Biometrika 89*, 491–512.

Philipov, A. and M. E. Glickman (2006). Multivariate stochastic volatility via Wishart processes. *Journal of Business and Economic Statistics 24*, 313–328.

Pitt, M. and N. Shephard (1999). Time varying covariances: A factor stochastic volatility approach (with discussion). In J. Bernardo, J. Berger, A. Dawid, and A. Smith (Eds.), *Bayesian statistics 6*, pp. 547–570. London: Oxford University Press.

Polasek, W. (1997). Factor analysis and outliers: A Bayesian approach. Discussion paper, University of Basel.

Press, S. (1982). *Applied Multivariate Analysis: Using Bayesian and Frequentist Methods of Inference (2nd edition).* New York: Krieger.

Press, S. and K. Shigemasu (1989). Bayesian inference in factor analysis. In *ASA Proceedings of Social Statistics Section*, pp. 292–294.

Press, S. and K. Shigemasu (1994). Posterior distribution for the number of factors. In *ASA Proceedings of Bayesian Statistical Science*, pp. 75–77.

Rosenberg, B. and W. McKibben (1973). The prediction of systematic and specific risk in common stocks. *The Journal of Financial and Quantitative Analysis 8*, 17–333.

Rubin, D. B. and D. T. Thayer (1982). EM algorithms for ML factor analysis. *Psychometrika 47*, 69–76.

Rubin, D. B. and D. T. Thayer (1983). More on the EM for factor analysis. *Psychometrika 48*, 253–257.

Sahu, S. K. and K. V. Mardia (2005). A Bayesian kriged-Kalman model for short-term forecasting of air pollution levels. *Applied Statistics 54*, 223–44.

Sansó, B., A. M. Schmidt, and A. A. Nobre (2008). Bayesian spatio-temporal models based on discrete convolutions. *Canadian Journal of Statistics 36*, 239–58.

Sentana, E. (1998). The relation between conditionally heteroskedastic factor models and factor GARCH models. *Econometrics Journal 1*, 1–9.

Sentana, E., G. Calzolari, and G. Fiorentini (2008). Indirect estimation of large conditionally heteroskedastic factor models, with an application to the dow 30 stocks. *Journal of Econometrics 146*, 10–25.

So, M. K. P., K. Lam, and W. K. Li (1998). A stochastic volatility model with Markov switching. *Journal of Business and Economic Statistics 16*, 244–253.

Song, X. Y. and S. Y. Lee (2001). Bayesian estimation and test for factor analysis model with continuous and polytomous data in several populations. *British Journal of Mathematical & Statistical Psychology 54*, 237–263.

Spearman (1904). General intelligence objectively determined and measured. *American Journal of Psychology 15*, 201–292.

Stein, M. L. (1999). *Statistical Interpolation of Spatial Data: Some Theory for Kriging.* New York: Springer.

Stock, J. H. and M. W. Watson (2002a). Forecasting using principal components from a large number of predictors. *Journal of the American Statistical Association 97*, 1167–1179.

Stock, J. H. and M. W. Watson (2002b). Macroeconomic forecasting using diffusion indexes. *Journal of Business and Economic Statistics 20*, 147–162.

Stock, J. H. and M. W. Watson (2005). Implications of dynamic factor models for var analysis. Working paper no. 11467, NBER.

Stock, J. H. and M. W. Watson (2006). Forecasting with many predictors. In G. Elliott, C. W. J. Granger, and A. Timmermann (Eds.), *Handbook of Economic Forecasting*, Volume 1, pp. 516–554. Elsevier.

Sun, D., R. Tsutakawa, and P. Speckman (1999). Posterior distribution of hierarchical models using CAR(1) distributions. *Biometrika 86*, 341–350.

Thomson, G. H. (1953). *The Factor Analysis of Human Ability.* University of London Press, London.

Thurstone, L. L. (1935). *Vectors of the Mind.* University of Chicago Press, Chicago.

Thurstone, L. L. (1947). *Multiple-Factor Analysis.* University of Chicago Press, Chicago.

Tsay, R. (2005). *Analysis of Financial Time Series.* Wiley.

Uhlig, H. and P. A. Ahmadi (2012). Measuring the dynamic effects of monetary policy shocks: A Bayesian FAVAR approach with sign restriction. 2012 Meeting Papers 1060, Society for Economic Dynamics.

Utsugi, A. and T. Kumagai (2001). Bayesian analysis of mixture of factor analysers. *Neural Computation 13*, 993–1002.

Vasicek, O. (1977). An equilibrium characterization of the term structure. *Journal of Financial Economics 5*, 177–188.

Vrontros, I. D., P. Dellaportas, and D. N. Politis (2003). A full-factor multivariate GARCH model. *Econometrics Journal 6*, 312–334.

Wang, F. and M. M. Wall (2003). Generalized common spatial factor model. *Biostatistics 4*, 569–582.

West, M. (2003). Bayesian factor regression models in the "large p, small n" paradigm. In J. M. Bernardo, M. J. Bayarri, J. O. Berger, A. P. Dawid, D. Heckerman, A. F. M. Smith, and M. West (Eds.), *Bayesian Statistics 7*, pp. 723–732. London: Oxford University Press.

West, M. and P. J. Harrison (1997). *Bayesian Forecasting and Dynamic Model.* New York: Springer-Verlag.

Wikle, C. K. and N. Cressie (1999). A dimension-reduced approach to space-time Kalman filtering. *Biometrika 86*, 815–829.

Yoshida, R. and M. West (2013). Bayesian learning in sparse graphical factor models via variational mean-field annealing. *Journal of Machine Learning Research 11*, 1771–1798.

Zellner, A. (1971). *An Introduction to Bayesian Inference in Econometrics.* New York: Wiley.

Zhang, Z. and J. R. Nesselroade (2007). Bayesian estimation of categorical dynamic factor models. *Multivariate Behavioral Research 42*, 729.

Zhou, X., J. Nakajima, and M. West (2012). Bayesian forecasting and portfolio decisions using dynamic dependent factor models. Discussion Paper 2012-09, Department of Statistical Science, Duke University.

CHAPTER 6

ESTIMATION OF STOCHASTIC VOLATILITY MODELS WITH HEAVY TAILS AND SERIAL DEPENDENCE

JOSHUA C.C. CHAN[1] AND CODY Y.L. HSIAO[2]

[1]Research School of Economics, Australian National University, Australia.
[2]Crawford School of Public Policy, Australian National University, Australia.

6.1 Introduction

Financial time series often exhibit properties that depart from the usual assumptions of serial independence and normality. These include volatility clustering, heavy-tailedness and serial dependence. A voluminous literature on different approaches for modeling these empirical regularities has emerged in the last decade. In this chapter we review the estimation of a variety of highly flexible stochastic volatility models, and introduce some efficient algorithms based on recent advances in state space simulation techniques. These estimation methods are illustrated via empirical examples involving precious

Bayesian Inference in the Social Sciences.

metal and foreign exchange returns. The corresponding MATLAB code is also provided.[1]

The remaining of the chapter is structured as follows. Section 6.2 first discusses the basic stochastic volatility model and its estimation. In particular, we provide details of the auxiliary mixture sampler and the precision sampler for linear Gaussian state space models. In Section 6.3 we extend the basic stochastic volatility model to allow for moving average errors. We then discuss an efficient estimation method based on fast band matrix routines.

Lastly, Section 6.4 considers another extension—instead of the conventional assumption of a Gaussian error distribution, we discuss some heavy-tailed distributions that can be written as scale mixtures of Gaussian distributions. We demonstrate the relevance of these heavy-tailed stochastic volatility models through an empirical example.

6.2 Stochastic Volatility Model

In financial time series, it is often observed that large changes in observations tend to be followed by large changes, while small changes are followed by small changes—the phenomenon that is referred to as **volatility clustering**. For instance, during financial crises movements in financial asset returns tend be large (of either sign), whereas in "normal periods" the same asset returns might exhibit little time variation.

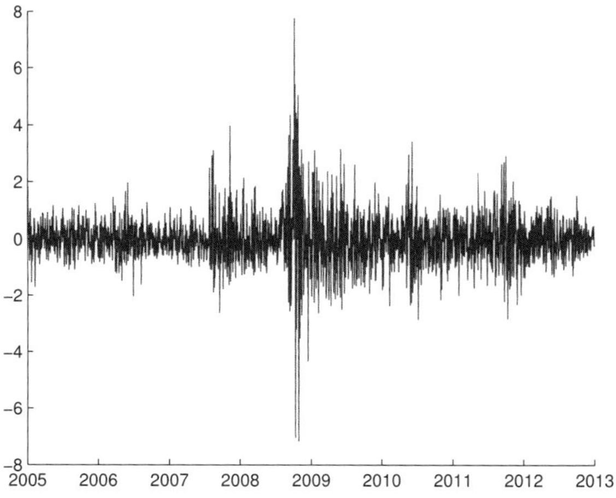

Figure 6.1: AUD/USD daily returns from January 2005 to December 2012.

[1] MATLAB code is available at `http://people.anu.edu.au/joshua.chan`.

As an example, Figure 6.1 depicts the AUD/USD daily returns from January 2005 to December 2012. For a long stretch of time from early 2005 to mid-2007, the daily returns mostly fluctuate between $\pm 2\%$. However, during the global financial crisis of 2007-2008, the volatility of the daily returns increases dramatically—often reaching as high as $\pm 4\%$, and sometimes even larger.

Models that assume constant variance, by definition, cannot accommodate time-varying volatility, and therefore cannot model volatility clustering that is a prominent feature in a wide variety of financial data. In this section we introduce a class of state space models that allow for this important feature. Extensions of these simple models are discussed in later sections.

We focus our discussion on modeling the variance of the time series, and for the moment assume that the observations have a zero mean; a constant mean or a suitable conditional mean process such as an AR(p) component can be added later on, as is done in the empirical example in Section 6.2.3. Specifically, under the **stochastic volatility** model, the observation at time t is given by

$$y_t = e^{\frac{1}{2}h_t}\epsilon_t \tag{6.1}$$

for $t = 1, \ldots, T$, where $\epsilon_t \sim \mathcal{N}(0, 1)$. Hence, the conditional variance of y_t is $Var(y_t \,|\, h_t) = e^{h_t}$, and the state h_t is often called the **log-volatility**. The states are assumed to evolve according to a stationary process

$$h_t = \mu_h + \phi_h(h_{t-1} - \mu_h) + \zeta_t \tag{6.2}$$

for $t = 2, \ldots, T$, where $\zeta_t \sim \mathcal{N}(0, \sigma_h^2)$ and is independent of ϵ_t at all leads and lags. Here we assume that $|\phi_h| < 1$, and the states are initialized with $h_1 \sim \mathcal{N}(\mu_h, \sigma_h^2/(1 - \phi_h^2))$, which is the stationary distribution of the process.

To complete the model specification, we assume independent prior distributions for μ_h, ϕ_h and σ_h^2, i.e., $p(\mu_h, \phi_h, \sigma_h^2) = p(\mu_h)p(\phi_h)p(\sigma_h^2)$. In particular, consider the following independent prior distributions:

$$\mu_h \sim \mathcal{N}(\mu_{h0}, V_{\mu_h}), \quad \phi_h \sim \mathcal{N}(\phi_{h0}, V_{\phi_h})1(|\phi_h| < 1), \quad \sigma_h^2 \sim \mathcal{IG}(\nu_h, S_h), \tag{6.3}$$

where $1(\cdot)$ denotes the indicator function and $\mathcal{IG}(\cdot, \cdot)$ represents the inverse-gamma distribution. Note that we impose the stationarity condition $|\phi_h| < 1$ through the prior distribution of ϕ_h.

The stochastic volatility model is an example of a *nonlinear* state space model where the measurement equation (6.1) is nonlinear in the state. One major challenge of estimating this nonlinear model is that the joint conditional density of the states $\mathbf{h} = (h_1, \ldots, h_T)'$ given the model parameters and the data is high-dimensional and nonstandard—in contrast to a linear Gaussian state space model where the conditional density of the states is Gaussian. Consequently, Bayesian estimation using MCMC becomes more difficult.

6.2.1 Auxiliary Mixture Sampler

In this subsection we discuss a popular approach—the so-called **auxiliary mixture sampler** (Kim et al., 1998)—for estimating the stochastic volatility model (6.1)–(6.2). In a nutshell, the idea is to approximate the nonlinear stochastic volatility model using a mixture of linear Gaussian models, where the estimation of the latter models is standard. Specifically, we first transform the observation y_t so that the measurement equation becomes linear in the log-volatility h_t. More precisely, we square both sides of (6.1) and take the logarithm:

$$y_t^* = h_t + \epsilon_t^*, \tag{6.4}$$

where $y_t^* = \log y_t^2$ and $\epsilon_t^* = \log \epsilon_t^2$. In practice, we often set $y_t^* = \log(y_t^2 + c)$ for some small constant c, say, $c = 10^{-4}$, to avoid numerical problems when y_t is close to zero. Then, (6.2) and (6.4) define a linear state space model in h_t. Note, however, that the error ϵ_t^* no longer has a Gaussian distribution—in fact, it follows a log-χ_1^2 distribution—and the machinery for fitting linear Gaussian state space models cannot be directly applied.

To tackle this difficulty, the second step of the auxiliary mixture sampling approach is to obtain an appropriate Gaussian mixture that well approximates $f(\epsilon_t^*)$, the density function of ϵ_t^*. More precisely,

$$f(\epsilon_t^*) \approx \sum_{i=1}^{n} p_i \, \varphi(\epsilon_t^* \,;\, \mu_i, \sigma_i^2), \tag{6.5}$$

where $\varphi(x\,;\,\mu, \sigma^2)$ denotes the Gaussian density with mean μ and variance σ^2, p_i is the probability of the ith mixture component, and n is the number of components.

We can equivalently write the mixture density in (6.5) in terms of an auxiliary random variable $s_t \in \{1, \ldots, n\}$ that serves as the mixture component indicator (hence, the name of the approach):

$$(\epsilon_t^* \mid s_t = i) \sim \mathcal{N}(\mu_i, \sigma_i^2), \tag{6.6}$$

$$P(s_t = i) = p_i. \tag{6.7}$$

Using this representation, we have a linear Gaussian model conditional on the component indicator s_t and the simulation techniques for estimating such a model can be applied.

The only missing piece is a suitable Gaussian mixture. By matching the moments of the log-χ_1^2 distribution, Kim et al. (1998) propose a seven-component Gaussian mixture

$$f(x) = \sum_{i=1}^{7} p_i \, \varphi(x\,;\,\mu_i - 1.2704, \sigma_i^2),$$

where the values of the parameters are given in Table 6.1. We emphasize that these values are fixed and do not depend on any unknown parameters. Hence,

Table 6.1: A seven-component Gaussian mixture for approximating the log-χ_1^2 distribution.

Component	p_i	μ_i	σ_i^2
1	0.00730	−10.12999	5.79596
2	0.10556	−3.97281	2.61369
3	0.00002	−8.56686	5.17950
4	0.04395	2.77786	0.16735
5	0.34001	0.61942	0.64009
6	0.24566	1.79518	0.34023
7	0.25750	−1.08819	1.26261

this approximation does not require any additional computation time in the estimation.

In summary, using the Gaussian mixture approximation in (6.6) and (6.7), the model (6.2) and (6.4) is now conditionally linear Gaussian given the component indicators $\mathbf{s} = (s_1, \ldots, s_T)'$. For later reference, let $\mathbf{y}^* = (y_1^*, \ldots, y_T^*)'$, and similarly define $\mathbf{h}$, $\boldsymbol{\zeta}$ and $\boldsymbol{\epsilon}^*$. Bayesian analysis can be performed using a sample from the joint posterior distribution $p(\mathbf{h}, \mathbf{s}, \mu_h, \phi_h, \sigma_h^2 \,|\, \mathbf{y})$. Posterior draws can be obtained via a Gibbs sampler that cycles through

1. $p(\mathbf{h} \,|\, \mathbf{y}^*, \mathbf{s}, \mu_h, \phi_h, \sigma_h^2)$;
2. $p(\mathbf{s} \,|\, \mathbf{y}^*, \mathbf{h}, \mu_h, \phi_h, \sigma_h^2) = p(\mathbf{s} \,|\, \mathbf{y}^*, \mathbf{h})$;
3. $p(\mu_h \,|\, \mathbf{y}, \mathbf{h}, \mathbf{s}, \phi_h, \sigma_h^2) = p(\mu_h \,|\, \mathbf{h}, \phi_h, \sigma_h^2)$;
4. $p(\phi_h \,|\, \mathbf{y}, \mathbf{h}, \mathbf{s}, \mu_h, \sigma_h^2) = p(\phi_h \,|\, \mathbf{h}, \mu_h, \sigma_h^2)$;
5. $p(\sigma_h^2 \,|\, \mathbf{y}, \mathbf{h}, \mathbf{s}, \mu_h, \phi_h) = p(\sigma_h^2 \,|\, \mathbf{h}, \mu_h, \phi_h)$.

These MCMC draws are from the approximate model, and one can reweight the draws using importance sampling weights to get the exact posterior moments under the original model. However, this step is often skipped in practice as reweighting makes little difference; see also the discussion in Kim et al. (1998).

6.2.2 Precision Sampler for Linear Gaussian State Space Models

Given the prior distributions in (6.3), the conditionally linear Gaussian model in (6.2), (6.4), (6.6) and (6.7) can be estimated using standard MCMC techniques. In particular, one can simulate from the joint distribution of the log-volatilities $p(\mathbf{h} \,|\, \mathbf{y}^*, \mathbf{s}, \mu_h, \phi_h, \sigma_h^2)$ using Kalman filter-based algorithms such as those discussed in Carter and Kohn (1994), Früwirth-Schnatter (1994), de Jong and Shephard (1995), Durbin and Koopman (2002).

Instead, here we apply a new algorithm that exploits the special structure of the model. Specifically, we will show below that the precision matrix—inverse of the covariance matrix—of $p(\mathbf{h} \mid \mathbf{y}^*, \mathbf{s}, \mu_h, \phi_h, \sigma_h^2)$ is a band matrix. That is, it contains only a small number of nonzero elements along a diagonal band. As such, computations involving band matrices can be done much faster compared to those involving full matrices.

For instance, obtaining the Cholesky decomposition of a band $T \times T$ matrix with fixed bandwidth involves only $\mathcal{O}(T)$ operations (see, e.g., Golub and van Loan, 1983, p. 156) as opposed to $\mathcal{O}(T^3)$ for a full matrix of the same size. A similar reduction in computations holds for other operations such as multiplication and forward-backward substitution for solving linear systems. These band matrix algorithms are implemented in standard packages such as MATLAB, GAUSS and R.

The idea of exploiting band precision matrices to speed up computations can be traced back to Rue (2001), who discusses simulation of Gaussian Markov random fields; Chan and Jeliazkov (2009) and McCausland et al. (2011) propose similar algorithms for linear Gaussian state space models, which are later used in various applications in, e.g., Belmonte and Koop (2013), Chan et al. (2012), Chan et al. (2013), and Jeliazkov and Liu (2010).

We first derive the joint distribution of the states $p(\mathbf{h} \mid \mathbf{y}^*, \mathbf{s}, \mu_h, \phi_h, \sigma_h^2)$. To that end, we rewrite (6.4) in matrix notation:

$$\mathbf{y}^* = \mathbf{h} + \boldsymbol{\epsilon}^*,$$

where $(\boldsymbol{\epsilon}^* \mid \mathbf{s}) \sim \mathcal{N}(\mathbf{d}, \Sigma_{\mathbf{y}^*})$, with $\mathbf{d} = (\mu_{s_1} - 1.2704, \ldots, \mu_{s_T} - 1.2704)'$, $\Sigma_{\mathbf{y}^*} = \operatorname{diag}(\sigma_{s_1}^2, \ldots, \sigma_{s_T}^2)$, and the fixed parameters $\mu_1, \ldots, \mu_7$ and $\sigma_1^2, \ldots, \sigma_7^2$ are given in Table 6.1. By a simple change of variable, we have $(\mathbf{y}^* \mid \mathbf{s}, \mathbf{h}) \sim \mathcal{N}(\mathbf{h} + \mathbf{d}, \Sigma_{\mathbf{y}^*})$. It follows that

$$\log p(\mathbf{y}^* \mid \mathbf{s}, \mathbf{h}) = -\frac{1}{2}(\mathbf{y}^* - \mathbf{h} - \mathbf{d})'\Sigma_{\mathbf{y}^*}^{-1}(\mathbf{y}^* - \mathbf{h} - \mathbf{d}) + c_1, \tag{6.8}$$

where c_1 is a normalization constant.

Next, rewrite the state equation (6.2) in matrix form:

$$\mathbf{H}_{\phi_h}\mathbf{h} = \widetilde{\boldsymbol{\alpha}} + \boldsymbol{\zeta},$$

where $\boldsymbol{\zeta} \sim \mathcal{N}(\mathbf{0}, \Sigma_{\mathbf{h}})$, $\Sigma_{\mathbf{h}} = \operatorname{diag}(\sigma_h^2/(1-\phi_h^2), \sigma_h^2, \ldots, \sigma_h^2)$, and

$$\widetilde{\boldsymbol{\alpha}} = \begin{pmatrix} \mu_h \\ (1-\phi_h)\mu_h \\ \vdots \\ (1-\phi_h)\mu_h \end{pmatrix}, \quad \mathbf{H}_{\phi_h} = \begin{pmatrix} 1 & 0 & 0 & \cdots & 0 \\ -\phi_h & 1 & 0 & \cdots & 0 \\ 0 & -\phi_h & 1 & \cdots & 0 \\ \vdots & \ddots & \ddots & \ddots & \vdots \\ 0 & \cdots & 0 & -\phi_h & 1 \end{pmatrix}.$$

Here $\mathbf{H}_{\phi_h}$ is a band matrix with determinant $|\mathbf{H}_{\phi_h}| = 1$ for all values of ϕ_h. By a change of variable, we have

$$(\mathbf{h} \,|\, \mu_h, \phi_h, \sigma_h^2) \sim \mathcal{N}(\boldsymbol{\alpha}, (\mathbf{H}_{\phi_h}' \Sigma_{\mathbf{h}}^{-1} \mathbf{H}_{\phi_h})^{-1}),$$

where $\boldsymbol{\alpha} = \mathbf{H}_{\phi_h}^{-1} \widetilde{\boldsymbol{\alpha}}$. Noting that $|(\mathbf{H}_{\phi_h}' \Sigma_{\mathbf{h}}^{-1} \mathbf{H}_{\phi_h})^{-1}| = |\Sigma_{\mathbf{h}}| = (\sigma_h^2)^T/(1-\phi_h^2)$, we have

$$\log p(\mathbf{h} \,|\, \mu_h, \phi_h, \sigma_h^2) = -\frac{1}{2} \log \left(\frac{(2\pi\sigma_h^2)^T}{1-\phi_h^2} \right) - \frac{1}{2} (\mathbf{h} - \boldsymbol{\alpha})' \mathbf{H}_{\phi_h}' \Sigma_{\mathbf{h}}^{-1} \mathbf{H}_{\phi_h} (\mathbf{h} - \boldsymbol{\alpha}). \tag{6.9}$$

Notice that in this case the precision matrix $\mathbf{H}_{\phi_h}' \Sigma_{\mathbf{h}}^{-1} \mathbf{H}_{\phi_h}$ is also band; in fact, it is tridiagonal.

Since $p(\mathbf{h} \,|\, \mathbf{y}^*, \mathbf{s}, \mu_h, \phi_h, \sigma_h^2) \propto p(\mathbf{y}^* \,|\, \mathbf{s}, \mathbf{h}) p(\mathbf{h} \,|\, \mu_h, \phi_h, \sigma_h^2)$, using the expressions in (6.8) and (6.9), we have

$$\begin{aligned}
&\log p(\mathbf{h} \,|\, \mathbf{y}^*, \mathbf{s}, \mu_h, \phi_h, \sigma_h^2) \\
&\quad = -\frac{1}{2} (\mathbf{y}^* - \mathbf{h} - \mathbf{d})' \Sigma_{\mathbf{y}^*}^{-1} (\mathbf{y}^* - \mathbf{h} - \mathbf{d}) \\
&\qquad - \frac{1}{2} (\mathbf{h} - \boldsymbol{\alpha})' \mathbf{H}_{\phi_h}' \Sigma_{\mathbf{h}}^{-1} \mathbf{H}_{\phi_h} (\mathbf{h} - \boldsymbol{\alpha}) + c_2 \qquad (6.10) \\
&\quad = -\frac{1}{2} \left[\mathbf{h}' (\mathbf{H}_{\phi_h}' \Sigma_{\mathbf{h}}^{-1} \mathbf{H}_{\phi_h} + \Sigma_{\mathbf{y}^*}^{-1}) \mathbf{h} \right. \\
&\qquad \left. -2\mathbf{h}' (\mathbf{H}_{\phi_h}' \Sigma_{\mathbf{h}}^{-1} \mathbf{H}_{\phi_h} \boldsymbol{\alpha} + \Sigma_{\mathbf{y}^*}^{-1} (\mathbf{y}^* - \mathbf{d})) \right] + c_3,
\end{aligned}$$

where c_2 and c_3 are constants not depending on $\mathbf{h}$. Since this log-density is quadratic in $\mathbf{h}$, it is Gaussian and we have

$$(\mathbf{h} \,|\, \mathbf{y}^*, \mathbf{s}, \mu_h, \phi_h, \sigma_h^2) \sim \mathcal{N}(\widehat{\mathbf{h}}, \mathbf{K}_{\mathbf{h}}^{-1})$$

for some mean vector $\widehat{\mathbf{h}}$ and precision matrix $\mathbf{K}_{\mathbf{h}}$. To determine $\widehat{\mathbf{h}}$ and $\mathbf{K}_{\mathbf{h}}$, compare (6.10) with the log-density

$$\log p(\mathbf{h} \,|\, \mathbf{y}^*, \mathbf{s}, \mu_h, \phi_h, \sigma_h^2) = -\frac{1}{2} (\mathbf{h}' \mathbf{K}_{\mathbf{h}} \mathbf{h} - 2\mathbf{h}' \mathbf{K}_{\mathbf{h}} \widehat{\mathbf{h}}) + c_4$$

for some normalization constant c_4. Hence, we have

$$\mathbf{K}_{\mathbf{h}} = \mathbf{H}_{\phi_h}' \Sigma_{\mathbf{h}}^{-1} \mathbf{H}_{\phi_h} + \Sigma_{\mathbf{y}^*}^{-1}, \quad \widehat{\mathbf{h}} = \mathbf{K}_{\mathbf{h}}^{-1} (\mathbf{H}_{\phi_h}' \Sigma_{\mathbf{h}}^{-1} \mathbf{H}_{\phi_h} \boldsymbol{\alpha} + \Sigma_{\mathbf{y}^*}^{-1} (\mathbf{y}^* - \mathbf{d})).$$

Next, we discuss an efficient method to sample from the high-dimensional Gaussian distribution $\mathcal{N}(\widehat{\mathbf{h}}, \mathbf{K}_{\mathbf{h}}^{-1})$ by exploiting the special structure of the precision matrix $\mathbf{K}_{\mathbf{h}}$. More specifically, since $\mathbf{K}_{\mathbf{h}}$ is a band matrix, $\widehat{\mathbf{h}}$ can be obtained quickly by solving the linear system

$$\mathbf{K}_{\mathbf{h}} \mathbf{x} = \mathbf{H}_{\phi_h}' \Sigma_{\mathbf{h}}^{-1} \mathbf{H}_{\phi_h} \boldsymbol{\alpha} + \Sigma_{\mathbf{y}^*}^{-1} (\mathbf{y}^* - \mathbf{d})$$

for $\mathbf{x}$, which avoids computing the inverse $\mathbf{K}_{\mathbf{h}}^{-1}$.

To get a draw from $\mathcal{N}(\widehat{\mathbf{h}}, \mathbf{K}_{\mathbf{h}}^{-1})$, we first obtain the Cholesky factor $\mathbf{C}_{\mathbf{h}}$ of $\mathbf{K}_{\mathbf{h}}$, i.e., $\mathbf{K}_{\mathbf{h}} = \mathbf{C}_{\mathbf{h}}\mathbf{C}_{\mathbf{h}}'$. Again, this operation can be done quickly as $\mathbf{K}_{\mathbf{h}}$ is a band matrix. Next, we obtain T independent $\mathcal{N}(0,1)$ draws $\mathbf{U} = (U_1, \ldots, U_T)'$, and return $\mathbf{V} = \widehat{\mathbf{h}} + (\mathbf{C}_{\mathbf{h}}')^{-1}\mathbf{U}$. It is easy to check that the mean vector of $\mathbf{V}$ is $\widehat{\mathbf{h}}$ and its covariance matrix is

$$(\mathbf{C}_{\mathbf{h}}')^{-1}((\mathbf{C}_{\mathbf{h}}')^{-1})' = (\mathbf{C}_{\mathbf{h}}')^{-1}(\mathbf{C}_{\mathbf{h}})^{-1} = (\mathbf{C}_{\mathbf{h}}\mathbf{C}_{\mathbf{h}}')^{-1} = \mathbf{K}_{\mathbf{h}}^{-1}.$$

We note that once again $(\mathbf{C}_{\mathbf{h}}')^{-1}\mathbf{U}$ can be computed quickly by solving a suitable linear system without computing the inverse $(\mathbf{C}_{\mathbf{h}}')^{-1}$. This completes Step 1 of the Gibbs sampler for the stochastic volatility model.

To implement Step 2, note that $p(\mathbf{s} \,|\, \mathbf{y}^*, \mathbf{h}) = \prod_{t=1}^T p(s_t \,|\, y_t^*, h_t)$, and therefore we can sample each s_t independently conditional on $\mathbf{y}^*$ and $\mathbf{h}$. Since s_t is a discrete random variable that follows a seven-point distribution, it can be easily sampled via the inverse-transform method (see, e.g., Algorithm 3.2 in Kroese et al., 2011), provided we can compute $P(s_t = i \,|\, y_t^*, h_t)$ for $i = 1, \ldots, 7$. In fact, we have

$$P(s_t = i \,|\, y_t^*, h_t) = \frac{1}{c_t} p_i \, \varphi(y_t^* \,;\, h_t + \mu_i - 1.2704, \sigma_i^2),$$

where $c_t = \sum_{j=1}^7 p_j \, \varphi(y_t^* \,;\, h_t + \mu_j - 1.2704, \sigma_j^2)$ is a normalization constant.

Steps 3 and 5 can be done easily, as both conditional distributions are standard. In particular,

$$(\mu_h \,|\, \mathbf{h}, \phi_h, \sigma_h^2) \sim \mathcal{N}(\widehat{\mu}_h, D_{\mu_h}),$$

where $D_{\mu_h} = (V_{\mu_h}^{-1} + \mathbf{X}_{\mu_h}' \Sigma_{\mathbf{h}}^{-1} \mathbf{X}_{\mu_h})^{-1}$ and $\widehat{\mu}_h = D_{\mu_h}(V_{\mu_h}^{-1}\mu_{h0} + \mathbf{X}_{\mu_h}'\Sigma_{\mathbf{h}}^{-1}\mathbf{z}_{\mu_h})$ with $\mathbf{X}_{\mu_h} = (1, 1-\phi_h, \ldots, 1-\phi_h)'$, $\mathbf{z}_{\mu_h} = (h_1, h_2 - \phi_h h_1, \ldots, h_T - \phi_h h_{T-1})'$ and $\Sigma_{\mathbf{h}} = \mathrm{diag}(\sigma_h^2/(1-\phi_h^2), \sigma_h^2, \ldots, \sigma_h^2)$. In addition, we have

$$(\sigma_h^2 \,|\, \mathbf{h}, \mu_h, \phi_h) \sim \mathcal{IG}(\nu_h + T/2, \widetilde{S}_h),$$

where $\widetilde{S}_h = S_h + [(1-\phi_h^2)(h_1 - \mu_h)^2 + \sum_{t=2}^T (h_t - \mu_h - \phi_h(h_{t-1} - \mu_h))^2]/2$.

Finally, it follows from (6.2) and (6.3) that

$$p(\phi_h \,|\, \mathbf{h}, \mu_h, \sigma_h^2) \propto p(\phi_h) g(\phi_h) e^{-\frac{1}{2\sigma_h^2}\sum_{t=2}^T (h_t - \mu_h - \phi_h(h_{t-1}-\mu_h))^2},$$

where $g(\phi_h) = (1-\phi_h^2)^{1/2}\exp(-\frac{1}{2\sigma_h^2}(1-\phi_h^2)(h_1-\mu_h)^2)$ and $p(\phi_h)$ is the truncated normal prior given in (6.3). The conditional density $p(\phi_h \,|\, \mathbf{h}, \mu_h, \sigma_h^2)$ is nonstandard, and we implement an independence-chain Metropolis-Hastings step with proposal distribution $\mathcal{N}(\widehat{\phi}_h, D_{\phi_h})1(|\phi_h| < 1)$, where $D_{\phi_h} = (V_{\phi_h}^{-1} + \mathbf{X}_{\phi_h}'\mathbf{X}_{\phi_h}/\sigma_h^2)^{-1}$ and $\widehat{\phi}_h = D_{\phi_h}(V_{\phi_h}^{-1}\phi_{h0} + \mathbf{X}_{\phi_h}'\mathbf{z}_{\phi_h}/\sigma_h^2)$, with $\mathbf{X}_{\phi_h} = (h_1 - \mu_h, \ldots, h_{T-1} - \mu_h)'$ and $\mathbf{z}_{\phi_h} = (h_2 - \mu_h, \ldots, h_T - \mu_h)'$. Then, given the current draw ϕ_h, a proposal ϕ_h^* is accepted with probability $\min(1, g(\phi_h^*)/g(\phi_h))$; otherwise the Markov chain stays at the current state ϕ_h.

6.2.3 Empirical Example: Modeling AUD/USD Returns

We illustrate the estimation of the stochastic volatility model using the data depicted in Figure 6.1—AUD/USD daily returns from January 2005 to December 2012. More specifically, we allow for a constant conditional mean in the measurement equation

$$y_t = \mu + e^{\frac{1}{2}h_t}\epsilon_t$$

for $t = 1, \ldots, T$, where $\epsilon_t \sim \mathcal{N}(0, 1)$. The state equation is again given in (6.2).

We assume the same independent prior distributions for μ_h, ϕ_h and σ_h^2 as in (6.3). In particular, we set $\mu_{h0} = 0, V_{\mu_h} = 5, \phi_{h0} = 0.95, V_{\phi_h} = 1$, $\nu_h = 10$ and $S_h = 0.19$. These values imply relatively diffuse priors with prior means $E(\mu_h) = 0$, $E(\phi_h) = 0.95$ and $E(\sigma_h^2) = 0.02$. For μ we assume the normal prior distribution $\mathcal{N}(\mu_0, V_\mu)$, where $\mu_0 = 0$ and $V_\mu = 5$. Again this prior is relatively diffuse with prior mean $E(\mu) = 0$.

To estimate the model with the constant conditional mean μ, we simply need an extra block to sample from the conditional distribution $p(\mu \,|\, \mathbf{y}, \mathbf{h})$, and modify the Gibbs sampler discussed in Section 6.2.2 by replacing y_t with $y_t - \mu$. In particular, it can be checked that

$$(\mu \,|\, \mathbf{y}, \mathbf{h}) \sim \mathcal{N}(\widehat{\mu}, D_\mu),$$

where $D_\mu = (V_\mu^{-1} + \sum_{t=1}^T e^{-h_t})^{-1}$ and $\widehat{\mu} = D_\mu(V_\mu^{-1}\mu_0 + \sum_{t=1}^T e^{-h_t}y_t)$.

We obtain 20000 draws from the posterior distribution using the Gibbs sampler outlined above, after a burn-in period of 1000. Figure 6.2 depicts the posterior means and quantiles of the time-varying standard deviation $\exp(h_t/2)$. As the figure shows, there is substantial time-variation in the volatility. In particular, between 2005 and early 2007, the estimated standard deviation mostly fluctuates around 0.5%. It increases to about 1% in mid-2007 and peaks at 3% during the global financial crisis. Although it goes down substantially after 2009, it is still much higher than the pre-crisis level.

Table 6.2: Posterior means, standard deviations and quantiles of model parameters; AUD/USD daily returns data.

Parameter	Posterior mean	Posterior std. dev.	5%-tile	95%-tile
μ	−0.029	0.013	−0.051	−0.006
μ_h	−0.748	0.351	−1.275	−0.229
ϕ_h	0.989	0.004	0.982	0.995
σ_h^2	0.017	0.003	0.012	0.023

The posterior means, standard deviations and quantiles of the model parameters are reported in Table 6.2. For instance, over the period from January 2005 to December 2012, the average daily return is estimated to be −0.029%.

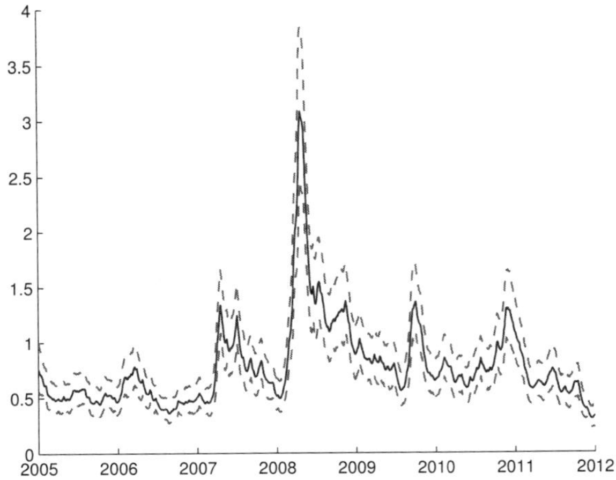

Figure 6.2: Posterior means (solid line) and 90% credible intervals (dash lines) of the time-varying standard deviation $\exp(h_t/2)$; AUD/USD daily returns data.

The posterior mean of the AR(1) coefficient of the state equation is 0.989, indicating a very high level of persistence.

6.3 Moving Average Stochastic Volatility Model

We now discuss an extension of the plain vanilla stochastic volatility model introduced in Section 6.2. Specifically, the model (6.1)–(6.2) assumes the errors in the measurement equation are serially independent given the log-volatilities. This is often an appropriate assumption for modeling financial data if we expect that the market is in some sense efficient. Of course, this assumption can be tested, for example, by building a model that allows for persistence via moving average errors.

To that end, we consider a variant of the moving average stochastic volatility models proposed in Chan and Jeliazkov (2009). More precisely, we expand the stochastic volatility model (6.1)–(6.2) by allowing the errors in the measurement equation to follow an MA(q) process:

$$y_t = \mu + u_t, \tag{6.11}$$

$$u_t = \epsilon_t + \psi_1 \epsilon_{t-1} + \cdots + \psi_q \epsilon_{t-q}, \tag{6.12}$$

$$h_t = \mu_h + \phi_h (h_{t-1} - \mu_h) + \zeta_t, \tag{6.13}$$

where $\epsilon_t \sim \mathcal{N}(0, e^{h_t})$ and $\zeta_t \sim \mathcal{N}(0, \sigma_h^2)$ are independent of each other, and $\epsilon_0 = \epsilon_{-1} = \cdots = \epsilon_{-q+1} = 0$. For identification, we impose the invertiblilty conditions—the roots of the characteristic polynomial associated with the MA coefficients $\boldsymbol{\psi} = (\psi_1, \ldots, \psi_q)'$ are all outside the unit circle. As before, we assume that $|\phi_h| < 1$ and the states are initialized with $h_1 \sim \mathcal{N}(\mu_h, \sigma_h^2/(1-\phi_h^2))$. It is obvious that if $\psi_1 = \cdots = \psi_q = 0$, the model (6.11)–(6.13) reduces to the standard stochastic volatility model discussed in Section 6.2.

Recall that under the standard stochastic volatility model (6.1)–(6.2), the conditional variance of the observation y_t is simply e^{h_t}. Here, under the moving average variant, the conditional variance of y_t is given by

$$Var(y_t \,|\, \mu, \boldsymbol{\psi}, \mathbf{h}) = e^{h_t} + \psi_1^2 e^{h_{t-1}} + \cdots + \psi_q^2 e^{h_{t-q}}.$$

That is, the conditional variance is time-varying through two channels: first, it is a moving average of the $q+1$ most recent variances $e^{h_t}, \ldots, e^{h_{t-q}}$; and second, the log-volatilities also evolve according to the stationary AR(1) process in (6.13).

Compared to the standard stochastic volatility model, another difference is that y_t is no longer serially independent (even after conditioning on the log-volatilities). In fact, its conditional autocovariances are given by

$$Cov(y_t, y_{t-j} \,|\, \mu, \boldsymbol{\psi}, \mathbf{h}) = \sum_{i=0}^{q-j} \psi_{i+j}\psi_i e^{h_{t-i}}$$

for $j = 1, \ldots, q$ and 0 for $j > q$, where $\psi_0 = 1$. Due to the presence of the time-varying log-volatilities, the autocovariances of y_t are also time-varying.

To complete the model specification, we assume independent prior distributions for μ, $\boldsymbol{\psi}$, μ_h, ϕ_h and σ_h^2, i.e., $p(\mu, \boldsymbol{\psi}, \mu_h, \phi_h, \sigma_h^2) = p(\mu)\, p(\boldsymbol{\psi})\, p(\mu_h)\, p(\phi_h)$ $p(\sigma_h^2)$. For $\boldsymbol{\psi}$, we assume a multivariate normal prior with support in the region where the invertibility conditions on $\boldsymbol{\psi}$ hold. For other model parameters, we assume the following independent prior distributions: $\mu \sim \mathcal{N}(\mu_0, V_\mu)$, $\mu_h \sim \mathcal{N}(\mu_{h0}, V_{\mu_h})$, $\phi_h \sim \mathcal{N}(\phi_{h0}, V_{\phi_h})1(|\phi_h| < 1)$ and $\sigma_h^2 \sim \mathcal{IG}(\nu_h, S_h)$.

The next section discusses an efficient algorithm to estimate this moving average model that exploits the band structure of covariance matrix of $\mathbf{y}$. Then in Section 6.3.2 we illustrate the algorithm and the relevance of the extension through an empirical example that involves Philippine peso returns during the global financial crisis.

6.3.1 Estimation

We consider an efficient sampler—that builds upon band matrix algorithms—for estimating the moving average stochastic volatility model in (6.11)–(6.13). A key ingredient of this sampler is a quick way to evaluate the likelihood function that exploits the band structure of the covariance matrix of $\mathbf{y}$—instead of using conventional methods based on the Kalman filter (see, e.g., Gardner et al., 1980).

To obtain the likelihood function, we derive the joint distribution of the observations $\mathbf{y}$. First rewrite (6.12) as

$$\mathbf{u} = \mathbf{H}_{\boldsymbol{\psi}}\boldsymbol{\epsilon}, \tag{6.14}$$

where $\mathbf{H}_{\boldsymbol{\psi}}$ is a $T \times T$ lower triangular matrix with ones on the main diagonal, ψ_1 on first lower diagonal, ψ_2 on second lower diagonal, and so forth. For example, for $q = 2$, we have

$$\mathbf{H}_{\boldsymbol{\psi}} = \begin{pmatrix} 1 & 0 & 0 & 0 & \cdots & 0 \\ \psi_1 & 1 & 0 & 0 & \cdots & 0 \\ \psi_2 & \psi_1 & 1 & 0 & \cdots & 0 \\ 0 & \psi_2 & \psi_1 & 1 & \cdots & 0 \\ \vdots & & \ddots & \ddots & \ddots & \vdots \\ 0 & 0 & \cdots & \psi_2 & \psi_1 & 1 \end{pmatrix}.$$

Since $\boldsymbol{\epsilon} \sim \mathcal{N}(\mathbf{0}, \mathbf{S_y})$ with $\mathbf{S_y} = \text{diag}(e^{h_1}, \ldots, e^{h_T})$, we have $\mathbf{u} \sim \mathcal{N}(\mathbf{0}, \boldsymbol{\Sigma}_\mathbf{y})$, where $\boldsymbol{\Sigma}_\mathbf{y} = \mathbf{H}_{\boldsymbol{\psi}}\mathbf{S_y}\mathbf{H}'_{\boldsymbol{\psi}}$. It follows from (6.11) and (6.14) that

$$(\mathbf{y} \,|\, \mu, \boldsymbol{\psi}, \mathbf{h}) \sim \mathcal{N}(\mu\mathbf{1}, \boldsymbol{\Sigma}_\mathbf{y}),$$

where $\mathbf{1}$ is a $T \times 1$ column of ones.

Note that even though $\mathbf{H}_{\boldsymbol{\psi}}$ is a $T \times T$ matrix that contains T^2 elements, only $(T - q/2)(q+1) < T(q+1)$ of those elements are nonzero. In addition, since $\mathbf{S_y}$ is diagonal, the covariance matrix $\boldsymbol{\Sigma}_\mathbf{y}$ is a band matrix. This band structure is again exploited to speed up computations by using band matrix routines.

Furthermore, noting that $|\mathbf{H}_{\boldsymbol{\psi}}| = 1$ for any $\boldsymbol{\psi}$, we have $|\boldsymbol{\Sigma}_\mathbf{y}| = |\mathbf{S_y}| = \exp\left(\sum_{t=1}^T h_t\right)$. The log-density of $\mathbf{y}$ is therefore given by

$$\log p(\mathbf{y} \,|\, \mu, \boldsymbol{\psi}, \mathbf{h}) = -\frac{T}{2}\log(2\pi) - \frac{1}{2}\sum_{t=1}^T h_t - \frac{1}{2}(\mathbf{y} - \mu\mathbf{1})'\boldsymbol{\Sigma}_\mathbf{y}^{-1}(\mathbf{y} - \mu\mathbf{1}). \tag{6.15}$$

Once again, the product $\boldsymbol{\Sigma}_\mathbf{y}^{-1}(\mathbf{y} - \mu\mathbf{1})$ can be obtained by solving a linear system of equations without computing the inverse of $\boldsymbol{\Sigma}_\mathbf{y}^{-1}$. Therefore, the log-likelihood function (6.15) can be evaluated quickly.

Next, posterior draws can be obtained by sequentially sampling from:

1. $p(\mu \,|\, \mathbf{y}, \mathbf{h}, \boldsymbol{\psi}, \mu_h, \phi_h, \sigma_h^2) = p(\mu \,|\, \mathbf{y}, \mathbf{h}, \boldsymbol{\psi})$;

2. $p(\mathbf{h} \,|\, \mathbf{y}, \mu, \boldsymbol{\psi}, \mu_h, \phi_h, \sigma_h^2)$;

3. $p(\boldsymbol{\psi} \,|\, \mathbf{y}, \mathbf{h}, \mu, \mu_h, \phi_h, \sigma_h^2) = p(\boldsymbol{\psi} \,|\, \mathbf{y}, \mathbf{h}, \mu)$;

4. $p(\mu_h \,|\, \mathbf{y}, \mathbf{h}, \mu, \boldsymbol{\psi}, \phi_h, \sigma_h^2) = p(\mu_h \,|\, \mathbf{h}, \phi_h, \sigma_h^2)$;
5. $p(\phi_h \,|\, \mathbf{y}, \mathbf{h}, \mu, \boldsymbol{\psi}, \mu_h, \sigma_h^2) = p(\phi_h \,|\, \mathbf{h}, \mu_h, \sigma_h^2)$;
6. $p(\sigma_h^2 \,|\, \mathbf{y}, \mathbf{h}, \mu, \boldsymbol{\psi}, \mu_h, \phi_h) = p(\sigma_h^2 \,|\, \mathbf{h}, \mu_h, \phi_h)$.

Step 1 can be easily done by drawing from

$$(\mu \,|\, \mathbf{y}, \mathbf{h}, \boldsymbol{\psi}) \sim \mathcal{N}(\widehat{\mu}, D_\mu),$$

where $D_\mu = (V_\mu^{-1} + \mathbf{1}'\mathbf{\Sigma}_\mathbf{y}^{-1}\mathbf{1})$ and $\widehat{\mu} = D_\mu(V_\mu^{-1}\mu_0 + \mathbf{1}'\mathbf{\Sigma}_\mathbf{y}^{-1}\mathbf{y})$. For Step 2, first note that if we define $\widetilde{\mathbf{y}} = \mathbf{H}_{\boldsymbol{\psi}}^{-1}(\mathbf{y} - \mu\mathbf{1})$, then

$$(\widetilde{\mathbf{y}} \,|\, \mu, \boldsymbol{\psi}, \mathbf{h}) \sim \mathcal{N}(\mathbf{0}, \mathbf{S}_\mathbf{y}),$$

with $\mathbf{S}_\mathbf{y} = \mathrm{diag}(e^{h_1}, \ldots, e^{h_T})$. Hence, the auxiliary mixture sampler discussed in Section 6.2.1 can be directly applied to sample from $p(\mathbf{h} \,|\, \mathbf{y}, \mu, \boldsymbol{\psi}, \mu_h, \phi_h, \sigma_h^2)$.

To implement Step 3, observe that given the prior $p(\boldsymbol{\psi})$, the conditional posterior density is given by $p(\boldsymbol{\psi} \,|\, \mathbf{y}, \mu, \mathbf{h}) \propto p(\boldsymbol{\psi})\, p(\mathbf{y} \,|\, \mu, \boldsymbol{\psi}, \mathbf{h})$, where both densities on the right-hand side can be evaluated quickly at any $\boldsymbol{\psi}$. Here we maximize $\log p(\boldsymbol{\psi} \,|\, \mathbf{y}, \mu, \mathbf{h})$ numerically and obtain the mode and the negative Hessian evaluated at the mode, denoted as $\widehat{\boldsymbol{\psi}}$ and $\mathbf{K}_{\boldsymbol{\psi}}$, respectively. Then, draws from $p(\boldsymbol{\psi} \,|\, \mathbf{y}, \mu, \mathbf{h})$ can be obtained using an independence-chain Metropolis-Hastings step with proposal density $\mathcal{N}(\widehat{\boldsymbol{\psi}}, \mathbf{K}_{\boldsymbol{\psi}}^{-1})$.

Finally, we can sample from the densities $p(\mu_h \,|\, \mathbf{h}, \sigma_h^2, \phi_h)$, $p(\phi_h \,|\, \mathbf{h}, \mu_h, \sigma_h^2)$ and $p(\sigma_h^2 \,|\, \mathbf{h}, \mu_h, \phi_h)$ in the same way as discussed in Section 6.2.2.

6.3.2 Empirical Example: Modeling PHP/USD Returns During Crisis

Figure 6.3 depicts the daily returns of the Philippine peso against the US dollar from the beginning of the global financial crisis, specifically from July 2007 to December 2012. We fit the data using the moving average stochastic volatility model in (6.11)–(6.13). In particular, we consider the MA(1) process $u_t = \epsilon_t + \psi\epsilon_{t-1}$. Recall that this moving average variant reduces to the standard stochastic volatility model when $\psi = 0$. Hence, one main focus is to investigate if the posterior density of ψ has mass concentrated around zero.

For the prior hyperparameters, we set $\mu_0 = 0$, $V_\mu = 5$, $\mu_{h0} = 0, V_{\mu_h} = 5, \phi_{h0} = 0.95, V_{\phi_h} = 1$, $\nu_h = 10$ and $S_h = 0.19$. These values imply relatively diffuse priors with prior means $E(\mu) = 0$, $E(\mu_h) = 0$, $E(\phi_h) = 0.95$ and $E(\sigma_h^2) = 0.02$. For ψ we assume the truncated normal prior distribution $\mathcal{N}(\psi_0, V_\psi)1(|\psi| < 1)$ with $\psi_0 = 0$ and $V_\psi = 1$.

Posterior analysis is based on 20000 draws from the posterior distribution obtained via the sampler described in Section 6.3.1, after a burn-in period of 1000. We report the posterior means, standard deviations and quantiles of the model parameters in Table 6.3. The average daily return over the sample period is estimated to be -0.009%. Compared to the results for AUD/USD

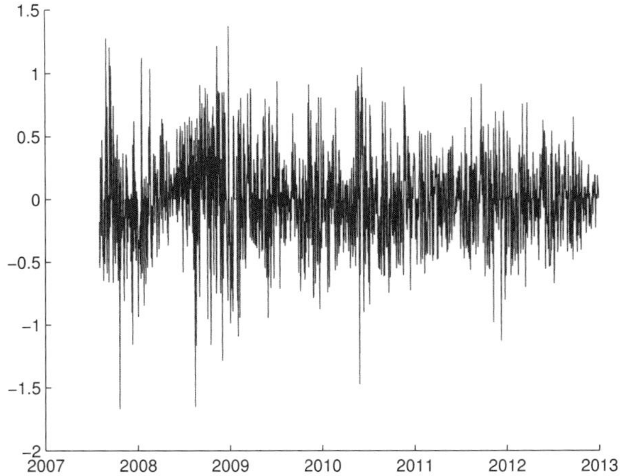

Figure 6.3: PHP/USD daily returns from July 2007 to December 2012.

Table 6.3: Posterior means, standard deviations and quantiles of model parameters; PHP/USD daily returns data.

Parameter	Posterior mean	Posterior std. dev.	5%-tile	95%-tile
μ	−0.009	0.010	−0.024	0.007
μ_h	−2.187	0.290	−2.589	−1.793
ϕ_h	0.983	0.008	0.970	0.994
σ_h^2	0.018	0.004	0.012	0.025
ψ	0.141	0.026	0.098	0.184

daily returns data (see Table 6.2), PHP/USD daily returns exhibit much lower volatility, although ϕ_h is estimated to be about the same magnitude.

The MA(1) coefficient ψ is quite precisely estimated with posterior mean around 0.141. In addition, a 90% credible interval is (0.098, 0.184)—which excludes zero—indicating that ψ is highly unlikely to be around zero. To investigate this issue further, we compute the posterior density $p(\psi \,|\, \mathbf{y})$ as follows. Since this density has support in $(-1, 1)$ and is known up to a normalization constant, it can be evaluated on a grid and renormalized such that the area under the curve is one. We can then estimate $p(\psi \,|\, \mathbf{y})$ using the Monte Carlo average

$$\widehat{p(\psi \,|\, \mathbf{y})} = \frac{1}{R}\sum_{i=1}^{R} p(\psi \,|\, \mathbf{y}, \mu^{(i)}, \mathbf{h}^{(i)})$$

by summing over the posterior draws $\mu^{(i)}$ and $\mathbf{h}^{(i)}$ for $i = 1, \ldots, R$.

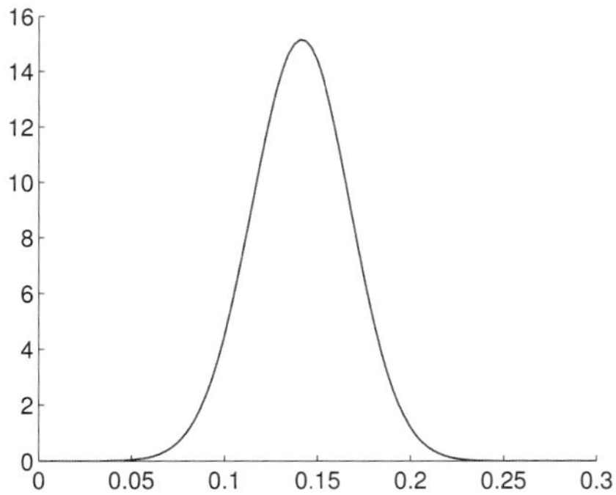

Figure 6.4: Estimate of $p(\psi \mid \mathbf{y})$; PHP/USD daily returns data.

The result is reported in Figure 6.4. The plot shows that most of the mass for ψ is concentrated between 0.05 and 0.25, with virtually no mass around zero. This highlights the relevance of extending the standard stochastic volatility model to allow for moving average errors. To formally compare the moving average model with the standard variant, we perform a Bayesian model comparison exercise by computing the Bayes factor (see, e.g., Koop, 2003, pp. 3–4) of the two models.

Since the models are nested—the standard stochastic volatility model can be recovered by setting $\psi = 0$—the Bayes factor in favor of the moving average model against the standard variant can be obtained using the Savage-Dickey density ratio (see, e.g., Verdinelli and Wasserman, 1995): $BF = p(\psi = 0)/p(\psi = 0 \mid \mathbf{y})$. That is, we evaluate the marginal prior and posterior densities for ψ at zero, and their ratio gives the relevant Bayes factor. The numerator density is a univariate truncated normal density that can be easily evaluated. The denominator density is nonstandard form, but it can be estimated using $\widehat{p(\psi = 0 \mid \mathbf{y})}$. Using this approach, the Bayes factor is estimated to be 49600, indicating overwhelming evidence in favor of the moving average stochastic volatility model.

6.4 Stochastic Volatility Models with Heavy-Tailed Error Distributions

In addition to volatility clustering, another prominent feature in typical financial data is the presence of outliers. The conventional assumption of a Gaussian error distribution might be inappropriate in this context as Gaussian distributions have exponentially decaying tails. Consequently they have little mass for more extreme values. In this section, we discuss some heavy-tailed distributions that can be used to address this issue. In particular, these distributions can be written as scale mixtures of Gaussian distributions, which facilitates estimation via data augmentation. Specifically, consider the model

$$y_t = \mu + e^{\frac{1}{2}h_t}\lambda_t^{\frac{1}{2}}\epsilon_t, \quad (6.16)$$

$$h_t = \mu_h + \phi_h(h_{t-1} - \mu_h) + \zeta_t, \quad (6.17)$$

where $\epsilon_t \sim \mathcal{N}(0,1)$, $\zeta_t \sim \mathcal{N}(0,\sigma_h^2)$ and λ_t are independent of each other. As before, we assume $|\phi_h| < 1$ and $h_1 \sim \mathcal{N}(\mu_h, \sigma_h^2/(1-\phi_h^2))$.

Different distributional assumptions of the scale mixture variable λ_t leads to different error distributions. For example, if we assume that λ_t has an exponential distribution with mean parameter 2, i.e., $\lambda_t \sim \mathcal{E}(1/2)$, then the error $\widetilde{\epsilon}_t = \lambda_t^{1/2}\epsilon_t$ follows the double-exponential distribution with density function (see Kroese et al., 2011, pp. 118–119)

$$f(x) = \frac{1}{2}e^{-|x|}.$$

Note that even though the tails of this density function are exponentially decaying, they decay at a rate that is slower than those of Gaussian distributions.

Another example is the normal log-normal distribution. More precisely, if

$$(\log\lambda_t \,|\, \tau) \sim \mathcal{N}(-0.5\tau^2, \tau^2),$$

then $\widetilde{\epsilon}_t = \lambda_t^{1/2}\epsilon_t$ follows the normal log-normal distribution; for more details see Omori et al. (2007).

A third example is the well-known Student-t distribution: if

$$(\lambda_t \,|\, \nu) \sim \mathcal{IG}(\nu/2, \nu/2), \quad (6.18)$$

then $\widetilde{\epsilon}_t = \lambda_t^{1/2}\epsilon_t$ has a standard Student-t distribution with degree of freedom parameter ν. For a textbook treatment, see Koop et al. (2007) pp. 254–255; for applications involving stochastic volatility models, see Nakajima and Omori (2009) and Wang et al. (2011).

Hence, the stochastic volatility model with Student-t errors are defined by (6.16), (6.17) and (6.18). To complete the model specification, we assume independent prior distributions for μ, ν, μ_h, ϕ_h and σ_h^2. More specifically, consider: $\mu \sim \mathcal{N}(\mu_0, V_\mu)$, $\nu \sim \mathcal{U}(0,\bar{\nu})$, $\mu_h \sim \mathcal{N}(\mu_{h0}, V_{\mu_h})$, $\phi_h \sim \mathcal{N}(\phi_{h0}, V_{\phi_h})1(|\phi_h| < 1)$ and $\sigma_h^2 \sim \mathcal{IG}(\nu_h, S_h)$. In particular, ν follows a uniform distribution with mean $\bar{\nu}/2$ and support $(0,\bar{\nu})$.

6.4.1 Estimation

We now discuss the estimation of the stochastic volatility model with a Student-t error distribution in (6.16), (6.17) and (6.18). The algorithm is illustrated in Section 6.4.2 via an empirical example involving daily returns on the silver spot price.

Let $\boldsymbol{\lambda} = (\lambda_1, \ldots, \lambda_T)'$. Posterior draws can be obtained via the following sampler that cycles through:

1. $p(\mu \mid \mathbf{y}, \mathbf{h}, \boldsymbol{\lambda}, \nu, \mu_h, \phi_h, \sigma_h^2) = p(\mu \mid \mathbf{y}, \mathbf{h}, \boldsymbol{\lambda})$;
2. $p(\mathbf{h} \mid \mathbf{y}, \boldsymbol{\lambda}, \mu, \nu, \mu_h, \phi_h, \sigma_h^2) = p(\mathbf{h} \mid \mathbf{y}, \boldsymbol{\lambda}, \mu, \mu_h, \phi_h, \sigma_h^2)$;
3. $p(\boldsymbol{\lambda} \mid \mathbf{y}, \mathbf{h}, \mu, \nu, \mu_h, \phi_h, \sigma_h^2) = \prod_{t=1}^T p(\lambda_t \mid y_t, h_t, \mu, \nu)$;
4. $p(\nu \mid \mathbf{y}, \mathbf{h}, \boldsymbol{\lambda}, \mu, \mu_h, \phi_h, \sigma_h^2) = p(\nu \mid \boldsymbol{\lambda})$;
5. $p(\sigma_h^2 \mid \mathbf{y}, \mathbf{h}, \boldsymbol{\lambda}, \mu, \nu, \mu_h, \phi_h) = p(\sigma_h^2 \mid \mathbf{h}, \mu_h, \phi_h)$;
6. $p(\mu_h \mid \mathbf{y}, \mathbf{h}, \boldsymbol{\lambda}, \mu, \nu, \phi_h, \sigma_h^2) = p(\mu_h \mid \mathbf{h}, \phi_h, \sigma_h^2)$;
7. $p(\phi_h \mid \mathbf{y}, \mathbf{h}, \boldsymbol{\lambda}, \mu, \nu\mu_h, \sigma_h^2) = p(\phi_h \mid \mathbf{h}, \mu_h, \sigma_h^2)$.

For Step 1, we only need to modify the corresponding step in the standard stochastic volatility model. More precisely, we have

$$(\mu \mid \mathbf{y}, \mathbf{h}, \boldsymbol{\lambda}) \sim \mathcal{N}(\widehat{\mu}, D_\mu),$$

where $D_\mu = (V_\mu^{-1} + \sum_{t=1}^T e^{-h_t}\lambda_t^{-1})^{-1}$ and $\widehat{\mu} = D_\mu(V_\mu^{-1}\mu_0 + \sum_{t=1}^T e^{-h_t}\lambda_t^{-1}y_t)$.

To implement Step 2, we first transform $\widetilde{\widetilde{\mathbf{y}}} = \boldsymbol{\Lambda}^{-\frac{1}{2}}(\mathbf{y} - \mu\mathbf{1})$, where $\boldsymbol{\Lambda}^{-\frac{1}{2}} = \text{diag}(\lambda_1^{-\frac{1}{2}}, \ldots, \lambda_T^{-\frac{1}{2}})$. Then,

$$(\widetilde{\widetilde{\mathbf{y}}} \mid \mu, \mathbf{h}, \boldsymbol{\lambda}) \sim \mathcal{N}(\mathbf{0}, \mathbf{S}_\mathbf{y})$$

with $\mathbf{S}_\mathbf{y} = \text{diag}(e^{h_1}, \ldots, e^{h_T})$. Therefore, we can directly apply the auxiliary mixture sampler in Section 6.2.1 to simulate $(\mathbf{h} \mid \mathbf{y}, \boldsymbol{\lambda}, \mu, \mu_h, \phi_h, \sigma_h^2)$.

Next, since $\lambda_1, \ldots, \lambda_T$ are conditionally independent given the model parameters and the data, we can sample each of them sequentially. In fact, we have

$$(\lambda_t \mid y_t, h_t, \mu, \nu) \sim \mathcal{IG}\left(\frac{\nu+1}{2}, \frac{\nu + e^{-h_t}(y_t - \mu)^2}{2}\right).$$

To implement Step 4, we first derive the log-density $\log p(\nu \mid \boldsymbol{\lambda})$: it follows from (6.18) and the prior assumption $\nu \sim \mathcal{U}(0, \bar{\nu})$ that

$$\log p(\nu \mid \boldsymbol{\lambda}) = \frac{T\nu}{2}\log(\nu/2) - T\log\Gamma(\nu/2) - \left(\frac{\nu}{2} + 1\right)\sum_{t=1}^T \log\lambda_t - \frac{\nu}{2}\sum_{t=1}^T \lambda_t^{-1} + c_5$$

for $0 < \nu < \bar{\nu}$, where c_5 is a normalization constant. It is easy to check that the first and second derivatives of the log-density with respect to ν are given by

$$\frac{d\log p(\nu \mid \boldsymbol{\lambda})}{d\nu} = \frac{T}{2}\log(\nu/2) + \frac{T}{2} - \frac{T}{2}\Psi(\nu/2) - \frac{1}{2}\sum_{t=1}^T \log\lambda_t - \frac{1}{2}\sum_{t=1}^T \lambda_t^{-1}$$

$$\frac{d^2\log p(\nu \mid \boldsymbol{\lambda})}{d\nu^2} = \frac{T}{2\nu} - \frac{T}{4}\Psi'(\nu/2),$$

where $\Psi(x) = \frac{d}{dx}\log\Gamma(x)$ and $\Psi'(x) = \frac{d}{dx}\Psi(x)$ are respectively the digamma and trigamma functions.

Since the first and second derivatives can be evaluated quickly, we can maximize $\log p(\nu \,|\, \boldsymbol{\lambda})$ using, e.g., Newton-Raphson method and obtain the mode and the negative Hessian evaluated at the mode, denoted as $\widehat{\nu}$ and K_ν, respectively. Then, we implement an independence-chain Metropolis-Hastings step with proposal distribution $\mathcal{N}(\widehat{\nu}, K_\nu^{-1})$.

Finally, sampling from the densities $p(\mu_h \,|\, \mathbf{h}, \phi_h, \sigma_h^2)$, $p(\phi_h \,|\, \mathbf{h}, \mu_h, \sigma_h^2)$ and $p(\sigma_h^2 \,|\, \mathbf{h}, \mu_h, \phi_h)$ can be done as discussed in Section 6.2.2.

6.4.2 Empirical Example: Modeling Daily Returns on the Silver Spot Price

In Figure 6.5 we plot the daily returns on the silver spot price from January 2005 to December 2012. The data exhibit volatility clustering as expected from financial data. In addition, there are numerous more extreme returns—as large as ±15%—indicating that even after allowing for stochastic volatility, the assumption of a Gaussian error distribution might not be appropriate.

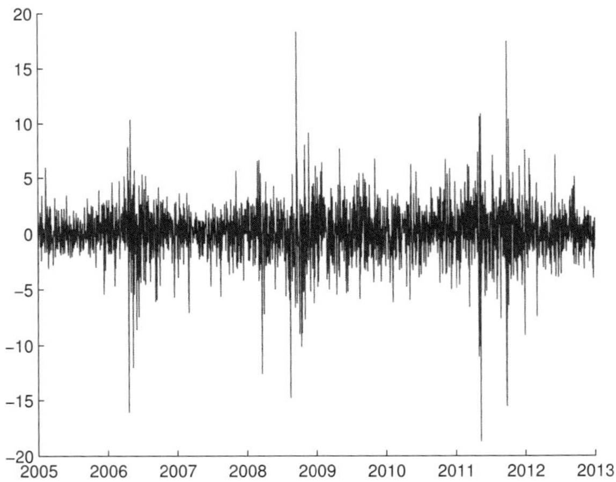

Figure 6.5: Daily returns on the silver spot price from January 2005 to December 2012.

To accommodate the presence of extreme returns, we fit the data using a stochastic volatility model with Student-t errors. In addition, we also allow for persistence through an MA(1) error process:

$$\begin{aligned} y_t &= \mu + u_t, \\ u_t &= \epsilon_t + \psi\epsilon_{t-1}, \end{aligned}$$

where $\epsilon_t \sim \mathcal{N}(0, \lambda_t e^{h_t})$, $\epsilon_0 = 0$ and $|\psi| < 1$. The log-volatilities evolve according to the state equation (6.17), and the distribution for the scale mixture variables is given in (6.18).

For the prior hyperparameters, we set $\mu_0 = 0$, $V_\mu = 5$, $\psi_0 = 0$ and $V_\psi = 1$, $\mu_{h0} = 0, V_{\mu_h} = 5, \phi_{h0} = 0.95, V_{\phi_h} = 1$, $\nu_h = 10$ and $S_h = 0.19$. These values imply relatively diffuse priors with prior means $E(\mu) = 0$, $E(\psi) = 0$, $E(\mu_h) = 0$, $E(\phi_h) = 0.95$ and $E(\sigma_h^2) = 0.02$. The prior distribution for ν is $\mathcal{U}(0, \bar{\nu})$ with $\bar{\nu} = 50$, implying a prior mean of 25.

Estimation of the model can proceed by combining the samplers in Sections 6.3.1 and 6.4.1. More specifically, if we define $\mathbf{z} = \mathbf{H}_\psi^{-1}\mathbf{y}$, then

$$(\mathbf{z} \,|\, \mathbf{h}, \boldsymbol{\lambda}, \boldsymbol{\psi}, \mu) \sim \mathcal{N}(\mu \mathbf{H}_\psi^{-1}\mathbf{1}, \boldsymbol{\Sigma}_\mathbf{z}),$$

where $\boldsymbol{\Sigma}_\mathbf{z} = \text{diag}(\lambda_1 e^{h_1}, \ldots, \lambda_T e^{h_T})$. Hence, the sampler in Section 6.4.1 can be applied to $\mathbf{z}$ to sample from the full conditional distributions of $\mathbf{h}$, $\boldsymbol{\lambda}$, μ, ν, μ_h, ϕ_h and σ_h^2. Moreover, we can quickly evaluate the density of $\mathbf{z}$, which can be used for a Metropolis-Hastings step for simulating from the full conditional distribution of ψ.

Table 6.4: Posterior means, standard deviations and quantiles of model parameters; silver spot price daily returns data.

Parameter	Posterior mean	Posterior std. dev.	5%-tile	95%-tile
μ	0.125	0.037	0.064	0.186
μ_h	1.102	0.232	0.744	1.436
ϕ_h	0.984	0.006	0.974	0.993
σ_h^2	0.017	0.005	0.011	0.024
ψ	−0.088	0.022	−0.120	−0.050
ν	6.675	1.235	5.071	8.894

We obtain 20000 draws from the posterior distribution, after a burn-in period of 1000. Table 6.4 reports the posterior means, standard deviations and quantiles of the model parameters. For example, the average daily return over the sample period is estimated to be 0.125%. The estimates for μ_h, ϕ_h and σ_h^2 are similar to those obtained in the previous two empirical examples.

The posterior mean of the MA(1) coefficient ψ is −0.088 with a 90% credible interval (−0.12, −0.05)—which excludes 0—indicating some persistence in the errors. Moreover, the degree of freedom parameter ν is estimated to be about 6.7. This shows that the error distribution seems to have heavier tails than those of a Gaussian distribution.

The right panel of Figure 6.6 depicts the estimate of the marginal density $p(\psi \,|\, \mathbf{y})$. The density plot indicates that the value 0 is highly unlikely given the data. The left panel shows the density plot of $p(\nu \,|\, \mathbf{y})$, in which most of

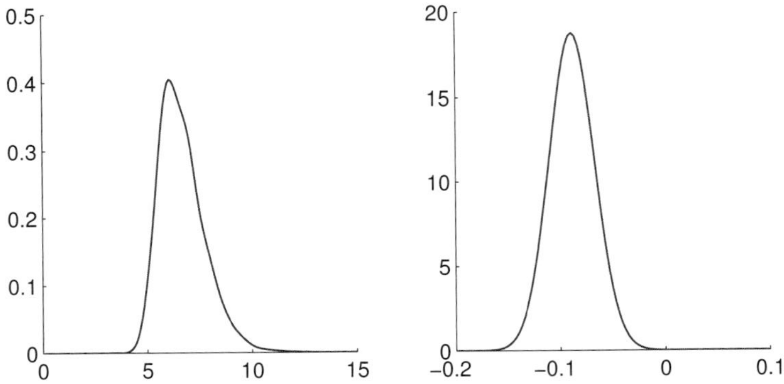

Figure 6.6: Estimates of $p(\nu \,|\, \mathbf{y})$ (left panel) and $p(\psi \,|\, \mathbf{y})$ (right panel); silver spot price daily returns data.

its mass is concentrated between 5 and 10, highlighting the relevance of the extension to a Student-t error distribution.

Bibliography

Belmonte, M. and G. Koop (2013). Model switching and model averaging in time-varying parameter regression models. *Strathclyde Discussion Papers in Economics*.

Carter, C. K. and R. Kohn (1994). On Gibbs sampling for state space models. *Biometrika 81*(3), 541–553.

Chan, J. C. C. and I. Jeliazkov (2009). Efficient simulation and integrated likelihood estimation in state space models. *International Journal of Mathematical Modelling and Numerical Optimisation 1*, 101–120.

Chan, J. C. C., G. Koop, R. Leon-Gonzalez, and R. Strachan (2012). Time varying dimension models. *Journal of Business and Economic Statistics 30*(3), 358–367.

Chan, J. C. C., G. Koop, and S. M. Potter (2013). A new model of trend inflation. *Journal of Business and Economic Statistics 31*(1), 94–106.

de Jong, P. and N. Shephard (1995). The simulation smoother for time series models. *Biometrika 82*, 339–350.

Durbin, J. and S. J. Koopman (2002). A simple and efficient simulation smoother for state space time series analysis. *Biometrika 89*, 603–615.

Früwirth-Schnatter, S. (1994). Data augmentation and dynamic linear models. *Journal of Time Series Analysis 15*, 183–202.

Gardner, G., A. Harvey, and G. Phillips (1980). Algorithm AS 154: An algorithm for exact maximum likelihood estimation of autoregressive-moving average models by means of Kalman filtering. *Journal of the Royal Statistical Society Series C 29*(3), 311–322.

Golub, G. H. and C. F. van Loan (1983). *Matrix computations.* Baltimore: Johns Hopkins University Press.

Jeliazkov, I. and R. Liu (2010). A model-based ranking of U.S. recessions. *Economics Bulletin 30*(3), 2289–2296.

Kim, S., N. Shepherd, and S. Chib (1998). Stochastic volatility: Likelihood inference and comparison with ARCH models. *Review of Economic Studies 65*(3), 361–393.

Koop, G. (2003). *Bayesian Econometrics.* New York: Wiley & Sons.

Koop, G., D. J. Poirier, and J. L. Tobias (2007). *Bayesian Econometric Methods.* Cambridge University Press.

Kroese, D. P., T. Taimre, and Z. I. Botev (2011). *Handbook of Monte Carlo Methods.* New York: John Wiley & Sons.

McCausland, W. J., S. Miller, and D. Pelletier (2011). Simulation smoothing for state-space models: A computational efficiency analysis. *Computational Statistics and Data Analysis 55*(1), 199–212.

Nakajima, J. and Y. Omori (2009). Leverage, heavy-tails and correlated jumps in stochastic volatility models. *Computational Statistics and Data Analysis 53*(6), 2335–2353.

Omori, Y., S. Chib, N. Shephard, and J. Nakajima (2007). Stochastic volatility with leverage: Fast and efficient likelihood inference. *Journal of Econometrics 140*(2), 425–449.

Rue, H. (2001). Fast sampling of Gaussian Markov random fields with applications. *Journal of the Royal Statistical Society Series B 63*(2), 325–338.

Verdinelli, I. and L. Wasserman (1995). Computing Bayes factors using a generalization of the Savage-Dickey density ratio. *Journal of the American Statistical Association 90*(430), 614–618.

Wang, J. J. J., J. S. K. Chan, and S. T. B. Choy (2011). Stochastic volatility models with leverage and heavy-tailed distributions: A Bayesian approach using scale mixtures. *Computational Statistics and Data Analysis 55*(1), 852–862.

CHAPTER 7

FROM THE GREAT DEPRESSION TO THE GREAT RECESSION: A MODEL-BASED RANKING OF U.S. RECESSIONS

RUI LIU[1] AND IVAN JELIAZKOV[2]

[1]San Jose State University, San Jose, USA.
[2]University of California, Irvine, USA.

7.1 Introduction

The seminal work by Burns and Mitchell (1946) characterizes business cycles as the cyclical comovement of many macroeconomic activities. A number of recent studies such as Gregory and Head (1999), Kose et al. (2003), Kose et al. (2008), Bernanke and Boivin (2003), Bernanke et al. (2005), Stock and Watson (2005), and Belviso and Milani (2006), have utilized dynamic factor models to capture the comovment and linkages among macroeconomic variables. The intuition is that information contained in very large data sets of economic variables can be characterized by a relatively small group of latent factors. Extracting such factors can yield information on the behavior of business cycles and benefit future policy analysis.

Due to their desirable features, dynamic factor models have been widely employed in constructing economic indicators (e.g., Burns and Mitchell, 1946; Sargent and Sims, 1977; Geweke, 1977) and characterizing and monitoring regional, country-specific and global business cycles (e.g., Gregory and Head, 1999; Kose et al., 2003, 2008). Factor models have also provided a useful analytical framework for forecasting with many macroeconomic predictors (Koop and Potter, 2004; Stock and Watson, 2002, 2005). Chauvet (1998) and Chauvet and Hamilton (2006) use dynamic factor models with regime switching to examine business cycle turning points and compute recession probabilities. Additionally, factor models are broadly applied in finance and exchange rate analysis (e.g., Geweke and Zhou, 1996; Kim et al., 1998; Aguilar and West, 2000; Chib et al., 2006).

This chapter extends the methodology of Jeliazkov and Liu (2010) and implements a dynamic factor model with time-varying parameters to study the ranking of 14 U.S. recessions starting with data from the Great Depression. In prior work, recession ranks have not been produced by formal econometric techniques and comparisons have instead relied on single-variable metrics (e.g., peak-to-trough changes in output growth or employment), which can suffer from important drawbacks. One obvious deficiency is that a single macroeconomic aggregate, treated in isolation, necessarily ignores information about business conditions that is contained in the trajectories of other variables. It is well known that for this reason the National Bureau of Economic Research (NBER) relies on a variety of data series, rather than on individual macroeconomic variables, in determining the timing of recessions. This is desirable because recessions do not always manifest themselves the same way through the same variable and any comparisons must accommodate potential discrepancies. For instance, the beginning of the March–November 2001 recession was driven by a rise in unemployment, while its end was announced as soon as growth increased. However, other measures of economic activity do not support the March–November 2001 dates – industrial production had been falling since October 2000 whereas unemployment continued to increase in the jobless recovery that followed the announced end of the recession. Similarly, labor market conditions and output growth provided divergent evidence in the case of the latest recession. Even though unemployment deteriorated in December 2007, it was not until the third quarter of 2008 that GDP growth turned negative. Moreover, during the post-war period, the 1973-1975 recession ranks fourth by GDP decline, but second worst by increase in unemployment. A less obvious problem with univariate comparisons is that they typically do not distinguish between idiosyncratic and economy-wide shocks, so that it is not possible to disentangle variable-specific from common shocks. In addition, single-variable methods also ignore the dynamic interdependence in the series. In contrast, our dynamic factor model resolves potential conflicts and produces recession rankings that fully account for information in multiple data series, e.g., output growth, unemployment, interest rates and inflation.

Another important consideration in our setting is that, given the long span of our data (1929:Q2 to 2012:Q4), the model must be suitably specified to accommodate possible structural instability. We do this by augmenting the standard dynamic factor specification with time-varying parameters. The resulting model is meant to allow for possible non-linearity through gradual change in the model parameters, which is an important disadvantage of fixed coefficient models. The appeal of our flexible model specification is that it does not require sub-sample estimates to study the evolution of the economic system, does not impose simultaneous structural breaks across all sectors, and naturally incorporates heterogeneity in the changes occurring across series. Our simulation-based estimation approach, based on recently developed Markov chain Monte Carlo (MCMC) algorithms, also allows us to make probabilistic statements about recession ranks (which are difficult to obtain from point estimates and point-wise confidence bands alone), and allows us to detect business cycle turning points, which is of central importance in policy analysis. In particular, since business cycles are characterized by phases of contraction and expansion, turning points are extremely important for monitoring economic conditions. The current NBER consensus estimates on the timing of turning points are (i) announced long after their occurrence and (ii) employ dating techniques that are highly subjective and hard to reproduce and replicate. For these reasons, there is a literature dealing with the dating of turning points (Kim and Nelson, 1988; Chauvet and Piger, 2003; Chauvet and Hamilton, 2006). In this chapter, we utilize the convenient simulation output from our MCMC sampler to generate turning point probabilities and pinpoint turning point dates. The methodology is useful because it can be used in both historical and real-time scenarios.

In our analysis, we utilize quarterly data spanning 1929:Q2 to 2012:Q4, which is preferable to the commonly used annual data on business cycle fluctuations prior to World War II. This benefits our analysis by improving the estimation precision including information on within-year variation. The significant range of the data makes it possible to compare the relative severity of post-war and interwar recessions, enabling comparisons between the Great Depression (comprising the 1929-33 and 1937-38 recessions) and the Great Recession of 2007-2009. We estimate a dynamic factor common to all macroeconomic variables within the model and an idiosyncratic component that captures variable-specific characteristics, and use the former to gauge downturns and trace turning points. We present evidence on the relative ranks of recessions as well as the possible turning point locations, both of which account for information in all of our data series and account for parameter uncertainty. Our results confirm the traditional belief that the Great Depression was by far the worst U.S. recession, and provide evidence that the 2007-09 economic downturn, also known as the Great Recession, ranks fourth since the Depression and first worst since World War II. Our turning point estimates are in close agreement with those announced by the NBER.

The remainder of the paper is organized as follows. In Section 7.2, we formalize the econometric model and present our estimation methodology. Section 7.3 presents our main results, whereas Section 7.4 offers concluding remarks.

7.2 Methodology

7.2.1 Model

We apply a dynamic factor model with time-varying parameters (DF-TVP) to analyze macroeconomic data for the United States which consists of $T = 335$ quarterly observations (1929:Q2 to 2012:Q4) on $n = 4$ general macroeconomic variable series: output growth, unemployment rate, interest rate, and inflation rate. For $t = 1, \ldots, T$, the DF-TVP model is given by

$$y_t = \mu_t + \Gamma_t y_{t-1} + A f_t + \varepsilon_t, \qquad \varepsilon_t \sim \mathcal{N}(0, \Omega), \tag{7.1}$$

where y_t is a $n \times 1$ vector of observables, Γ_t is a diagonal matrix capturing intertemporal relations between observables and their lagged values, A is a vector of factor loadings, f_t is the common factor at time t, and $\Omega = \text{diag}(\omega_{11}, \ldots, \omega_{nn})$.

For the purposes of estimation, the model is written in the form of a seemingly unrelated regression model (Zellner, 1962) as

$$y_t = X_t \beta_t + A f_t + \varepsilon_t,$$

where $X_t \equiv I_n \otimes (1, y'_{t-1})$, $\beta_t \equiv \text{vecr}\left([\mu_t : \text{diag}(\Gamma_t)]'\right)$, where the vecr$(\cdot)$ operator stacks the rows of a matrix into a column vector and the diag$(\cdot)$ operator extracts the main diagonal from a matrix. The next level of the hierarchical model is concerned with the appropriate specification of the evolution process of β_t and f_t, where we let

$$\beta_t = \beta_{t-1} + \eta_t, \qquad \eta_t \sim \mathcal{N}(0, \Psi), \tag{7.2}$$

and

$$f_t = \gamma_t f_{t-1} + \nu_t, \qquad \nu_t \sim \mathcal{N}(0, \sigma^2). \tag{7.3}$$

The specification in (7.2) captures the time-varying aspect of the coefficients of the model through a random walk process that allows them to vary from one period to the next. The conditional variance of the process is captured through a diagonal matrix Ψ. The specification in (7.3) guides the evolution of the dynamic economy-wide factor f_t, which reflects unobserved sources of macroeconomic volatility. Due to the diagonal structure of Ω and Γ_t, the comovement among the macro variables is fully reflected in the latent factor. The model is capable of reproducing complex dynamic behavior because intertemporal cross-correlations can be captured through the unobserved factor, whereas the potential for structural instability is embedded in the time-varying parameters and shocks. Similarly to equations (7.2) and (7.3), the time-varying persistence of the common factor, captured by γ_t, is modelled as

$$\gamma_t = \gamma_{t-1} + u_t, \qquad u_t \sim \mathcal{N}(0, \tau^2). \tag{7.4}$$

Note that both the factor loadings A and factors f_t are unknown and must be estimated. Despite the high dimensionality of the quantities of interest in this case, Bayesian MCMC estimation methods can be easily implemented. Note that the full-conditional posterior distribution of the latent factors given all parameters and observations is tractable and readily available. At the same time, the posterior distributions of each parameter conditioning on other parameters and observations can also be conveniently obtained. Thus a joint posterior distribution of all the model unknowns can be sampled using an MCMC procedure that iteratively simulates from the set of full-conditional distributions, as will be discussed in Section 7.2.2.

Since the factor loadings A and the latent factors f_t are both unknown and enter the model multiplicatively, we must address identification issues relating to the sign and scale of these quantities. This can be seen by recognizing that Af_t is equivalent to $(cA)(f_t/c)$, for any $c \neq 0$ (note that c can be either positive or negative). To resolve the identification problem, we fix the first component of A to unity, which removes the possibility for simultaneously rescaling the two quantities without changing the magnitude of the product. Note that this normalization also identifies the sign of f_t and implies that the in our particular setting, the magnitudes of the factor and loadings is normalized on the basis of the output growth equation (the first variable in y_t). Unlike Kose et al. (2008), in our model the factor loadings do not drift over time for several reasons. First, the presence of varying intercepts and lag coefficients in the observation equation (7.1) enables rich and sophisticated dynamic behavior of the system. Second, our model incorporates varying persistence in the dynamics of the factor. This is an important feature of the model that requires a restriction on the factor loadings in order to identify the time variation in the factor, as can be seen by substituting the transition equation for the factor into the observation equation. Moreover, normalizing on the loadings not only resolves identification problems when facing several time-varying unobserved components, but also leads to a relatively straightforward estimation algorithm as discussed next.

7.2.2 Estimation Framework

Equation (7.3) is initialized with the distribution $f_1 \sim N(0, D_F)$, so that the joint distribution for $f = (f_1, \ldots, f_T)'$ becomes $[f|y, \sigma^2] \sim \mathcal{N}(0, \sigma^2 F_0^{-1})$, where the precision F_0 is given by

$$F_0 = \begin{pmatrix} \sigma^2/D_F + \gamma_2^2\sigma^2/D_F & -\gamma_2 & & & \\ -\gamma_2 & \gamma_3^2+1 & -\gamma_3 & & \\ & \ddots & \ddots & \ddots & \\ & & -\gamma_{T-1} & \gamma_T^2+1 & -\gamma_T \\ & & & -\gamma_T & 1 \end{pmatrix}. \quad (7.5)$$

Under the priors $\beta \sim \mathcal{N}(\beta_0, B_0)$, $a \sim \mathcal{N}(a_0, A_0)$, $\omega_{ii} \sim \mathcal{IG}(\nu_{i0}/2, r_{i0}/2)$, $\gamma \sim \mathcal{N}(g_0, G_0)$, $\psi_{ii} \sim \mathcal{IG}(s_{i0}/2, \delta_{i0}/2)$, $\tau^2 \sim \mathcal{IG}(\nu_0, r_0)$ and $\sigma \sim \mathcal{IG}(s_0, \delta_0)$, and initial condition distributions $\beta_1 \sim \mathcal{N}(\beta_{10}, B_{10})$,$\gamma_1 \sim \mathcal{N}(\gamma_{10}, G_{10})$ and $f_1 \sim N(0, D_F)$, and upon stacking $y = (y_1', \ldots, y_T')'$, $X = (X_1', \ldots, X_T')'$, $\beta = (\beta_1, \ldots, \beta_T)'$, and $\gamma = (\gamma_2, \ldots, \gamma_T)'$, the model can be estimated efficiently by precision-based MCMC samplers for state space models (see, e.g., Chan and Jeliazkov, 2009). Specifically, estimation proceeds as follows.

Algorithm: MCMC Estimation of DF-TVP Model

1. Sample $[\beta|y, A, f, \Omega, \sigma^2, \Psi] \sim \mathcal{N}\left(\hat{\beta}, B^{-1}\right)$, where $B = B_0 + X'(I_T \otimes \Omega^{-1})X$, and $\hat{\beta} = B^{-1}[B_0\beta_0 + X'(I_T \otimes \Omega^{-1})(y - Af)]$. B_0 and β_0 are defined as

$$\beta_0 = \begin{pmatrix} I & & & & \\ -I & I & & & \\ & \ddots & \ddots & & \\ & & \ddots & I & \\ & & & -I & I \end{pmatrix}^{-1} \times \begin{pmatrix} \beta_{10} \\ 0 \\ 0 \\ \vdots \\ 0 \end{pmatrix}_{(n+n)T\times 1}, \quad (7.6)$$

$$B_0 = \begin{pmatrix} \Psi^{-1} + B_{10}^{-1} & -\Psi^{-1} & & & \\ \Psi^{-1} & 2\Psi^{-1} & & & \\ & \ddots & \ddots & \ddots & \\ & & -\Psi^{-1} & 2\Psi^{-1} & -\Psi^{-1} \\ & & & -\Psi^{-1} & \Psi^{-1} \end{pmatrix}. \quad (7.7)$$

2. Sample $[a, f|y, \beta, \Omega, \gamma, \sigma^2]$ in one block as follows

   (a) Sample $[a|y, \beta, \Omega, \gamma, \sigma^2]$ marginally of f by a Metropolis-Hastings step with tailored proposal $a^\dagger \sim q(\hat{a}, V)$, and accept the proposed draw $a^\dagger$ with probability

$$\alpha_{MH}(a, a^\dagger) = \min\left\{1, \frac{\pi(a^\dagger|y, \beta, A, \Omega, \gamma, \sigma^2)q(a|\hat{a}, V)}{\pi(a|y, \beta, A, \Omega, \gamma, \sigma^2)q(a^\dagger|\hat{a}, V)}\right\} \quad (7.8)$$

(b) Sample $[f|y,\beta,A,\Omega,\gamma,\sigma^2] \sim \mathcal{N}(\hat{f},F)$, where $\hat{f} = F(I_T \otimes A)'(I_T \otimes \Omega^{-1})(y - X\beta)$ and $F = \left(F_0/\sigma^2 + (I_T \otimes A)'(I_T \otimes \Omega^{-1})(I_T \otimes A)\right)^{-1}$.

3. Sample $[\gamma|f,\sigma^2,\tau^2] \sim \mathcal{N}\left(\hat{\gamma},G^{-1}\right)$, where $G = G_0/\tau^2 + \sigma^2\hat{X}'\hat{X}$, and $\hat{\gamma} = G^{-1}[G_0\gamma_0/\tau^2 + \sigma^2\hat{X}'\hat{y}]$. $\hat{X}$ is a $(T-1)\times(T-1)$ diagonal matrix with $(f_1,\ldots,f_{T-1})'$ on its diagonal and zeros otherwise. $\hat{y}$ is a $(T-1)\times 1$ vector, $(f_2,\ldots,f_T)\prime$. G_0 and γ_0 are defined as

$$\gamma_0 = \begin{pmatrix} 1 & & & & \\ -1 & 1 & & & \\ & \ddots & \ddots & & \\ & & \ddots & 1 & \\ & & & -1 & 1 \end{pmatrix}^{-1} \times \begin{pmatrix} \gamma_{10} \\ 0 \\ 0 \\ \vdots \\ 0 \end{pmatrix}_{(T-1)\times 1}, \quad (7.9)$$

$$G_0 = \begin{pmatrix} 1+\tau^2 G_{10}^{-1} & -1 & & & \\ -1 & 2 & & & \\ & \ddots & \ddots & \ddots & \\ & & -1^{-1} & 2 & -1 \\ & & & -1 & 1 \end{pmatrix}. \quad (7.10)$$

4. Sample $[\Omega|y,\beta,A,f]$ by drawing $\omega_{ii} \sim \mathcal{IG}((\nu_{i0}+T)/2,(r_{i0}+e_i'e_i)/2)$ for $i=1,\ldots,n$, where e_i is a T-vector of residuals from the ith observation equation.

5. Sample $[\sigma^2|f,\gamma] \sim \mathcal{IG}\left(\frac{s_0+T-1}{2},\frac{\delta_0+(f^*-\hat{f})'(f^*-\hat{f})}{2}\right)$, where $f^* = (f_2,...,f_T)'$ and $\hat{f} = (\gamma_2 f_1,\ldots,\gamma_T f_{T-1})'$.

6. Sample $[\tau^2|\gamma] \sim \mathcal{IG}\left(\frac{\nu_0+T-1}{2},\frac{r_0+(\gamma^*-\hat{\gamma})'(\gamma^*-\hat{\gamma})}{2}\right)$, where $\gamma^* = (\gamma_2,\ldots,\gamma_T)'$ and $\hat{\gamma} = (\gamma_1,\ldots,\gamma_{T-1})'$.

7. Sample $[\Psi|\beta]$ by drawing $\psi_{ii} \sim \mathcal{IG}\left(\frac{s_{i0}+T-1}{2},\frac{\delta_{i0}+(\psi_i^*-\hat{\psi}_i)'(\psi_i^*-\hat{\psi}_i)}{2}\right)$ for $i=1,\ldots,2n$, where $\psi_i^* = (\beta_{i2},\ldots,\beta_{iT})'$ and $\hat{\psi}_i = (\beta_{i1},\ldots,\beta_{iT-1})'$.

It is important to note that because this algorithm produces draws from the joint posterior distribution of all model unknowns, subsequent inferences based on the simulated factors fully account for all parameter uncertainty (unlike plug-in approaches). Moreover, the framework is quite flexible and can easily accommodate systems with variable-specific lag lengths, more complex factor dynamics, or many macroeconomic predictors (e.g., Stock and Watson, 2002).

In our application, the steps in the Algorithm are iterated 11,000 times—the first 1,000 draws are discarded as burn-in and inference is based on the remaining 10,000 draws.

7.3 Results

Our data set starts with the second half of the interwar period (1929-1938), covers World War II, and includes the postwar period (1948-2012) which is most frequently studied in recent empirical work. The postwar period can be divided into the Bretton Woods system (1949-1972) and the post-Bretton Woods regime (1973-2012). We will pay special attention to these episodes in the following analysis.

The estimated dynamic factor over the sample period is shown in Figure 7.1, together with the timing of NBER-announced recessions. A striking similarity is visible between the NBER reference cycles dates and the pattern of the estimated factor, which captures these economic downturns quite well. What is more, the behavior of the factor also gives an idea of the relative severity and dynamic evolution of each recession. Figure 7.1 shows that the factor volatility under the interwar regime is much higher than that during the postwar period, which is consistent with the notion that that U.S business cycles have dampened after World War II. In addition, the relative decline in macroeconomic volatility after the mid-1980s—the so-called Great Moderation—is also well depicted in figure. However, an interesting observation is that the Great Moderation stage seems to have come to an end with the arrival of the most recent recession. Table 7.1 reports the factor volatility (measured by the variance of the estimated factor) under different regimes. The two Great Moderation periods summarized in Table 7.1, one including and the other excluding the latest recession, clearly point to the higher factor volatility in the former case and indicate presence of volatility growth since 2008. Table 7.2 documents the evolution of the factor volatility over the sample period. It also confirms that the 2008 recession breaks the trend of business cycle dampening, with the most recent recession inducing volatility increases from 0.28 to 0.63.

To better understand the information captured by the factor, we present the correlations between the estimated factor and the four macroeconomic series in Table 7.3. Several points emerge from this part of the analysis. First, inflation is much more highly correlated with the estimated factor during the interwar period, when the deflationary pressure presents itself together with the government pursuit of contractionary policies. The high correlation dropped almost to zero after World War II. In addition, the signs of correlations are opposite under the Bretton Woods and post-Bretton Woods regimes, along with a higher absolute value in the former period. Second, the interest rate exhibits a higher correlation with the factor during the interwar stage. However, the correlation has switched signs with the advent of the Great

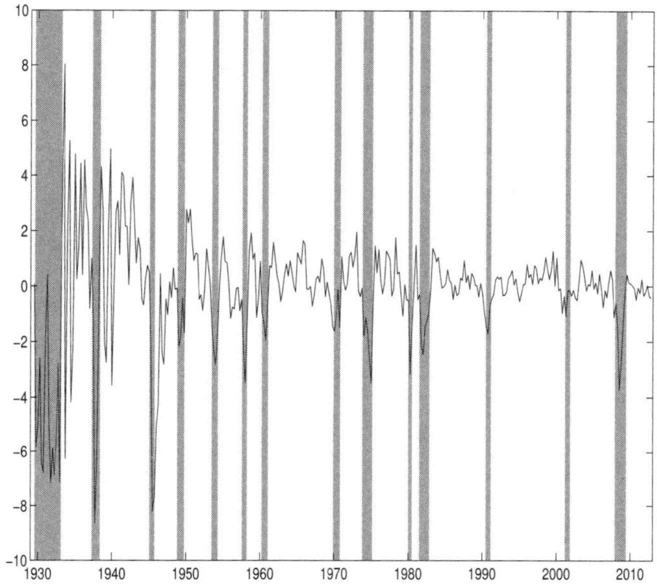

Figure 7.1: Estimated factor in the DF-TVP model of the U.S. economy along with the NBER recession dates (shaded regions).

Table 7.1: Volatility of the common factor under different regimes.

Period	Factor Volatility
Interwar(1929-1938)	19.99
Postwar (1948-2012)	1.04
Pre Great Moderation(1948-1984)	1.48
Great Moderation (1984-2012)	0.46
Great Moderation (1984-2007)	0.28
Bretton Woods (1949-1972)	1.45
Post-Bretton Woods (1973-2012)	0.81

Moderation. Third, and quite surprisingly, the interwar unemployment has an extremely low correlation of 0.121 with the estimated factor, even though the unemployment rate in fact shot up to 25% concurrently with the sharp fall in the national output. However, we note that unemployment is a lagging macroeconomic indicator, and hence contemporaneous correlation does not always reveal the true dynamics of the economy. A recalculation of the cor-

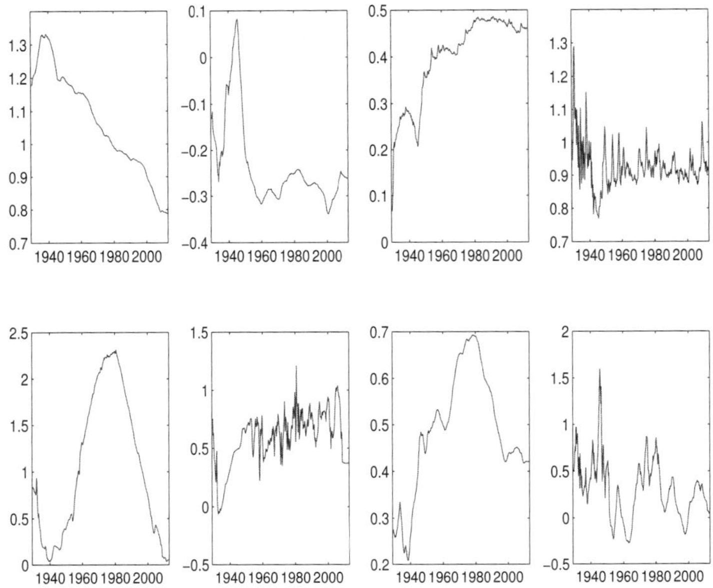

Figure 7.2: Evolution of the parameters.

1. First row displays estimated intercepts of macro series; second row displays the associated autoregressive coefficients.

2. Each column shows estimates for individual series, from left to right: output, unemployment, interest rate and inflation.

relation between the factor and the four-period lagged unemployment during the interwar period shows a stronger connection with a value of -0.35.[1]

The estimated factor loadings are reported in Table 7.4. Inflation loads positively on the latent factor. As expected, unemployment has a negative loading due to its counter-cyclical nature.

Figure 7.2 reports the time-varying plot of the model parameters. The first row displays estimated intercepts for each macro series; the second row shows the associated autoregressive coefficients. We see a very strong pattern of changing behavior in these parameters over time, indicating potential structural changes in the economy. Whether the change has been due to luck or policy (which has been a debateable issue in the literature) is not revealed by the estimates, although the evidence is clearly there. Importantly, Figure 7.2 serves as a warning against adopting dynamic factor models with fixed pa-

[1]The correlations between the factor and the two-, three- and four-period lagged unemployment are -0.11, -0.29, and -0.35, respectively.

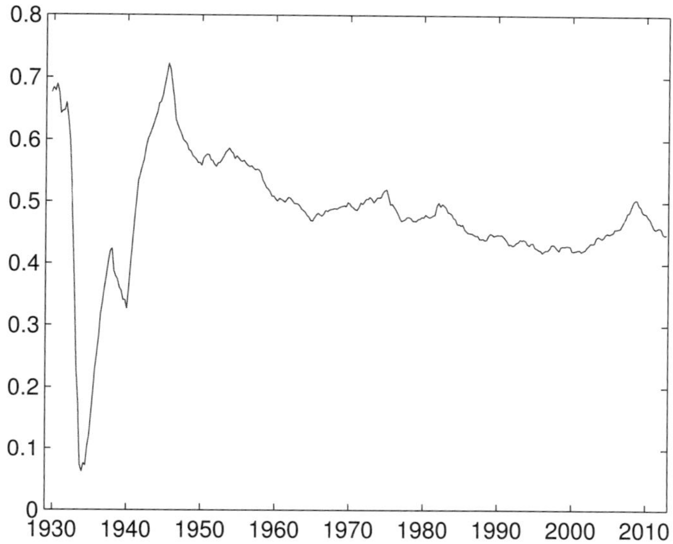

Figure 7.3: Estimated persistence in the evolution of the common factor.

Table 7.2: Evolution of the volatility of the estimated factor.

1930s	**1970s**
19.70	1.40
1940s	**1980s**
7.94	0.96
1950s	**1990s**
2.23	0.36
1960s	**2000–12**
0.69	0.63
Full Sample	**2000–08 (excl. 2008-12)**
4.21	0.28

rameters especially when dealing with a time scale such as that in our data set, suggesting that time-varying specifications may be more appropriate. Finally, it is revealing to note that conventional methods of testing for breaks or forming sub-sample estimates typically impose simultaneous structural breaks across all series and may fail to adequately incorporate heterogeneity in the occurrence of those changes occurring across series.

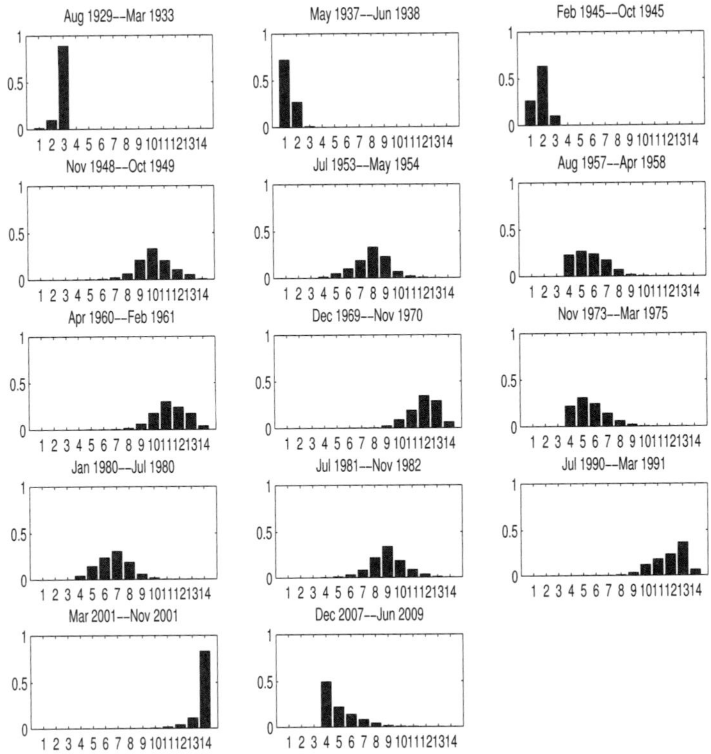

Figure 7.4: Recession rank probabilities (1 = Most Severe, 14 = Mildest).

The evolution of γ is illustrated in Figure 7.3. The overall message is that of a positive serial correlation in the dynamics of the latent factor. However, we do see drastic fluctuation in the persistence of the factor during the Great Depression period (1929-39). The factor persistence plummets from 0.7 to less than 0.1, followed by a rapid increase back to the pre-depression level by the late 1940's. From then on, persistence has gradually declined, with fluctuation ranging between 0.4 and 0.5 in the last five decades.

Taking a second look at Figure 7.1, we note another interesting finding, namely, that the latest downturn that started in December 2007 and known as the Great Recession appears to be the most severe in the post-war sample. Figure 7.1 also displays some potential contenders for different severity ranks. For example, the two recessions during the Great Depression period display comparably sharp drops; two other recessions (those in 1957-58 and 1973-75) exhibit similar severity scale; 1990-91 and 1969-70 downturns are difficult to

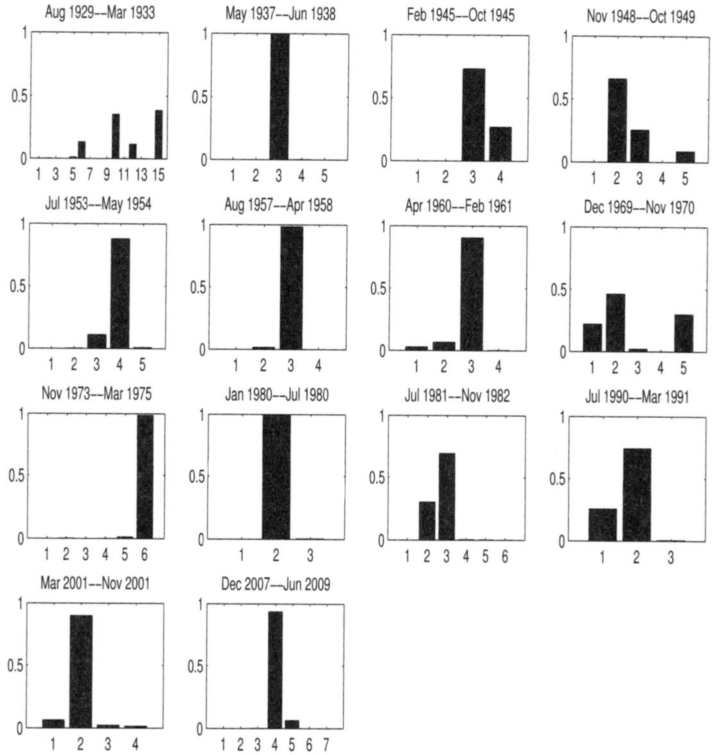

Figure 7.5: Turning point probabilities (1 = the 1st quarter of a recession as dated by the NBER).

Note: the estimated trough coincides with the NBER trough when the tallest bar is located at the end quarter of a recession.

rank as well. For this reason, in comparing these recessions to each other, we must account for the uncertainty in the estimated factor. This, however, is not a straightforward issue that can be addressed by looking at point estimates and point-wise confidence bands. Moreover, due to the estimation uncertainty, it is quite possible that at each draw, the latent factor could actually achieve its local minima at points other than the ones depicted in Figure 7.1. To address these concerns and provide a ranking of the 14 recessions, we proceed as follows. For each MCMC draw of f, we determine the minima in the factor corresponding to each recession. We compare and rank these 14 extrema, and record the number of times each recession achieves a given rank over the course of MCMC sampling. Doing so allows us to make comparisons and probabilistic

Table 7.3: Correlations between the common factor and the four macroeconomic series in different regimes.

	Correlation b/n factor & series			
Periods	GNP	Unemp	Int	Inf
Interwar (1929-1938)	0.980	0.121	−0.361	0.656
Postwar (1948-2012)	0.965	−0.074	−0.086	−0.005
Pre Great Moderation(1948-1984)	0.970	−0.072	−0.158	−0.011
Great Moderation (1984-2012)	0.964	−0.055	0.162	0.026
Great Moderation (1984-2007)	0.949	0.018	−0.066	−0.274
Bretton Woods (1949-1972)	0.969	0.004	−0.042	0.284
Post-Bretton Woods (1973-2012)	0.968	−0.074	−0.082	−0.200

Table 7.4: Factor loadings.

Observables	Factor loadings	Standard deviation
Output Growth	1	-
Unemployment Rate	−0.178	0.017
Interest Rate	−0.007	0.008
Inflation	0.162	0.019

statements about recession ranks while simultaneously producing estimates about the location and uncertainty associated with each trough. This is an important benefit of the simulation-based estimation framework that is quite non-trivial in other approaches.

Before continuing with the discussion of recession ranks, we mention an interesting observation based on Figure 7.1. In particular, the figure shows that while the officially announced recessions of 1990-91 and 2001 were shorter, shallower, and further apart than recessions earlier in the sample, their ends might have been announced prematurely. The ability of the model to capture these features is notable because the 1990-91 recession was followed by a period of jobless growth when unemployment actually increased and remained elevated for several years. Recovery from the 2001 recession was stymied by the immediate aftermath of the 9/11 attacks, which affected consumer behavior, investor sentiment, and demand in key industries.

We now turn our attention to a discussion of the relative recession ranks obtained by the proposed approach. Our main results, based on a sample of 10,000 MCMC draws, are given in Figure 7.4. Figure 7.4 illustrates the marginal rank probabilities associated with each recession. The second recession (1937-38) during the Great Depression period (1929-39) has a probability of 72% being the deepest recession. The two closest contenders (the

Table 7.5: Severity Rank Probabilities of U.S. Recessions (1 = Most Severe, 14 = Mildest)

	Recession Ranks													
Recession Timing	1	2	3	4	5	6	7	8	9	10	11	12	13	14
Aug 1929–Mar 1933	0.01	0.10	0.89	0	0	0	0	0	0	0	0	0	0	0
May 1937–Jun 1938	0.72	0.27	0.01	0	0	0	0	0	0	0	0	0	0	0
Feb 1945–Oct 1945	0.27	0.63	0.10	0	0	0	0	0	0	0	0	0	0	0
Nov 1948–Oct 1949	0	0	0	0	0	0.01	0.02	0.06	0.21	0.33	0.20	0.10	0.05	0.01
Jul 1953–May 1954	0	0	0	0.01	0.05	0.10	0.19	0.33	0.23	0.07	0.02	0.01	0	0
Aug 1957–Apr 1958	0	0	0	0.23	0.27	0.24	0.17	0.07	0.02	0	0	0	0	0
Apr 1960–Feb 1961	0	0	0	0	0	0	0	0.01	0.06	0.17	0.30	0.24	0.17	0.04
Dec 1969–Nov 1970	0	0	0	0	0	0	0	0	0.02	0.09	0.19	0.34	0.29	0.06
Nov 1973–Mar 1975	0	0	0	0.22	0.31	0.25	0.14	0.06	0.02	0	0	0	0	0
Jan 1980–Jul 1980	0	0	0	0.04	0.14	0.24	0.31	0.19	0.06	0.02	0	0	0	0
Jul 1981–Nov 1982	0	0	0	0	0.01	0.03	0.08	0.22	0.34	0.19	0.09	0.03	0.01	0
Jul 1990–Mar 1991	0	0	0	0	0	0	0	0.01	0.03	0.12	0.18	0.23	0.37	0.06
Mar 2001–Nov 2001	0	0	0	0	0	0	0	0	0	0	0.01	0.04	0.11	0.83
Dec 2007–Jun 2009	0	0	0	0.50	0.22	0.14	0.08	0.04	0.01	0.01	0	0	0	0

Table 7.6: Turning Point Probabilities of U.S. Recessions (1 = the 1st Quarter of a Recession Dated by the NBER)

	Turning Point Probabilities														
Recession Timing	1	2	3	4	5	6	7	8	9	10	11	12	13	14	15
Aug 1929–Mar 1933	0	0	0	0	0.01	0.13	0	0	0	0.35	0	0.12	0	0	0.38
May 1937–Jun 1938	0	0	1.00	0	0										
Feb 1945–Oct 1945	0	0	0.73	0.27											
Nov 1948–Oct 1949	0	0.66	0.26	0	0.09										
Jul 1953–May 1954	0	0	0.11	0.88	0.01										
Aug 1957–Apr 1958	0	0.02	0.98	0											
Apr 1960–Feb 1961	0.03	0.07	0.90	0											
Dec 1969–Nov 1970	0.22	0.46	0.02	0	0.30										
Nov 1973–Mar 1975	0	0	0	0	0.01	0.99									
Jan 1980–Jul 1980	0	1.00	0												
Jul 1981–Nov 1982	0	0.30	0.70	0	0	0									
Jul 1990–Mar 1991	0.25	0.74	0.01												
Mar 2001–Nov 2001	0.06	0.90	0.02	0.02											
Dec 2007–Jun 2009	0	0	0	0.95	0.05	0	0								

Table 7.7: A comparison between model estimated trough dates and the NBER reference dates.

NBER Trough	Model Trough	Discrepancy	NBER Trough	Model Trough	Discrepancy
1933Q1	1933Q1	0	1970Q4	1970Q1	3Q
1938Q2	1937Q4	2Q	1975Q1	1975Q1	0
1945Q4	1945Q3	1Q	1980Q3	1980Q2	1Q
1949Q4	1949Q1	3Q	1982Q4	1982Q1	3Q
1954Q2	1954Q1	1Q	1991Q1	1990Q4	1Q
1958Q2	1958Q1	1Q	2001Q4	2001Q2	2Q
1961Q1	1960Q4	1Q	2009Q2	2008Q3	3Q

1929-34 and 1945 recessions) rank first with lower probability (1% and 27%, respectively). The figure also confirms that the latest recession (2007-09) is indeed the worst post-war recession with probability 63%, see Jeliazkov and Liu (2010). There is a 53% probability that the 1945 recession is the second deepest and that the 1929-33 recession ranks third with probability 89%. The rank probabilities for the full set of U.S. recessions is given in Table 7.5. To rank two recessions with similar severity and simultaneously account for the uncertainty of the troughs, we adopt a pairwise comparison which records the number of times during the MCMC sampling that one recession attains a deeper trough than the other. Take the 1937-38 recession and the 1945 recession, for example, where with probability 86% the late 1930s recession is deeper and more severe than the ensuing downturn. For the recessions during 1957-58 and 1973-75, the former is more severe than the latter with probability 45%.

As argued earlier, an important by-product of the simulation approach is the probability distribution of turning points for each recession. Business cycle turning points, as part of an economic warning system (Chauvet and Yu, 2006), have gained more and more popularity since the seminal work of Stock and Watson (2002, 2005) on dynamic factor models and the works by Hamilton (1989) on regime switching models. In this chapter, we adopt a relatively agnostic approach to produce turning points (of which of particular interest are recession troughs, although expansions are handled similarly). Turning point probabilities are obtained from the simulation output by simply documenting the number of times during a given recession where the estimated factor achieves a local minimum in a given quarter. The outcomes from this exercise are recorded in Table 7.6 along with corresponding histograms in Figure 7.5. For instance, for the 1929-33 recession during the Great Depression period, the most likely turning point when the economy switches from contraction to expansion, takes place 14 quarters after the start of the recession with a probability of 38%, which coincides with the NBER official end date of the recession. Table 7.7 compares the NBER trough dates with the corresponding dates obtained from our proposed model. It is interesting to note that out of the 14 estimated turning points, two match exactly, six are off by one quarter, and the rest differ by two or more quarters. Finally, while we have demonstrated how such analysis can be performed to analyze the historical record, it is worth mentioning that "now-casting" and forecasting can be done identically without any methodological differences.

7.4 Conclusions

This chapter has developed a dynamic factor model with time-varying parameters that aims to capture possible structural instability or gradual evolution in macroeconomic series, while simultaneously capturing the overall state of the economy in a single dynamic factor. An advantage of this approach is that it allows for flexible modeling that does not require arbitrary sub-sample estimation and instead employs time-variation in the parameters to mirror the underlying structural change. Unlike models with sharp breaks, which impose simultaneous structural breaks across all sectors, this model allows changes to differ across series.

We have employed the methodology to produce a model-based approach for ranking the relative severity and locating the possible turning points of recessions. From a statistical standpoint, the method is desirable because it involves joint modeling and estimation of all model unknowns, thereby producing estimates that fully reflect parameter uncertainty, while also accounting for the co-movement and dynamic interdependence in the series. From a practical perspective, the method is useful because it accommodates uncertainty about the location of business cycle turning points. The simulation-based methodology provides rank probabilities and probabilistic depictions of events which are difficult to obtain from point estimates and point-wise confidence bands. Our results indicate that the the Great Depression is indeed the deepest in the sample period, whereas the latest recession is the worst post-war downturn.

In future research we intend to apply the proposed techniques to models containing additional macroeconomic indicators in order to capture a broader range of measures of output, income and employment. Moreover, the adequacy of more general dynamic specifications, such as models with variable-specific lags, will also be examined using formal model choice techniques.

Appendix: Data

`Output growth:` 1929:Q2-1983:Q4 real GNP, Appendix B, "The American Business Cycle: Continuity and Change" edited by Robert J. Gordon, National Bureau of Economic Research Studies in Business Cycles Volume 25,

University of Chicago Press 1986. 1984:Q1-2012:Q4 real GNP, Federal Reserve Economic Database (FRED).
`Unemployment rate:` 1929:Q2-2012:Q4 unemployment rate, Global Financial Data, UNUSAM.
`Interest rate:` 1929:Q2-2012:Q4 USA Government 90-day T-Bills Secondary Market, Global Financial Data, ITUSA3D.
`Inflation rate:` 1929:Q2-1983:Q4 GNP deflator, Appendix B, "The American Business Cycle: Continuity and Change" edited by Robert J. Gordon, National Bureau of Economic Research Studies in Business Cycles Volume 25, University of Chicago Press 1986. 1984:Q1-2010:Q3 GNP deflator, FRED.

Bibliography

Aguilar, O. and M. West (2000). Bayesian dynamic factor models and portfolio allocation. *Journal of Business and Economic Statistics 18*, 338–357.

Belviso, F. and F. Milani (2006). Structural factor-augmented VARs (SF-VARs) and the effects of monetary policy. *Topics in Macroeconomics 6*(3), Iss. 3, Article 2.

Bernanke, B. and J. Boivin (2003). Bayesian dynamic factor models and portfolio allocation. *Journal of Monetary Economics 50*, 3525–546.

Bernanke, B., J. Boivin, and P. Eliasz (2005). Measuring the effects of monetary policy: a factor-augmented vector autoregressive (FAVAR) approach. *Quarterly Journal of Economics 120*, 387–422.

Burns, A. and W. Mitchell (1946). *Measuring Business Cycles.* New York: National Bureau of Economic Research.

Chan, J. and I. Jeliazkov (2009). Efficient simulation and integrated likelihood estimation in state space models. *International Journal of Mathematical Modelling and Numerical Optimisation 1*, 101–120.

Chauvet, M. (1998). An econometric characterization of business cycle dynamics with factor structure and regime switching. *International Economic Review 39*(4), 969–996.

Chauvet, M. and J. D. Hamilton (2006). Dating business cycle turning points in real time. In V. Dijk, Milas, and Rothman (Eds.), *Nonlinear Time Series Analysis of Business Cycles*, Elsevier's Contributions to Economic Analysis, pp. 1–54. Federal Reserve Bank of Minneapolis.

Chauvet, M. and J. Piger (2003). Identifying business cycle turning points in real time. *Federal Reserve Bank of St. Louis Review 85*, 47–61.

Chauvet, M. and C. Yu (2006). International business cycles: G7 and OECD countries. *Economic Review, Federal Reserve Bank of Atlanta*, 43–54.

Chib, S., F. Nardari, and N. Shephard (2006). Analysis of high dimensional multivariate stochastic volatility models. *Journal of Econometrics 134*, 341–371.

Geweke, J. (1977). Wage and price dynamics in U.S. manufacturing. In *New Methods in Business Cycle Research: Proceedings from a Conference*, pp. 111–156. Federal Reserve Bank of Minneapolis.

Geweke, J. and G. Zhou (1996). Measuring the pricing error of the arbitrage pricing theory. *The Review of Financial Studies 9*(2), 557–587.

Gregory, A. and A. Head (1999). Common and country-specific fluctuations in productivity, investment, and the current account. *Journal of Monetary Economics 44*, 423–451.

Hamilton, J. D. (1989). A new approach to the economic analysis of nonstationary time series and the business cycle. *Econometrica 57*, 357–384.

Jeliazkov, I. and R. Liu (2010). A model-based ranking of U.S. recessions. *Economics Bulletin 30*(2), 2289–2296.

Kim, C. J. and C. Nelson (1988). Business cycle turning points, a new coincident index, and tests of duration dependence based on a dynamic factor model with regime switching. *Review of Economics and Statistics 80*, 188–201.

Kim, S., N. Shephard, and S. Chib (1998). Stochastic volatility: Likelihood inference and comparison with ARCH models. *The Review of Economic Studies 65*, 361–393.

Koop, G. and S. Potter (2004). Forecasting in dynamic factor models using Bayesian model averaging. *Econometrics Journal 7*, 550–565.

Kose, M. A., C. Otrok, and C. H. Whiteman (2003). International business cycles: World, region and country specific factors. *American Economic Review 93*, 1216–1239.

Kose, M. A., C. Otrok, and C. H. Whiteman (2008). Understanding the evolution of world business cycles. *Journal of International Economics 75*, 110–130.

Sargent, T. and C. Sims (1977). Business cycle modeling without pretending to have too much *a priori* economic theory. In *New Methods in Business Cycle Research: Proceedings from a Conference*, pp. 45–109. Federal Reserve Bank of Minneapolis.

Stock, J. H. and M. W. Watson (2002). Forecasting using principal components from a large number of predictors. *Journal of the American Statistical Association 97*, 1167–1179.

Stock, J. H. and M. W. Watson (2005). Implications of dynamic factor models for VAR analysis. `NBER working paper 11467`.

Zellner, A. (1962). An efficient method of estimating seemingly unrelated regressions and tests for aggregation bias. *Journal of the American Statistical Association 57*(298), 348–368.

CHAPTER 8

WHAT DIFFERENCE FAT TAILS MAKE: A BAYESIAN MCMC ESTIMATION OF EMPIRICAL ASSET PRICING MODELS

PASKALIS GLABADANIDIS

Finance Discipline, University of Adelaide Business School, Adelaide SA 5005, Australia.

This chapter presents an empirical application of Bayesian MCMC estimation to the three main asset pricing models in use in the financial econometrics literature, namely, the Capital Asset Pricing Model (CAPM), the Fama and French (1992) three-factor model, and the Carhart (1997) four-factor model for decile portfolios sorted on market capitalization and book-to-market ratios. I use a SUR model similar in spirit to Zellner (1976) with a Student t-link to allow for fat tails in the distribution of the return innovations. The empirical findings are encouraging in that fat tails lead to posterior distributions of the asset pricing errors that encompass zero and indicate that previous findings of statistically and economically significant intercepts might be an artifact of using a Gaussian error distribution.

8.1 Introduction

The paramount purpose of financial asset pricing theory is to propose an equilibrium expected return on a risky asset as a function of pervasive systematic risk factors or asset characteristics. The equilibrium expected return is usually also a functional of the adopted risk measure. Following the early ground-breaking work of Markowitz (1952), the portfolio risk measure in the framework of the first two moments of asset expected returns has been the dominant approach to defining the contribution of a risky asset to the overall riskiness of a portfolio of risky assets. As shown in Sharpe (1964) among others, this leads to a very elegant expression for the equilibrium expected return equaling the risk-free plus a risk premium which has been referred to ever since as the Capital Asset Pricing Model (CAPM). The risk premium is proportional to the risk premium on the entire portfolio of risky assets and the standardized covariance of the financial asset return with the return on this large portfolio. Unfortunately, the entire portfolio of risky assets is unobserved as many of them do not trade in liquid markets, like human capital or Manhattan skyscrapers, for example. Hence, most of the empirical tests of the CAPM use narrower aggregate portfolios limited to the entire universe of publicly listed stocks and treat that portfolio as the market portfolio. The second complication arising from attempts at testing the CAPM is due to the fact that it has something to say about *ex ante* expected returns which are not directly observable and researcher frequently resort to using time series of realized asset and portfolio *ex post* returns as noisy proxies for the true *ex ante* expected returns. The primary predictive implication of the CAPM is that the so-called market portfolio is mean-variance efficient. Early empirical tests of this implication met with some qualified successes. However, further studies began documenting regularities showing consistent deviations from the equilibrium CAPM relation. In particular, Ball (1978) showed persistent differences between the average returns of portfolios of stocks with high dividend yields relative to a portfolio of stocks with low or zero dividend yields. In a similar fashion, Banz (1981) discovered a significant difference between the average returns of a portfolio consisting of small market capitalization stocks and a portfolio constructed with large market capitalization stocks. Reinganum (1981), Basu (1983), and Rosenberg et al. (1985) revisited the evidence against the CAPM with portfolios sorted on both market capitalization and dividend yields and documenting a positive relationship between size and yield as well as size and the book-to-market ratio. In light of this mounting evidence against the CAPM, Fama and French (1992) proposed a new empirically motivated asset pricing model which included a market capitalization and a book-to-market risk factor in addition to the standard stock market factor. Further empirical studies uncovered persistent and systematic inability of the Fama-French three-factor model to price portfolio sorted on past price performance or the so-called momentum effect. In a study of the performance of US mutual funds, Carhart (1997) proposed the addition of a fourth risk factor related to the cross-section difference between the performance of winner stocks and loser stocks which have experience large run-ups and run-downs, respectively, in the prior year.

The applications of Bayesian estimation and inference in the field of empirical asset pricing and investments are too numerous to list here. I will briefly discuss and mention a few studies that are a reflection of my own personal tastes and are relevant to the study undertaken in this chapter. Early applications of Bayesian inference in the empirical finance literature include Winkler (1973) and Winkler and Barry (1975). More recently, Kandel et al. (1995) apply Bayesian inference to the empirical question regarding the mean-variance efficiency of the market portfolio. Pastor (2000) developed a framework which allows the researcher to provide various degrees of belief in or against the validity of any asset pricing model which then combined with a sample of asset returns supplies a posterior belief in the validity of the model under investigation. This idea along with the exploitation cross-asset error correlations has been extended and applied to tests of mutual fund performance in Pastor and Stambaugh (2002). Avramov (2002) uses closed-form Bayes factors to perform pair-wise tests between a multitude of predictability models for asset prices. Tu and Zhou (2004) use a very similar model to the one in this chapter but their research question is regarding the impact of fat tails in portfolio returns on the portfolio choice decision of a Bayesian investor. Jostova and Philipov (2005) apply Bayesian inference to an application of time-varying factor loadings. Avramov and Chao (2006) present an exact Bayes test in the framework of the empirical Bayes approach in testing the value-added of a portfolio of international stocks to a US-based equity investor. Kan and Zhou (2007) explore the impact on portfolio choice of uncertainty about the true values of the parameters driving the securities return-generating process. Building on the work of Pastor (2000), Tu and Zhou (2010) incorporate investor's beliefs about the validity of the asset pricing model into the optimal portfolio choice problem while allowing for a non-Gaussian distribution for the return innovations. A very nice survey of these and other studies employing Bayesian techniques to empirical asset pricing tests, problems of asset allocation and portfolio choice as well as model and parameter uncertainty is provided in Avramov and Zhou (2010).

The goals of this chapter are several. First, I apply a familiar Bayesian MCMC setup to empirical asset pricing tests in finance under both a Gaussian and a Student-t distribution. I investigate three popular asset pricing models using two sets of portfolio returns sorted on market capitalization and the book-to-market ratio. Second, I investigate several hypotheses regarding these portfolios regarding the distribution of the errors, the mean absolute pricing error across the ten portfolios and whether a subset of the asset pricing factors is sufficient in describing the actual realized portfolio decile returns. Thirdly, I offer a few concluding thoughts and discuss possible future direction of work.

This chapter proceeds as follows. Section 8.2 present the empirical set-up and the Bayesian MCMC algorithm. Section 8.3 presents the details behind the construction of the returns of the test and factor portfolios. Section 8.4 presents the key findings while Section 8.5 offers a few concluding remarks.

8.2 Methodology

Let the realized excess returns $r_{i,t}$ of portfolio i at time period t be stacked in an $N \times 1$ vector y_t while a column vector of ones, 1_N, and the K factor returns and constant for all portfolios be stacked in the $N \times (K+1)$ matrix X_t with the unanticipated return innovations stacked in an $N \times 1$ column-vector ϵ_t. The parameter estimate for all intercepts and factor loadings is denoted as $\theta = [\alpha_1, \beta_1', \ldots, \alpha_N, \beta_N']'$. We can express the system of portfolio realized excess returns more compactly as follows:

$$y_t = X_t\theta + \epsilon_t. \tag{8.1}$$

In order to allow ϵ_t to have a non-Gaussian distribution, we can introduce a mixing variable λ_t which is driven by an inverse gamma (IG) distribution as follows:

$$\epsilon_t|\lambda_t \sim \mathcal{N}_p(0, \Omega), \tag{8.2}$$

$$\lambda_t \sim \mathcal{IG}\left(\frac{\nu}{2}, \frac{\nu}{2}\right), \tag{8.3}$$

where the conditional distribution of ϵ_t is Gaussian but the mixing variable is essentially following a χ distribution with a degrees of freedom coefficient ν taking on values from any in the following prior set of possible values:

$$\nu \in \{2.1, 3, 4, 5, \ldots, 30\}. \tag{8.4}$$

Prior densities for the parameters θ, Ω^{-1}, λ and ν are as follows: $\theta \sim \mathcal{N}_k(\theta_0, B_0^{-1})$, $\Omega^{-1} \sim \mathcal{W}_p(\rho_0, R_0)$, $\lambda_t \sim \mathcal{IG}\left(\frac{\nu_0}{2}, \frac{\nu_0}{2}\right)$, and $\nu \sim \frac{1}{29}$. It is straightforward to show that the conditional posterior distributions are given by $\theta|Y, \Omega^{-1}, \lambda \sim \mathcal{N}_k(\hat{\theta}, B_n^{-1})$, $\Omega^{-1}|Y, \theta, \lambda \sim \mathcal{W}_p(\nu_0 + n, R_n)$, and $\lambda_t|Y, \theta, \Omega^{-1} \sim \mathcal{IG}\left(\frac{\nu_0+p+k}{2}, \frac{\nu_0+S_n}{2}\right)$ where

$$\hat{\theta} = B_n^{-1}\left(B_0\theta_0 + \sum_{t=1}^{t=T}\lambda_t\left(X_t'\Omega^{-1}y_t\right)\right), \tag{8.5}$$

$$B_n = \left(B_0 + \sum_{t=1}^{t=T}\lambda_t\left(X_t'\Omega^{-1}X_t\right)\right), \tag{8.6}$$

$$R_n = \left[R_0^{-1} + \sum_{t=1}^{t=T}\lambda_t\left(y_t - X_t\theta\right)\left(y_t - X_t\theta\right)'\right]^{-1}, \tag{8.7}$$

$$S_n = \left(y_t - X_t\theta\right)'\Omega^{-1}\left(y_t - X_t\theta\right). \tag{8.8}$$

Pioneered by Hastings (1970), a Markov chain Monte Carlo (MCMC) sampling from the full conditional densities is initiated at a random point in the support of the prior densities of the parameters θ, λ, and ν. A total of subsequent 10000 parameter draws are taken and retained following an initial burn-in period of 2000 discarded parameter draws. This large sample of draws converges in distribution to the joint posterior parameter distribution. Following Chib (1995) and Chib and Jeliazkov (2001), marginal model likelihoods are computed by running the MCMC chain for a further 10000 iterations and evaluating the full conditional likelihoods of the additional draws.

8.3 Data

The excess value-weighted return on all CRSP firms incorporated in the United States and listed on NYSE, AMEX or NASDAQ is denoted by MKT. The description of the construction of the size, value and momentum factors are as follows. The six portfolios used to construct the size and value factors are based on a double-sort of all NYSE, AMEX, and NASDAQ stocks with available market capitalization and book value of equity based on size and the book-to-market ratio. The 50th percentile is used to separate small-cap from large-cap stocks while the top 70th percentile of the book-to-market ratio is used to separate value from value-neutral stocks and the bottom 30th percentile of the book-to-market ratio is the cut-off between growth and value-neutral stocks.

Small Value (SV)	Big Value (BV)
Small Neutral (SN)	Big Neutral (BN)
Small Growth (SG)	Big Growth (BG)

The value-weighted return of the small-cap factor SMB is calculated as the difference between the equal-weighted small-cap portfolio returns across the value spectrum and the equal-weighted large-cap portfolio returns for value, growth and value-neutral stocks as follows:

$$\mathrm{SMB} = \frac{1}{3}\left(\mathrm{SV} + \mathrm{SN} + \mathrm{SG}\right) - \frac{1}{3}\left(\mathrm{BV} + \mathrm{BN} + \mathrm{BG}\right). \tag{8.9}$$

The value-weighted return of the value factor HML is given by the difference between the value-weighted returns of the high book-to-market stocks across market caps and the value-weighted returns of the low book-to-market small-cap and large-cap stocks as follows:

$$\mathrm{HML} = \frac{1}{2}\left(\mathrm{SV} + \mathrm{BV}\right) - \frac{1}{2}\left(\mathrm{SG} + \mathrm{BG}\right). \tag{8.10}$$

The six portfolios used to construct the momentum factor are based on the following double-sort of all NYSE, AMEX and NASDAQ stocks with available prior monthly returns based on market capitalization using the 50th percentile as the cut-off point between small-cap and large-cap as well as prior 12 month return using the top 70th percentile to separate past winners from medium performing stocks and the bottom 30th percentile to distinguish losing stocks from medium past performers.

Small Up (SU)	Big Up (BU)
Small Medium (SM)	Big Medium (BM)
Small Down (SD)	Big Down (BD)

Value-weighted momentum portfolio returns UMD are calculated based on the spread between the value-weighted return of past winners and the value-weighted return of past losers as follows:

$$\text{UMD} = \frac{1}{2}\left(\text{SU} + \text{BU}\right) - \frac{1}{2}\left(\text{SD} + \text{BD}\right). \tag{8.11}$$

The size decile portfolio returns are value-weighted and each portfolio includes one tenth of the total security count on NYSE, AMEX, and NASDAQ. Similarly, the value decile portfolio returns are value-weighted and each portfolio includes one tenth of the number of securities with positive book-to-market ratios.[1]

8.4 Empirical Results

Table 8.1 presents summary statistics for the returns of the two sets of decile portfolios using monthly returns and covering 53 years of monthly returns between 1960:01 and 2012:12. Panel A of the table presents the summary statistics for decile portfolios sorted on the market capitalization of all stocks listed on NYSE, NASDAQ and AMEX. It is clear that the average historical returns of small-cap stocks (Low) tend to be larger than the average historical returns of large-cap stocks (High) which is commonly referred to in the finance literature as the small-firm effect. At the same time, the standard deviation of historical returns of the small-cap portfolio is substantially higher than the standard deviation of returns of the large-cap decile. At a first glance this might appear to be justified as stocks with more total risk appear to have higher average returns. All the decile returns exhibit a negatively skewed tail as well as excess kurtosis which is typically the case for individual stock and portfolio returns. The first-order serial autocorrelation is much higher for the small-cap portfolio returns than the one for large-cap returns which is a cause for concern related to low liquidity and stale prices of small stocks. The standard error for any of the ρ_i is approximately equal to 0.04[2] leading to a

[1] I am grateful to Ken French for providing the historical factor and portfolio return data.
[2] S.E.$= 1/\sqrt{12 \times 53} = 1/\sqrt{636} = 0.03965$.

substantial and significant persistence of size decile returns at the one month lag for all portfolios except the large-cap one.

Table 8.1: Summary statistics for the Fama and French size/book-to-market decile portfolios over the period 1960:01–2012:12. Returns are value-weighted and expressed in percent.

Panel A: Size-sorted decile portfolios.									
Portfolio	Mean	Std. Dev.	Skewness	Kurtosis	ρ_1	ρ_2	ρ_3	ρ_6	ρ_{12}
Low	1.133	6.356	−0.135	5.311	0.243	0.001	−0.022	−0.029	0.088
2	1.078	6.322	−0.220	5.214	0.158	−0.031	−0.043	−0.030	0.058
3	1.152	6.048	−0.407	4.907	0.144	−0.059	−0.038	−0.044	0.050
4	1.079	5.818	−0.461	4.989	0.142	−0.050	−0.037	−0.037	0.046
5	1.112	5.627	−0.477	5.031	0.131	−0.033	−0.021	−0.041	0.034
6	1.047	5.280	−0.497	4.882	0.139	−0.070	−0.014	−0.040	0.031
7	1.053	5.193	−0.455	5.287	0.135	−0.040	0.003	−0.045	0.019
8	1.012	5.052	−0.438	4.734	0.103	−0.047	−0.006	−0.052	0.022
9	0.954	4.648	−0.411	4.888	0.107	−0.043	0.002	−0.068	0.018
High	0.807	4.270	−0.334	4.574	0.031	−0.029	0.050	−0.058	0.055
Panel B: Book-to-market-sorted decile portfolios.									
Portfolio	Mean	Std. Dev.	Skewness	Kurtosis	ρ_1	ρ_2	ρ_3	ρ_6	ρ_{12}
Low	0.768	5.168	−0.202	4.319	0.093	−0.022	0.026	−0.041	0.021
2	0.884	4.723	−0.447	4.850	0.077	−0.048	0.003	−0.036	0.009
3	0.918	4.622	−0.469	5.219	0.078	−0.064	0.000	−0.057	0.015
4	0.916	4.742	−0.449	5.137	0.106	−0.045	0.002	−0.077	0.033
5	0.919	4.469	−0.408	5.411	0.076	−0.075	−0.007	−0.062	0.043
6	0.981	4.513	−0.403	5.572	0.067	−0.057	0.058	−0.050	0.051
7	1.037	4.458	−0.096	4.705	0.070	0.001	0.051	−0.091	0.032
8	1.076	4.552	−0.434	6.348	0.103	−0.009	0.028	−0.100	0.104
9	1.157	4.831	−0.297	4.870	0.098	−0.063	−0.000	−0.091	0.072
High	1.272	5.867	0.070	7.567	0.144	−0.044	0.014	−0.135	0.112

Panel B of Table 8.1 presents the summary statistics for decile portfolios sorted on the book-to-market ratio which is a way of sorting out value stocks from growth stocks. Stocks with the 10% lowest market-to-book ratios are grouped in the first decile (Low) and contain growth stocks while stocks with the 10% highest book-to-market ratios are grouped in the tenth decile (High) and contain value stocks. The phenomenon referred to as the value premium in the finance literature is immediately obvious from the reported historical average returns across the ten book-to-market decile portfolios. Value stocks have considerably outperformed growth stocks in the past more than half a century. In this case, there seems to be less of an obvious relationship between the average returns and the standard deviation of returns across the book-to-market decile portfolios. If anything, the risk profile appears to be U-shaped instead of monotonic. There is also some evidence of second-order stochastic dominance between Deciles 2 through 9 and Decile Low in the sense than each of the former portfolio has a larger average return and a smaller standard deviation of return than the growth decile portfolio. All of the deciles with the exception of the Value portfolio have negatively skewed returns and excess kurtosis that latter being the largest for the Value portfolio returns. Some of the autocorrelation coefficients appear statistically significant at considerable lag lengths of up to twelve months.

The Capital Asset Pricing Model of Sharpe (1964):

$$E(R_i) - R_f = \beta_{m,i}\left(E(R_m) - R_f\right),$$

where $E(R_i)$ and $E(R_m)$ are the expected returns on the individual security i and the overall stock market m while R_f is the risk-free rate of return. In practice, the expected returns are unobserved and one popular way to proxy for them is to use realized historical returns $R_{i,t}$ and $R_{m,t}$. Initially, empirical applications of the CAPM with portfolio returns met with some qualified successes. Later applications in Ball (1978), Banz (1981), Basu (1983), Reinganum (1981) and Rosenberg et al. (1985) presented convincing evidence that β_m has a hard time explaining average returns of small-cap and value and growth stock returns. To address this short-coming, Fama and French (1992) proposed their famous three-factor model:

$$E(R_i) - R_f = \beta_{m,i}(E(R_m) - R_f) + \beta_{s,i}\text{SMB} + \beta_{h,i}\text{HML},$$

where the additional variables are $\beta_{s,i}$ as the standardized covariance between asset i and portfolio SMB and $\beta_{h,i}$ is the standardized covariance between asset i and portfolio HML. Portfolio SMB (small minus big) consists of long positions in small-cap stocks and short positions in large-cap stocks while portfolio HML (high minus low) consists of long positions in value stocks (high book-to-market ratios) and short positions in growth stocks (low book-to-market ratios). This model has an easier job "explaining" average returns of portfolio sorted on market capitalization and book-to-market ratios. However, it has a much harder time with the returns of portfolios sorted on past performance or momentum. In his work on mutual fund performance, Carhart (1997) proposed taking momentum into account with the following extension to the Fama-French three-factor model:

$$E(R_i) - R_f = \beta_{m,i}(E(R_m) - R_f) + \beta_{s,i}\text{SMB} + \beta_{h,i}\text{HML} + \beta_{u,i}\text{WML},$$

where the additional variables are $\beta_{u,i}$ as the standardized covariance between asset i and portfolio WML. Portfolio WML (winners minus losers) consists of long positions in stocks that have increased significantly in value over the course of the prior year (winners) and short positions in stocks that have decreased significantly in value during the past year (losers).

Table 8.2 presents the classical estimates of the CAPM one-factor model, the Fama-French three-factor model, and the Carhart four-factor model for

Table 8.2: Classical factor regressions results.

Portfolio	α	β_m	$\bar{R}^2$	α	β_m	β_s	β_h	$\bar{R}^2$	α	β_m	β_s	β_h	β_u	$\bar{R}^2$
Panel A: Size-sorted decile portfolios.														
	CAPM			Fama-French					Carhart					
Low	2.58	1.11***	60.79	−0.74	0.90***	1.24***	0.26***	92.16	−0.68	0.90***	1.24***	0.26***	−0.01	92.15
2	1.42	1.20***	72.21	−1.25**	1.00***	1.09***	0.19***	97.05	−1.22*	1.00***	1.09***	0.18***	−0.00	97.05
3	2.33	1.20***	78.46	−0.16	1.04***	0.92***	0.20***	97.84	0.12	1.03***	0.92***	0.19***	−0.02**	97.86
4	1.62	1.16***	80.44	−0.55	1.02***	0.83***	0.17***	97.30	−0.56	1.02***	0.83***	0.17***	0.00	97.29
5	2.08*	1.15***	84.39	0.19	1.04***	0.69***	0.15***	96.83	0.40	1.03***	0.69***	0.15***	−0.02**	96.84
6	1.56	1.11***	87.93	0.02	1.03***	0.50***	0.14***	95.41	0.14	1.03***	0.50***	0.14***	−0.01	95.41
7	1.63*	1.11***	91.07	0.43	1.05***	0.38***	0.11***	95.57	0.41	1.05***	0.38***	0.11***	0.00	95.56
8	1.25	1.09***	92.80	0.14	1.05***	0.27***	0.12***	95.33	0.37	1.05***	0.27***	0.12***	−0.02	95.35
9	0.96*	1.01***	94.98	0.24	1.02***	0.08***	0.11***	95.49	0.34	1.01***	0.08***	0.10***	−0.01	95.49
High	−0.35	0.93***	95.19	0.41*	0.98***	−0.28***	−0.06***	98.78	0.46**	0.98***	−0.28***	−0.06***	−0.00	98.78
MAPE	1.58			0.41					0.47					
Panel B: Book-to-market sorted decile portfolios.														
Low	−1.61	1.07***	86.70	1.46**	1.00***	−0.12***	−0.51***	93.60	1.51**	1.00***	−0.12***	−0.51***	−0.00	93.59
2	0.14	1.01***	91.77	0.84	1.00***	−0.05***	−0.11***	92.18	1.06	1.00***	−0.05***	−0.12***	−0.02	92.20
3	0.71	0.98***	90.13	0.59	1.00***	−0.04**	0.03	90.23	0.64	0.99***	−0.05**	0.03	−0.00	90.22
4	0.64	0.99***	87.05	−0.68	1.04***	−0.03	0.24***	88.96	−0.61	1.04***	−0.03	0.24***	−0.01	88.94
5	1.11	0.91***	82.89	−0.57	0.98***	−0.04	0.31***	86.48	−0.70	0.98***	−0.04	0.31***	0.01	86.47
6	1.76*	0.93***	84.43	−0.37	1.00***	0.01	0.37***	89.27	−0.09	0.99***	0.01	0.36***	−0.03*	89.30
7	2.69**	0.88***	78.31	−0.29	0.98***	0.02	0.52***	87.99	−0.25	0.98***	0.02	0.51***	−0.00	87.97
8	3.13**	0.88***	76.04	−0.99	0.99***	0.13***	0.69***	92.34	−0.80	0.99***	0.12***	0.68***	−0.02	92.35
9	3.80***	0.94***	76.17	−0.45	1.04***	0.18***	0.70***	91.26	−0.16	1.03***	0.18***	0.69***	−0.03*	91.29
High	4.47**	1.07***	66.65	−1.57	1.17***	0.39***	0.95***	87.22	−0.71	1.16***	0.39***	0.93***	−0.08***	87.49
MAPE	2.01			0.78					0.65					

the two sets of decile portfolios (value-weighted portfolio excess returns, annualized and in percent, over the period 1960:01–2012:12). The table also presents adjusted R^2, mean absolute pricing error (MAPE), and statistical significance (two-sided test) at the 1% ('***'), 5% ('**'), and 10% ('*'), computed using Newey-West standard errors with 12 lags. Panel A presents the classical parameter estimates for the size-sorted decile portfolios. As expected, the CAPM has a hard time pricing most of these portfolios though only three of the annualized risk-adjusted returns (αs) are statistically significant. The small-cap decile (Low) and decile 3 also have large positive alphas but they are not statistically significant. The goodness-of-fit is decreasing as the market cap of the decile is decreasing. However, the measure of systematic risk, β, is higher for the small-cap portfolios than the beta of the large-cap portfolio. This could partially explain the small-firm effect though the presence of large statistically significant pricing errors is a concern. The Fama-French three-factor model performs considerably better in terms of having uniformly large values of the adjusted R^2 and only two statistically significant pricing errors (deciles 2 and High). Based on the loadings on HML we can tell that small-cap portfolios are largely value-tilted as well while large-cap decile has a negative loading on HML meaning that it has a distinct growth tilt. The factor loadings on SMB are as expected and simply confirm that as the market cap of the stocks in the decile declines the loading on SMB increases and vice versa. Finally, the lack of variation in the market loadings, β_m, has led some researchers to conclude that "(the CAPM) beta is dead." The success of the Fama and French model is also easily seen in the more than three-fold decline in MAPE. Finally, turning to the Carhart four-factor model we see little improvement in terms of goodness-of-fit, with the loadings on the first three factor changing very little while the loadings on the WML momentum factor are mostly negative with only two statistically significant (deciles 3 and 5). The latter finding suggests that all the size decile portfolios exhibit a mild contrarian tilt at the twelve month frequency. There is also a slight increase in the MAPE for the Carhart model relative to the Fama-French model.

Panel B presents the classical parameter estimates using the three asset pricing models with the book-to-market decile portfolio returns. The failure of the CAPM to explain the average returns of these portfolios is more pronounced. First, the annualized alphas are large and highly significant for deciles 6 through High. Second, market betas appear to have a U-shaped relation to the book-to-market ratio and are virtually the same for the two extreme deciles of growth (Low) and value (High). The only way for the CAPM to account for the value premium is then though the large difference in the alphas of the two extreme decile portfolios. Thirdly, the goodness-of-fit deteriorates markedly as we move towards decile with increasing value tilt. Turning to the parameter estimates for the Fama-French model we observe a considerable improvement in terms of lower pricing errors and substantial increase in the goodness-of-fit. Now, only one alpha is significant, namely the growth decile (Low). The MAPE has declined by almost a factor of three times while the pattern of factor loadings is more meaningful compared to the CAPM. Growth stocks (Low) have negative loadings on SMB and HML indicating that these stocks tend to have large market capitalizations and exhibit a growth tilt. At the other extreme, value stocks (High) have positive loadings on SMB and HML meaning that these stocks have mostly small market capitalizations and tend to be value stocks on average. Even the pattern of market betas is increasing as we move from growth to value stocks even though that is accounting for only a small portion of the average return of the value premium. Finally, turning towards the parameter estimates for the Carhart model we observe a small improvement on the Fama-French model in terms of the point estimates and statistical significance of the pricing errors, the MAPE. The goodness-of-fit is either virtually unchanged or deteriorates marginally reflecting the insignificant loadings on the momentum factor. Nevertheless, all the β_u point estimates are negative and only three are statistically significant (deciles 6, 9 and High) suggesting that portfolios sorted on book-to-market have a mild contrarian tilt consistent with the findings for portfolios sorted on size. This becomes more pronounced as we focus on portfolios that include more and more value stocks.

Table 8.3 presents the medians of the posterior distributions of the parameters of the CAPM, Fama-French and Carhart asset pricing models as well 95% Bayesian credibility intervals for the two sets of decile portfolio returns under investigation assuming a Gaussian distribution for the error term. Panel A reports the empirical results for the ten portfolios sorted by size. The Bayesian estimates of the parameter show a small improvement on the classical estimates reported in Table 8.2. The MAPE increases by about 20% (from 1.58% up to 1.91%) and only the alpha of decile 7 has the posterior credibility interval that excludes zero. Next, the Bayesian posterior estimates for the Fama-French model show an even greater improvement compared to classical estimates reported previously. The MAPE declines by approximately 20% (from 0.41% down to 0.34% per year) and all of the posterior credibility intervals for the alphas now encompass zero suggesting that the evidence appears in favor of the correct pricing of the returns of all ten portfolios. Finally, turning to the parameter estimates for the Carhart model we observe the same pattern again. Now, all of the posterior credibility intervals for the alphas include zero and the MAPE has decreased by about 10% (from 0.47% down to 0.42% per year) relative to the classical parameter estimates.

Panel B of Table 8.3 presents the findings for the ten deciles sorted on the book-to-market ratio. The posterior Bayesian credibility intervals for the alphas of deciles 5 through 9 are to the right of zero indicating that the CAPM beta substantially underprices these four portfolios consisting largely of value stocks. Nevertheless, the MAPE is now down by about 10% (from 2.01% down to 1.78% per year). Turning to the Fama-French model we see similar findings to the classical estimates. The posterior credibility interval for the alpha of the growth decile (Low) lies to the right of zero implying a significant underpricing. The credibility intervals for the alphas of all the other

Table 8.3: Bayesian factor regression results: Gaussian errors.

Portfolio	CAPM α	CAPM β_m	Fama-French α	Fama-French β_m	Fama-French β_s	Fama-French β_h	Carhart α	Carhart β_m	Carhart β_s	Carhart β_h	Carhart β_u
Panel A: Size-sorted decile portfolios.											
Low	0.68	1.12	−1.38	0.90	1.22	0.27	−0.69	0.90	1.24	0.26	−0.00
	(−0.79,2.25)	(1.09,1.16)	(−4.40,1.44)	(0.84,0.96)	(1.14,1.30)	(0.17,0.38)	(−2.93,1.54)	(0.85,0.94)	(1.17,1.30)	(0.19,0.33)	(−0.05,0.04)
2	1.39	1.21	−1.56	1.01	1.08	0.18	−1.18	1.00	1.09	0.18	−0.00
	(−3.44,6.20)	(1.11,1.31)	(−3.71,0.81)	(0.96,1.05)	(1.01,1.15)	(0.09,0.26)	(−2.90,0.53)	(0.97,1.04)	(1.05,1.14)	(0.13,0.24)	(−0.04,0.03)
3	2.45	1.21	0.04	1.04	0.92	0.18	0.14	1.03	0.92	0.19	−0.02
	(−0.74,5.80)	(1.15,1.28)	(−1.08,1.17)	(1.02,1.06)	(0.89,0.96)	(0.14,0.21)	(−0.90,1.19)	(1.01,1.05)	(0.89,0.95)	(0.16,0.22)	(−0.05,−0.00)
4	1.74	1.18	−0.35	1.02	0.84	0.13	−0.53	1.02	0.83	0.17	0.00
	(−1.05,4.60)	(1.12,1.24)	(−1.26,0.61)	(1.00,1.04)	(0.81,0.87)	(0.10,0.17)	(−1.56,0.48)	(1.00,1.04)	(0.80,0.86)	(0.14,0.20)	(−0.02,0.02)
5	2.51	1.17	0.58	1.04	0.70	0.13	0.42	1.03	0.69	0.15	−0.02
	(−0.00,4.86)	(1.11,1.22)	(−0.41,1.47)	(1.02,1.06)	(0.67,0.73)	(0.09,0.16)	(−0.58,1.42)	(1.02,1.05)	(0.66,0.72)	(0.12,0.18)	(−0.04,0.00)
6	2.10	1.11	0.42	1.02	0.52	0.12	0.16	1.03	0.50	0.14	−0.01
	(−0.04,4.35)	(1.07,1.16)	(−0.57,1.47)	(1.00,1.05)	(0.48,0.55)	(0.08,0.15)	(−0.86,1.20)	(1.01,1.05)	(0.47,0.53)	(0.11,0.17)	(−0.03,0.01)
7	1.46	1.11	0.44	1.05	0.38	0.10	0.43	1.05	0.38	0.11	0.00
	(0.32,2.62)	(1.09,1.13)	(−0.62,1.54)	(1.02,1.07)	(0.35,0.42)	(0.06,0.14)	(−0.68,1.53)	(1.03,1.07)	(0.35,0.41)	(0.08,0.15)	(−0.02,0.02)
8	1.72	1.09	0.20	1.06	0.26	0.12	0.39	1.05	0.27	0.12	−0.02
	(−0.56,3.92)	(1.04,1.13)	(−1.21,1.72)	(1.03,1.09)	(0.21,0.30)	(0.07,0.17)	(−0.82,1.61)	(1.03,1.07)	(0.23,0.30)	(0.08,0.16)	(−0.04,0.00)
9	1.06	1.01	0.34	1.02	0.07	0.10	0.36	1.01	0.08	0.10	−0.01
	(−0.36,2.63)	(0.98,1.04)	(−0.67,1.35)	(1.00,1.04)	(0.04,0.10)	(0.06,0.13)	(−0.66,1.40)	(0.99,1.03)	(0.05,0.11)	(0.07,0.13)	(−0.03,0.01)
High	−0.49	0.93	0.26	0.98	−0.29	−0.06	0.45	0.98	−0.28	−0.06	−0.00
	(−1.79,0.85)	(0.90,0.95)	(−0.58,1.07)	(0.96,1.00)	(−0.31,−0.26)	(−0.08,−0.03)	(−0.51,1.47)	(0.96,1.00)	(−0.31,−0.25)	(−0.09,−0.03)	(−0.02,0.01)
MAPE	1.91		0.34				0.42				
Panel B: Book-to-market sorted decile portfolios.											
Low	−1.78	1.07	1.51	1.00	−0.12	−0.50	1.57	1.00	−0.12	−0.51	−0.00
	(−4.25,0.71)	(1.03,1.12)	(0.21,2.83)	(0.98,1.03)	(−0.15,−0.08)	(−0.54,−0.46)	(0.19,2.91)	(0.97,1.03)	(−0.15,−0.08)	(−0.55,−0.47)	(−0.03,0.02)
2	0.13	1.01	0.85	1.00	−0.05	−0.11	1.06	1.00	−0.05	−0.12	−0.02
	(−0.83,1.09)	(0.99,1.03)	(−0.13,1.83)	(0.98,1.02)	(−0.08,−0.03)	(−0.14,−0.08)	(0.07,2.05)	(0.98,1.02)	(−0.08,−0.03)	(−0.14,−0.08)	(−0.04,−0.00)
3	0.74	0.98	0.61	1.00	−0.04	0.03	0.65	0.99	−0.04	0.03	−0.00
	(−0.19,1.66)	(0.96,1.00)	(−0.38,1.57)	(0.98,1.01)	(−0.07,−0.02)	(0.00,0.06)	(−0.35,1.64)	(0.98,1.01)	(−0.07,−0.02)	(0.00,0.06)	(−0.02,0.01)
4	0.61	0.99	−0.59	1.04	−0.02	0.24	−0.53	1.04	−0.02	0.24	−0.01
	(−0.58,1.77)	(0.96,1.01)	(−2.31,1.13)	(1.00,1.07)	(−0.07,0.02)	(0.19,0.29)	(−2.28,1.28)	(1.00,1.07)	(−0.07,0.03)	(0.18,0.29)	(−0.04,0.03)
5	1.18	0.91	−0.55	0.98	−0.04	0.30	−0.67	0.98	−0.04	0.31	0.01
	(0.03,2.33)	(0.89,0.93)	(−1.51,0.42)	(0.96,1.00)	(−0.07,−0.02)	(0.28,0.33)	(−1.65,0.30)	(0.96,1.00)	(−0.07,−0.02)	(0.28,0.34)	(−0.01,0.03)
6	1.78	0.93	−0.31	0.99	0.01	0.37	−0.01	0.99	0.01	0.36	−0.02
	(0.84,2.69)	(0.91,0.94)	(−1.80,1.20)	(0.97,1.02)	(−0.03,0.05)	(0.33,0.42)	(−1.52,1.57)	(0.96,1.02)	(−0.04,0.05)	(0.31,0.41)	(−0.06,0.01)
7	2.76	0.88	−0.25	0.98	0.02	0.52	−0.19	0.97	0.02	0.51	−0.00
	(1.57,3.92)	(0.86,0.90)	(−1.42,0.96)	(0.95,1.00)	(−0.02,0.05)	(0.48,0.55)	(−1.43,1.05)	(0.95,1.00)	(−0.02,0.05)	(0.48,0.55)	(−0.03,0.02)
8	3.25	0.88	−0.93	0.99	0.13	0.69	−0.74	0.99	0.13	0.68	−0.02
	(1.61,4.87)	(0.85,0.92)	(−2.13,0.25)	(0.97,1.01)	(0.09,0.16)	(0.65,0.72)	(−2.03,0.55)	(0.96,1.01)	(0.09,0.16)	(0.64,0.72)	(−0.04,0.01)
9	3.98	0.94	−0.42	1.04	0.18	0.69	−0.13	1.03	0.18	0.69	−0.03
	(1.66,6.26)	(0.90,0.98)	(−1.38,0.51)	(1.02,1.06)	(0.15,0.20)	(0.67,0.72)	(−1.09,0.84)	(1.01,1.05)	(0.15,0.20)	(0.66,0.72)	(−0.05,−0.01)
High	4.89	1.07	−1.66	1.17	0.39	0.95	−0.80	1.16	0.39	0.93	−0.08
	(−0.63,10.27)	(0.96,1.17)	(−4.27,0.95)	(1.12,1.22)	(0.32,0.47)	(0.87,1.03)	(−3.32,1.71)	(1.11,1.21)	(0.32,0.46)	(0.85,1.00)	(−0.13,−0.03)
MAPE	1.78		0.60				0.66				

portfolios comfortably include zero suggesting correct pricing by the Fama-French model. The MAPE declines by about a quarter (from 0.78% down to 0.60% per year) compared to the classical parameter estimates. Finally, as expected, the posterior medians for all of the factor loadings agree with the classical estimates reported before.

Next, I turn to the Bayesian MCMC results for the three asset pricing models and two sets of size and book-to-market decile portfolios under the assumption of a Student-t error distribution. Table 8.4 reports the posterior median alphas and factor loadings as well as 95% Bayesian credibility intervals. Panel A presents the findings for the size-sorted decile portfolio returns. For the CAPM model the evidence for the alphas is largely the same as the evidence for the Gaussian error distribution model. However, the MAPE is significantly reduced compared to the Gaussian model and the posterior degrees of freedom estimate is 7 suggesting a substantial deviation from Gaussianity. The posterior credibility intervals for the alphas in the Fama-French and Carhart models all include the zero suggesting that both models price all ten portfolios reasonably well. The posterior median degree of freedom is still 7 consistent it the findings for the CAPM model. The MAPE follow the same pattern as the classical estimates with Gaussian error structure.

Panel B presents the empirical estimates for the book-to-market sorted decile portfolios. We find again that the CAPM has a hard time pricing the value-growth decile portfolios as the posterior credibility intervals for the alphas of deciles 6 through 9 are all to the right of zero indicating a substantial underpricing. The MAPE increases relative to the Gaussian error model but that appears to be an artifact of the pricing model. For comparison, the MAPE in the Fama-French and Carhart models decreases relative to the Gaussian error model. Similarly to what we saw in Table 8.3, all the posterior credibility intervals for the alphas of all ten portfolios include zero suggesting that the two models do a good job at pricing the book-to-market portfolio deciles. The posterior medians and credibility intervals for the factor loadings indicate that the fat-tailed error distribution largely leaves these parameters unchanged relative to the Gaussian baseline model.

Table 8.5 reports the Bayesian marginal likelihoods for the three asset pricing models using a Gaussian and a Student-t error distribution with both unconstrained and constrained alphas for the two sets of decile portfolios. Several observations are in order from the values of these marginal likelihoods. First, the Fama-French model provides a substantial improvement on the CAPM while the Carhart model provides a relatively smaller improvement on the Fama-French model. Second, the Gaussian error models have significantly lower marginal likelihoods than the corresponding models with Student-t errors. Finally, the restriction of zero pricing errors leads to marginal likelihoods that are much lower than the models with unconstrained alphas. All of these findings are consistent for both sets of decile portfolios.

The marginal likelihoods for the various versions of the models provide the necessary means to calculate Bayes factors allowing us to more formally

Table 8.4: Bayesian factor regression results: Student-t errors.

Portfolio	α	β_m	α	β_m	β_s	β_h	α	β_m	β_s	β_h	β_u
Panel A: Size sorted decile portfolios.											
	CAPM (median d.f. = 7)		Fama-French (median d.f. = 7)				Carhart (median d.f. = 7)				
Low	0.68	1.12	−1.38	0.90	1.22	0.27	−1.38	0.90	1.22	0.27	0.00
	(−0.79,2.25)	(1.09,1.16)	(−4.40,1.44)	(0.84,0.96)	(1.14,1.30)	(0.17,0.38)	(−4.65,1.70)	(0.84,0.97)	(1.13,1.32)	(0.18,0.38)	(−0.07,0.07)
2	1.39	1.21	−1.56	1.01	1.08	0.18	−1.62	1.01	1.07	0.18	−0.00
	(−3.44,6.20)	(1.11,1.31)	(−3.71,0.81)	(0.96,1.05)	(1.01,1.15)	(0.09,0.26)	(−4.02,0.93)	(0.96,1.06)	(1.00,1.15)	(0.10,0.26)	(−0.06,0.05)
3	2.45	1.21	0.04	1.04	0.92	0.18	0.29	1.04	0.92	0.17	−0.02
	(−0.74,5.80)	(1.15,1.28)	(−1.08,1.17)	(1.02,1.06)	(0.89,0.96)	(0.14,0.21)	(−0.84,1.44)	(1.01,1.06)	(0.89,0.96)	(0.13,0.20)	(−0.05,−0.00)
4	1.74	1.18	−0.35	1.02	0.84	0.13	−0.44	1.02	0.84	0.13	0.01
	(−1.05,4.60)	(1.12,1.24)	(−1.26,0.61)	(1.00,1.04)	(0.81,0.87)	(0.10,0.17)	(−1.47,0.62)	(1.00,1.04)	(0.81,0.87)	(0.10,0.17)	(−0.02,0.03)
5	2.51	1.17	0.58	1.04	0.70	0.13	0.77	1.03	0.70	0.12	−0.02
	(−0.00,4.86)	(1.11,1.22)	(−0.41,1.47)	(1.02,1.06)	(0.67,0.73)	(0.09,0.16)	(−0.19,1.76)	(1.01,1.05)	(0.67,0.73)	(0.09,0.15)	(−0.04,0.00)
6	2.10	1.11	0.42	1.02	0.52	0.12	0.45	1.02	0.52	0.11	−0.00
	(−0.04,4.35)	(1.07,1.16)	(−0.57,1.47)	(1.00,1.05)	(0.48,0.55)	(0.08,0.15)	(−0.68,1.51)	(1.00,1.04)	(0.48,0.55)	(0.08,0.15)	(−0.03,0.02)
7	1.46	1.11	0.44	1.05	0.38	0.10	0.55	1.05	0.38	0.10	−0.01
	(0.32,2.62)	(1.09,1.13)	(−0.62,1.54)	(1.02,1.07)	(0.35,0.42)	(0.06,0.14)	(−0.63,1.71)	(1.02,1.07)	(0.35,0.42)	(0.06,0.14)	(−0.03,0.02)
8	1.72	1.09	0.20	1.06	0.26	0.12	0.46	1.05	0.25	0.12	−0.03
	(−0.56,3.92)	(1.04,1.13)	(−1.21,1.72)	(1.03,1.09)	(0.21,0.30)	(0.07,0.17)	(−0.83,1.87)	(1.02,1.08)	(0.21,0.30)	(0.07,0.16)	(−0.06,0.00)
9	1.06	1.01	0.34	1.02	0.07	0.10	0.52	1.01	0.06	0.09	−0.01
	(−0.36,2.63)	(0.98,1.04)	(−0.67,1.35)	(1.00,1.04)	(0.04,0.10)	(0.06,0.13)	(−0.52,1.47)	(0.99,1.04)	(0.03,0.10)	(0.06,0.13)	(−0.04,0.01)
High	−0.49	0.93	0.26	0.98	−0.29	−0.06	0.33	0.98	−0.28	−0.06	−0.00
	(−1.79,0.85)	(0.90,0.95)	(−0.58,1.07)	(0.96,1.00)	(−0.31,−0.26)	(−0.08,−0.03)	(−0.57,1.24)	(0.96,1.00)	(−0.31,−0.26)	(−0.08,−0.03)	(−0.02,0.02)
MAPE	1.59		0.38				0.49				
Panel B: Book-to-market sorted decile portfolios.											
	CAPM (median d.f. = 7)		Fama-French (median d.f. = 7)				Carhart (median d.f. = 8)				
Low	−2.21	1.07	1.37	1.00	−0.11	−0.54	1.36	1.00	−0.11	−0.54	0.00
	(−6.15,1.37)	(0.99,1.15)	(−0.11,2.90)	(0.96,1.03)	(−0.16,−0.06)	(−0.60,−0.48)	(−0.37,3.01)	(0.96,1.03)	(−0.16,−0.06)	(−0.60,−0.48)	(−0.03,0.04)
2	−0.36	1.02	0.69	1.00	−0.05	−0.15	0.89	1.00	−0.05	−0.16	−0.02
	(−1.53,0.74)	(1.00,1.05)	(−0.41,1.77)	(0.98,1.02)	(−0.08,−0.01)	(−0.20,−0.11)	(−0.15,2.05)	(0.97,1.02)	(−0.08,−0.01)	(−0.20,−0.11)	(−0.05,0.01)
3	0.45	1.00	0.44	1.00	−0.02	0.01	0.59	1.00	−0.02	0.01	−0.01
	(−0.61,1.53)	(0.98,1.02)	(−0.56,1.54)	(0.98,1.02)	(−0.06,0.01)	(−0.03,0.04)	(−0.49,1.48)	(0.98,1.02)	(−0.05,0.01)	(−0.03,0.05)	(−0.04,0.02)
4	0.56	1.00	−0.39	1.03	−0.02	0.20	−0.28	1.03	−0.02	0.20	−0.01
	(−1.10,2.20)	(0.97,1.03)	(−2.77,1.96)	(0.98,1.09)	(−0.10,0.06)	(0.11,0.28)	(−2.83,2.21)	(0.98,1.08)	(−0.10,0.06)	(0.11,0.28)	(−0.07,0.04)
5	1.21	0.92	−0.52	0.97	−0.01	0.28	−0.56	0.97	−0.01	0.28	−0.00
	(−0.35,2.71)	(0.89,0.96)	(−1.56,0.59)	(0.95,1.00)	(−0.05,0.02)	(0.24,0.32)	(−1.62,0.66)	(0.95,0.99)	(−0.05,0.02)	(0.24,0.32)	(−0.03,0.03)
6	1.63	0.93	−0.21	0.99	0.01	0.34	0.07	0.98	0.01	0.33	−0.02
	(0.58,2.71)	(0.91,0.96)	(−2.18,1.80)	(0.94,1.03)	(−0.05,0.08)	(0.26,0.41)	(−2.11,2.29)	(0.94,1.03)	(−0.06,0.08)	(0.25,0.41)	(−0.07,0.02)
7	2.55	0.91	−0.23	0.99	0.04	0.48	−0.15	0.99	0.04	0.48	−0.01
	(1.03,4.05)	(0.88,0.95)	(−1.62,1.34)	(0.95,1.02)	(−0.01,0.09)	(0.43,0.54)	(−1.77,1.46)	(0.96,1.02)	(−0.02,0.09)	(0.42,0.54)	(−0.04,0.03)
8	2.85	0.90	−1.19	0.99	0.11	0.65	−0.94	0.99	0.12	0.64	−0.02
	(0.42,5.20)	(0.84,0.95)	(−2.48,0.29)	(0.96,1.02)	(0.07,0.16)	(0.60,0.69)	(−2.40,0.56)	(0.96,1.02)	(0.07,0.16)	(0.59,0.70)	(−0.05,0.02)
9	3.84	0.96	−0.45	1.05	0.18	0.68	−0.13	1.04	0.18	0.67	−0.03
	(0.45,7.28)	(0.89,1.04)	(−1.49,0.63)	(1.02,1.07)	(0.15,0.22)	(0.64,0.72)	(−1.23,0.87)	(1.02,1.06)	(0.15,0.22)	(0.64,0.71)	(−0.05,−0.00)
High	4.64	1.07	−1.89	1.16	0.40	0.95	−1.36	1.15	0.40	0.94	−0.05
	(−3.38,13.55)	(0.89,1.25)	(−6.03,2.12)	(1.06,1.25)	(0.27,0.54)	(0.80,1.09)	(−5.43,2.98)	(1.06,1.24)	(0.27,0.52)	(0.79,1.07)	(−0.14,0.04)
MAPE	1.92		0.48				0.58				

Table 8.5: Bayesian log marginal likelihoods.

Panel A: Size sorted decile portfolios.				
Error	Restriction	CAPM	Fama-French	Carhart
Gaussian	$\alpha = 0$	−14455.47	−9711.01	−9704.40
Gaussian	–	−14016.64	−9690.79	−9676.50
Student-t	$\alpha = 0$	−11833.41	−9519.01	−9499.73
Student-t	–	−11810.73	−9479.47	−9467.05
Panel B: Book-to-market sorted decile portfolios.				
Error	Restriction	CAPM	Fama-French	Carhart
Gaussian	$\alpha = 0$	−15338.20	−12829.19	−12809.70
Gaussian	–	−15227.98	−12823.51	−12797.82
Student-t	$\alpha = 0$	−12858.35	−11765.49	−11763.52
Student-t	–	−12857.41	−11753.88	−11750.56

investigate the comparison between models and the various built-in restrictions. Table 8.6 reports the log Bayes factors of all the meaningful pairwise model comparisons that are of interest. Panel A presents the findings for the size-sorted decile portfolios. The first interesting question concerns the significance of including SMB and HML to the CAPM market factor as well as the addition of the WML momentum factor to the three Fama-French factors in explaining the size-sorted decile portfolio returns. All four versions of the asset pricing models concur that SMB and HML are incredibly important with the WML factor being necessary as well though the Bayes factors decrease dramatically for the latter model comparison. Similarly, it is worthwhile noting that the log Bayes factors for the Student-t model's inclusion of SMB and HML are half as big as the log Bayes factors for the Gaussian error distribution. The second model comparison tests the two nested versions of all three asset pricing models with and without zero pricing errors. Overwhelmingly, the data prefer non-zero pricing errors especially for the CAPM model with Gaussian errors. The Bayes factors are also in favor of non-zero pricing errors for the other two asset pricing models and the Student-t error distribution versions of all three models though the log factors are an order of magnitude smaller. The final model comparison is the one between the Gaussian and Student-t versions for all three asset pricing models. Once again, the data prefer the models with fat-tailed error distribution for all three asset pricing models and regardless of whether the pricing errors have been constrained to zero or not.

The same pattern of finding with one notable exception is reported in Panel B for the portfolios sorted on book-to-market. First, the inclusion of SMB and

Table 8.6: Log Bayes factors.

Panel A: Size-sorted decile portfolios.			
Error	Restriction	Are size and value priced?	Is momentum priced?
Gaussian	$\alpha = 0$	4744.46	6.61
Gaussian	–	4325.85	14.29
Student-t	$\alpha = 0$	2314.40	19.28
Student-t	–	2331.26	12.42

Error	Is $\alpha = 0$?	CAPM	Fama-French	Carhart
Gaussian		438.83	20.22	27.90
Student-t		22.68	39.54	32.68

Is the error Gaussian?	Restriction	CAPM	Fama-French	Carhart
	$\alpha = 0$	2622.06	192.00	204.67
	–	2205.91	211.32	209.45

Panel B: Book-to-market sorted decile portfolios.			
Error	Restriction	Are size and value priced?	Is momentum priced?
Gaussian	$\alpha = 0$	2509.01	19.49
Gaussian	–	2404.47	25.69
Student-t	$\alpha = 0$	1092.86	1.97
Student-t	–	1103.53	3.32

Error	Is $\alpha = 0$?	CAPM	Fama-French	Carhart
Gaussian		110.22	5.68	11.88
Student-t		0.94	11.61	12.96

Is the error Gaussian?	Restriction	CAPM	Fama-French	Carhart
	$\alpha = 0$	2479.85	1063.70	1046.18
	–	2370.57	1069.63	1047.26

HML is strongly preferred by the data yet the log Bayes factor decline by one-half when the error is Student-t just as for the size decile portfolios. This time the inclusion of the momentum factor is still preferred by the data but much less strongly for the Student-t error structure with the constrained version showing a log Bayes factor of 1.97 while the unconstrained version a log Bayes factor of 3.32. Secondly, the Student-t error distribution is strongly preferred by the data across all three asset pricing models and regardless of whether we constrain the pricing errors or not. Lastly and most importantly, the log Bayes factor in favor of unconstrained alphas in the CAPM framework is a very low 0.94 for Student-t errors and a very large 110.22 for Gaussian errors. This finding suggests that the classical estimates of large CAPM pricing errors with book-to-market portfolio returns could be just an artefact of using a Gaussian error structure instead of more fat-tailed distribution. The Bayes factors for the Fama-French and Carhart models still indicate a mis-pricing as the data prefer non-zero alphas for both models regardless of the error distribution.

8.5 Concluding Remarks

The empirical application of the Bayesian MCMC model to financial asset pricing tests in this chapter presents several key findings. First, it allows for the model comparison between non-nested models. The Student-t distribution tends to a Gaussian in the limit when the degrees of freedom increase to infinity and in practical applications the two densities appear almost identical for much lower values of the degrees of freedom. However, the densities do not nest for any finite value of the degrees of freedom. This makes comparisons between the two error structures using a likelihood ratio test very difficult in the classical frequentist framework of statistical inference. Second, the Bayesian framework allows the incorporation of prior information into the inference about the pricing errors and factor loadings of the pricing models. Though not explicitly addressed in this chapter, an empirical Bayesian approach could be used by taking an older sample of historical securities' returns, drawing a posterior parameter distribution and then applying it as a prior to a more recent historical sample of securities' returns. Third, the ease of determining Bayesian credibility intervals makes it relatively easier to draw inference about alphas and betas following a single glance at the data rather than having to rely on a re-sampling argument. Fourth, any useful function of interest of the posterior distribution of the parameters can be easily computed from the MCMC draws and its posterior distribution analyzed to inform the researcher on risk-reward ratios, optimal portfolio weights, mean-variance efficiency tests, etc.

Bibliography

Avramov, D. (2002). Stock return predictability and model uncertainty. *Journal of Financial Economics 64*, 423–458.

Avramov, D. and J. Chao (2006). An exact bayes test of asset pricing models with implications to international markets. *Journal of Business 79*, 293–323.

Avramov, D. and G. Zhou (2010). Bayesian portfolio analysis. *Annual Review of Financial Economics 2*, 25–47.

Ball, R. (1978). Anomalies in relationships between securities' yields and yield surrogates. *Journal of Financial Economics 6*, 103–126.

Banz, R. (1981). The relationship between return and market value of common stocks. *Journal of Financial Economics 9*, 3–18.

Basu, S. (1983). The relationship between earnings yield, market value and return for NYSE common stocks: Further evidence. *Journal of Financial Economics 12*, 129–156.

Carhart, M. M. (1997). On persistence in mutual fund performance. *Journal of Finance 52*, 57–82.

Chib, S. (1995). Marginal likelihood from the Gibbs output. *Journal of American Statistical Association 90*, 1313–1321.

Chib, S. and I. Jeliazkov (2001). Marginal likelihood from the Metropolis-Hastings output. *Journal of American Statistical Association 96*, 270–281.

Fama, E. F. and K. R. French (1992). The cross-section of expected stock returns. *Journal of Finance 47*, 427–465.

Hastings, W. K. (1970). Monte Carlo sampling methods using Markov chains and their applications. *Biometrika 57*, 97–109.

Jostova, G. and A. Philipov (2005). Bayesian analysis of stochastic betas. *Journal of Financial and Quantitative Analysis 40*, 747–778.

Kan, R. and G. Zhou (2007). Optimal portfolio choice with parameter uncertainty. *Journal of Financial and Quantitative Analysis 42*, 621–656.

Kandel, S., R. McCulloch, and R. F. Stambaugh (1995). Bayesian inference and portfolio efficiency. *Review of Financial Studies 9*, 1–53.

Markowitz, H. M. (1952). Portfolio selection. *Journal of Finance 7*(1), 77–91.

Pastor, L. (2000). Portfolio selection and asset pricing models. *Journal of Finance 55*, 179–223.

Pastor, L. and R. F. Stambaugh (2002). Mutual fund performance and seemingly unrelated assets. *Journal of Financial Economics 63*, 315–349.

Reinganum, M. (1981). Misspecification of capital asset pricing: Empirical anomalies, based on earnings' yields and market values. *Journal of Financial Economics 9*, 19–46.

Rosenberg, B., K. Reid, and R. Lanstein (1985). Persuasive evidence of market inefficiency. *Journal of Portfolio Management 11*, 9–17.

Sharpe, W. F. (1964). Capital asset prices: A theory of market equilibrium under conditions of risk. *Journal of Finance 19*, 425–442.

Tu, J. and G. Zhou (2004). Data-generating process uncertainty: What difference does it make in portfolio decisions? *Journal of Financial Economics 72*, 385–421.

Tu, J. and G. Zhou (2010). Incorporating economic objectives into Bayesian priors: Portfolio choice under parameter uncertainty. *Journal of Financial and Quantitative Analysis 45*, 959–986.

Winkler, R. L. (1973). Bayesian models for forecasting future security prices. *Journal of Financial and Quantitative Analysis 8*, 387–405.

Winkler, R. L. and C. B. Barry (1975). A Bayesian model for portfolio selection and revision. *Journal of Finance 30*, 179–192.

Zellner, A. (1976). Bayesian and non-Bayesian analysis of the regression model with multivariate student-t error terms. *Journal of the American Statistical Association 71*, 400–405.

CHAPTER 9

STOCHASTIC SEARCH FOR PRICE INSENSITIVE CONSUMERS

ERIC EISENSTAT

University of Bucharest and RIMIR, Bucharest, Romania.

This chapter outlines a straightforward Gibbs sampling algorithm that yields posterior probabilities of price indifference for individuals in the sample. The algorithm is founded on a random utility model of consumer choice, with a censored distribution placed on the heterogeneous price coefficient. Assuming normal disturbances for the indirect utilities, this yields a parsimonious mixed probit specification with the computational advantage of fully tractable conditional distributions. Simulating the insensitivity to price fluctuations over a given price range allows for a straightforward way to segment markets, and consequently, is of particular interest to economists and marketers. Moreover, it is shown that the proposed formulation can be easily adapted to accommodate reference price models. Therefore, the sampling algorithm can be used to estimate heterogeneous price thresholds in discontinuous demand curves with asymmetric price effects.

9.1 Introduction

The focus of this chapter is on modeling the effect of prices on consumer choice, and in particular the lack thereof. Without doubt, price is a crucial variable in choice analysis as it is closely related to most quantities of interest that are ultimately inferred. Although this is true for a much wider class of models, in this chapter we focus on discrete choice formulations based on random utility maximization (Marschak, 1960; McFadden, 1974, 1986). In this context, an econometric specification relates multinomial choice observations to a latent utility index, which is modeled as a (typically linear) function of prices. The effect of price changes on consumer choices then reduces to analyzing their effects on the utilities.

Price coefficients in the underlying utility specification consequently play a central role in inference. For example, a widespread interest in applied microeconomics lies in inferring cross price elasticities of market shares in differentiated product markets, using market level data (Berry, 1994; Berry et al., 1995; Nevo, 2000; Pinkse and Slade, 2004). The relevant quantities in this case are proportional to the price coefficients. Alternatively, random utility models are often used to construct valuation measures. This concept is closely related to measuring changes in consumer surplus, which was formalized for the discrete choice framework by Small and Rosen (1981) and involves transforming ordinal quantities to the money-metric scale. In linear utility specifications, this transformation reduces to division by the price coefficient.

Since their seminal work, related constructs have been proposed and analyzed in a wide variety of applications. For example, transportation studies often focus on estimating the *value of time* or *value of reliability* (e.g., Brownstone et al., 2003), which are constructed as the marginal rate of substitution between, say, reduced travel time and a higher cost (i.e., price) of travel for a commuter trip. With linear utility, this is easily computed as the ratio of two coefficients: the coefficient on time divided by the coefficient on price. The measure itself is in essence the compensating/equivalent variation of reducing travel time. In the marketing context, a closely related measure—the *market value for an attribute improvement* (Ofek and Srinivasan, 2002)—captures the monetary value that a particular consumer attaches to an improvement in a specific product attribute; the corresponding aggregate measure (over a sample of consumers) can then be compared to the cost of implementing the improvement.[1]

The concept has been further extended in marketing studies. Jedidi and Zhang (2002) use the same approach to measure the difference in utilities between the utility of acquiring a particular product and the baseline utility

[1] In general, such valuations are typically regarded as measuring consumers' *willingness-to-pay* (i.e., for improved attributes, improved reliability, reduced travel time, etc.).

associated with the "outside good" option. The price that equates the two utilities is termed the *reservation price* for that product. Swait et al. (1993) use an identical measure—under the heading *equalization price*—to construct consumer brand valuations. Its computation only involves dividing the sum of all non-price partworths by the price coefficient, and it is interpreted as the value (to a particular consumer) of all non-price attributes embodied by the product. It may also be viewed as a change in welfare—i.e., the change in surplus associated with forcing a consumer that would otherwise choose that product to not consume at all. A more coherent application of measuring welfare change in the marketing context is used by Kim et al. (2002) who applied compensating variation to assess the value of variety.

Price is also allotted special attention because of its role as a policy variable. This is especially true in marketing, where price is generally regarded as a primary marketing tool (e.g., Han et al., 2001). In this spirit, a particularly interesting application of random utility models is in terms of target marketing, where estimated heterogenous responses to prices (and other product characteristics) are used to segment markets and design optimal promotional strategies. Allenby and Rossi (1998) further demonstrate that the Bayesian approach is especially advantageous in this setting because it can be used to not only construct general consumer profiles, but to specifically predict the effect of a discount on the purchase behaviour of a particular consumer in the sample, given her purchase history.

The way that price is modeled to affect utility in discrete choice applications, therefore, warrants extra care as a poorly modeled price variable would clearly undermine all of the inferential constructs outlined above. Some aspects related to how price should be modeled are clear. For example, it is obvious that a prudent model should not allow consumers to *prefer* price increases. Such a relationship is illogical from the point of view of microeconomic theory and leads to anomalies such as negative valuations of product characteristics with positive marginal utilities. Accordingly, there are a number of ways to bound coefficients in random utility specifications (for discussion, see Train and Sonnier, 2005). Target marketing applications further underscore the importance of allowing for consumer specific price coefficients, an approach that has been standard in Bayesian implementations of random utility models since at least the work of Rossi et al. (1995).

A more subtle issue is related to price insensitivity, a phenomenon by which a consumer's purchase decisions is not affected by price changes over a finite interval. In marketing, this is a well accepted phenomenon. It has been justified on theoretical grounds with psychological foundations (for example, the Assimilation-Contrast theory developed by Sherif, 1963), and it has been closely tied to other marketing concepts, such as deal-proneness Gázquez-Abad and Sánchez-Pérez (2009). In practical terms, allowing for the possibility that a consumer is price insensitive facilitates the design of more effective marketing instruments. Additionally, academic marketing formulates price insensitivity as arising from price thresholds within a reference price framework.

A consumer's sensitivity to price in this setting is specified as a three-part discontinuous function of the relative price (i.e., sticker price minus the reference price) that takes on different (typically constant) slopes depending on whether the relative price is below a threshold, above a threshold, or within the two thresholds. In the latter case, the slope is presumed to be zero, and therefore, the interval between the two thresholds constitutes a *latitude of price acceptance*.

The main contribution of this chapter is to the empirical literature focused on identifying and characterize price insensitive consumer choice behavior. Specifically, we propose that price insensitivity be modeled through a censored distribution that allows for no sensitivity with some probability and bounded sensitivity otherwise. In the simplest case, the censored distribution can be applied to the price coefficient in the linear utility model to probabilistically identify which consumers in the sample exhibit price insensitive behaviour. In terms of target marketing applications, this could provide a straightforward way to identify an important market segment consisting of consumers that (with a significant probability) would not, for example, respond to price promotions, at least within the range of price values observed in their historic purchase decisions.

Although the idea of using the censored distribution to model coefficients in random utility models is not new (Train and Sonnier, 2005, for example, use it to bound marginal utilities), its full potential for modeling price effects has yet to be fully explored. Indeed, we argue that in addition to facilitating a non-trivial probability of price insensitivity, the censored distribution embodies a number of other desirable properties for modeling price effects. A particularly important advantage in this respect is that it leads to straightforward Bayesian computation: because it constitutes a conjugate prior under the probit specification, a fully tractable Gibbs sampler is readily available. This property of the censored distribution can be further exploited to construct more sophisticated price response models. We illustrate this point in the context of the reference price model with heterogeneous thresholds by using the censored distribution to model both the heterogenous price effects and the heterogeneous price thresholds. The result is still a fully tractable Gibbs sampler.

The remainder of the chapter proceeds as follows. Section 9.2 briefly reviews some conventional random utility specifications in marketing and demonstrates how the censored distribution fits easily within this framework. Section 9.3 explores in more depth some properties of the censored distribution in terms of modeling sensitivity to prices and derives a simple Gibbs sampling algorithm. Section 9.4 extends these ideas to the reference price model with heterogeneous effects and price thresholds. Finally, Section 9.5 offers some concluding remarks.

9.2 Random Utility Models in Marketing Applications

A typical marketing model of consumer choice posits a consumer i choosing a single product j from a set of $m+1$ options (including the outside good) on purchase occasion t. Assume that *relative* (to the outside good) utility u_{ijt} for each good is linear in observed and unobserved characteristics, such that

$$u_{ijt} = \mathbf{x}'_{ijt}\boldsymbol{\beta}_i - \alpha_i p_{ijt} + \epsilon_{ijt}, \qquad \begin{array}{l} i = 1, \ldots, n, \\ j = 1, \ldots, m, \\ t = 1, \ldots, T_i, \end{array} \tag{9.1}$$

where $\mathbf{x}_{ijt}$ is a $K_0 \times 1$ vector of observed product characteristics, p_{ijt} is the price and ϵ_{ijt} is an idiosyncratic error capturing the effects of all unobserved components. Note that in marketing applications, $\mathbf{x}_{ijt}$ typically includes dummies to capture brand-specific effects, and in general, (9.1) is flexible enough to capture a wide variety of specifications employed in marketing and economic applications.

To operationalize this setup, one typically proceeds by specifying distributions for α_i, $\boldsymbol{\beta}_i$ and ϵ_{ijt}. In the classical literature, α_i and $\boldsymbol{\beta}_i$ are then referred to as *random coefficients* and a standard approach is to assign logistic distributions to $\epsilon_{i1t}, \ldots, \epsilon_{imt}$ (see, for example, Train, 2003), such that the resulting choice probability is

$$\Pr(y_{it} = j) = \int_{(\alpha_i, \boldsymbol{\beta}_i)} \frac{\exp\{\mathbf{x}'_{ijt}\boldsymbol{\beta}_i - \alpha_i p_{ijt}\}}{1 + \sum_{l=1}^{m} \exp\{\mathbf{x}'_{ilt}\boldsymbol{\beta}_i - \alpha_i p_{ilt}\}} dF(\alpha_i, \boldsymbol{\beta}_i).$$

Such a model is commonly known as the *mixed logit* with mixing distributions $F(\alpha_i, \boldsymbol{\beta}_i)$.

Alternatively, the Bayesian approach simply assigns hierarchical priors to $\alpha_i, \boldsymbol{\beta}_i$ and treats u_{ijt} as a latent variable for estimation purposes. In this context, $\epsilon_{i1t}, \ldots, \epsilon_{imt}$ are usually given normal distributions, leading a to a multivariate probit specification (although Train, 2003, also derives a Bayesian routine based on the *mixed logit* specification). Specifically, rewrite (9.1) in stacked matrix form as

$$\mathbf{u}_{it} = \mathbf{X}_{it}\boldsymbol{\beta}_i - \alpha_i \mathbf{p}_{it} + \boldsymbol{\epsilon}_{it}, \tag{9.2}$$

where

$$\mathbf{u}_{it} = \begin{pmatrix} u_{i1t} \\ \vdots \\ u_{imt} \end{pmatrix}, \quad \mathbf{X}_{it} = \begin{pmatrix} \mathbf{x}'_{i1t} \\ \vdots \\ \mathbf{x}'_{imt} \end{pmatrix}, \quad \mathbf{p}_{it} = \begin{pmatrix} p_{i1t} \\ \vdots \\ p_{imt} \end{pmatrix}, \quad \boldsymbol{\epsilon}_{it} = \begin{pmatrix} \epsilon_{i1t} \\ \vdots \\ \epsilon_{imt} \end{pmatrix},$$

and assign the following distributions to $\alpha_i, \boldsymbol{\beta}_i, \boldsymbol{\epsilon}_{it}$:

$$\alpha_i \sim \mathcal{N}(\mathbf{z}_i'\boldsymbol{\psi}_\alpha, \omega_\alpha^2), \tag{9.3}$$

$$\boldsymbol{\beta}_i \sim \mathcal{N}((\mathrm{I}_{K_0} \otimes \mathbf{z}_i')\boldsymbol{\psi}_\beta, \boldsymbol{\Omega}_\beta), \tag{9.4}$$

$$\boldsymbol{\epsilon}_{it} \sim \mathcal{N}(0, \boldsymbol{\Sigma}), \tag{9.5}$$

where $\mathbf{z}_i$ is a $K_1 \times 1$ vector of personal characteristics describing individual i. Note that the independence between α_i and $\boldsymbol{\beta}_i$ is for notational convenience only, and all methods discussed in the remainder of the chapter are easily extendable if α_i and $\boldsymbol{\beta}_i$ are allowed to be correlated. Under these distributional assumptions and with conjugate priors on the free parameters $\boldsymbol{\psi}_\alpha, \boldsymbol{\psi}_\beta, \omega_\alpha^2, \boldsymbol{\Omega}_\beta, \boldsymbol{\Sigma}$ (i.e., with the first element of $\boldsymbol{\Sigma}$ restricted to be 1 for identification), it is straightforward to construct a Gibbs sampler for estimation purposes.

To make the model more parsimonious, the marketing literature (Rossi et al., 1995, 1996; Allenby and Rossi, 1998; Terui and Dahana, 2006) typically treats $\boldsymbol{\Sigma}$ as diagonal (with $\sigma_1^2 = 1$), which also simplifies computation as follows. Let $y_{it} \in \{1, \ldots, m\}$ denote the observed choice of consumer i on choice occasion t. If $y_{it} = j$ and $\boldsymbol{\Sigma}$ is diagonal, then conditional on u_{ijt}, the remaining $\{u_{ilt}\}_{l \neq j}$ are independent of each other. Rossi et al. (1996) provide the appropriate Gibbs sampling algorithm under these assumptions. We note in passing, however, that a more appropriate modelling choice may be to assume independent errors $\epsilon_{i1t}, \ldots, \epsilon_{imt}$ *conditional* on u_{i0t}, the utility of the outside good, rather than imposing marginal independence. This would bring the resulting probit specification in close alignment with the mixed logit and establish a more direct link with other widely-used consumer choice models, such as that of Berry (1994). In terms of computation, u_{i0t} can be treated as an additional latent variable and hence only a minor modification of the Rossi et al. (1996) algorithm would be required: add a step to generate draws of u_{i0t} conditional on remaining parameters and proceed with the Rossi et al. (1996) algorithm conditional on u_{i0t}.

The focus of this chapter is on α_i. As previously discussed, this quantity plays a central role in inference drawn from random utility models, and hence, it is prudent to devote particular attention to how it is modelled. For example, (9.3) would be inadequate in most consumer choice contexts since it assigns positive probabilities to negative values of α_i, and therefore, allows for negative marginal utility of income. To address this criticism, the restriction $\alpha_i > 0$ may be introduced, but care must be taken in how this is accomplished. For example, modifying (9.3) to be a *truncated normal*, i.e.,

$$\alpha_i \sim \mathcal{N}_{(0,\infty)}(\mathbf{z}_i'\boldsymbol{\psi}_\alpha, \omega_\alpha^2),$$

would break the conjugacy and greatly complicate computation since the truncated normal distribution takes the form

$$\phi_{(0,\infty)}(\alpha_i \,|\, \mathbf{z}_i'\boldsymbol{\psi}_\alpha, \omega_\alpha^2) = \frac{\phi((\alpha_i - \mathbf{z}_i'\boldsymbol{\psi}_\alpha)/\omega_\alpha)}{\omega_\alpha \Phi(\mathbf{z}_i'\boldsymbol{\psi}_\alpha/\omega_\alpha)} \mathbf{1}(\alpha_i > 0),$$

where $\phi(\cdot)$ and $\Phi(\cdot)$ denote the pdf and cdf of the *standard normal* distribution, respectively, and $\mathbf{1}(\cdot)$ is the indicator function that evaluates to 1 if its argument is true and 0 otherwise. Here, the computational difficulty arises from the fact that parameters $\boldsymbol{\psi}_\alpha, \omega_\alpha^2$ enter the integrating constant, and therefore, a complicated M-H step would be required to sample $\boldsymbol{\psi}_\alpha, \omega_\alpha^2$ (for an application of this approach to constrained hierarchical modeling, see Boatwright et al., 1999).

The computation difficulty described above may be circumvented by specifying a joint prior on $\boldsymbol{\psi}_\alpha, \omega_\alpha^2, \alpha_1, \ldots, \alpha_n$ as

$$p(\boldsymbol{\psi}_\alpha, \omega_\alpha^2, \alpha_1, \ldots, \alpha_n) \propto p(\boldsymbol{\psi}_\alpha, \omega_\alpha^2) \prod_{i=1}^{n} \omega_\alpha^{-1} \phi((\alpha_i - \mathbf{z}_i' \boldsymbol{\psi}_\alpha)/\omega_\alpha) \mathbf{1}(\alpha_i > 0).$$

In this case, the only additional complication relative to the standard Gibbs algorithm is that α_i must be sampled from a univariate truncated normal distribution—a straightforward extension. The implied prior on $\boldsymbol{\psi}_\alpha, \omega_\alpha^2$, however, is not intuitive and its properties may be difficult to ascertain. Alternatively, α_i may be modelled with *log-normal* distributions by letting $\alpha_i = \exp\{\alpha_i^*\}$ and specifying a normal distribution for α_i^*. Continuing to work directly with α_i^*, this approach admits conjugate priors for $\boldsymbol{\psi}_\alpha, \omega_\alpha^2$, but leads to a non-standard conditional distribution for α_i^*, and therefore, either the need to employ univariate M-H steps to generate samples in a probit specification, or a slight modification of the M-H algorithm in the mixed logit case, as described in Train and Sonnier (2005).

Neither the truncated normal nor the log-normal specifications for α_i, however, allow for a non-trivial probability that $\alpha_i = 0$. Thus, both approaches demand that all consumers be treated as sensitive to price changes, albeit to varying degrees (and some with arbitrarily low sensitivity). On the other hand, the marketing literature and practice tend to ascribe particular significance to the concept that consumers may be insensitive to price changes, at least over a finite interval (Gupta and Cooper, 1992; Kalwani and Yim, 1992; Kalyanaram and Little, 1994; Han et al., 2001). For example, a consumer that is deemed price insensitive with a significant probability would likely not respond to price promotions, and therefore, needs to be reached with alternative marketing instruments.

In other cases, a marketer may wish to devise a discriminatory pricing strategy that more efficiently allocates prices among consumers with heterogenous tastes. Ideally, such a strategy would result in customer-specific prices that are, all else constant, inversely proportional to price sensitivity, but this is obviously infeasible in practice. As a next-best alternative, one may consider discriminating prices across segments. To implement this, however, it is necessary to segment consumers based on their price sensitivities. Yet applying the truncated normal or the log-normal mixing distribution in this case leads to complications. For example, it may be tempting to segment consumers based on α_i by defining segments through thresholds $\bar{\alpha}_0, \bar{\alpha}_1, \ldots$ such that consumer

i belongs to a particular sensitivity segment if and only if α_i lies between certain thresholds. However, the scale of α_i is not identified empirically, and so such thresholds are devoid of economic meaning. Allowing for a non-trivial probability that $\alpha_i = 0$, on the other hand, provides a clear delineation for a segment marked by price insensitivity. Moreover, since the associated probability [i.e., $\Pr(\alpha_i = 0 \,|\, \text{Data})$] is identifiable and inherits a well defined scale, it can be further used in more sophisticated segmentation schemes as well.

A straightforward way of introducing price insensitivity into random utility model is by modeling α_i with a censored mixing distribution, i.e., $\alpha_i = \max\{0, \alpha_i^*\}$, where α_i^* follows an unrestricted normal. As with the log-normal, such a specification is easy to implement within a sampling routine. Train and Sonnier (2005) provide the details for mixed logits; again, only a minor modification of the M-H algorithm is required. In the following section, we further examine the properties of the censored distribution on α_i and derive a fully tractable Gibbs sampler for the probit specification. In either case, the MCMC routine leads to an estimate of $\Pr(\alpha_i = 0 \,|\, \text{Data})$, which can be directly exploited in marketing applications.

9.3 The Censored Mixing Distribution in Detail

To better appreciate the advantages of a censored mixing distribution over truncated distributions in modeling the price coefficient α_i, begin by rewriting the distribution in the following form:

$$\alpha_i \sim \pi_{\alpha_i} \mathbf{1}(\alpha_i \leq 0) + (1 - \pi_{\alpha_i}) \phi_{(0,\infty)}(\alpha_i \,|\, \mathbf{z}_i' \boldsymbol{\psi}_\alpha, \omega_\alpha^2). \tag{9.6}$$

Thus, it is a mixture between a degenerate distribution (at point $\alpha_i = 0$) and the truncated (above zero) normal distribution. The application of similar *spike and slab* distributions as priors in Bayesian variable and model selection problems can be traced to the work of Mitchell and Beauchamp (1988) and George and McCulloch (1993, 1997). Recently, Frühwirth-Schnatter and Wagner (2011) demonstrated their usefulness in designing *stochastic model selection search* algorithms, and Reichl and Frühwirth-Schnatter (2012) use a censored distribution of the type discussed in this section to stochastically identify zero willingness to pay measures. In the present context, (9.6) may be regarded as facilitating a stochastic search algorithm for price insensitive consumers, where π_{α_i} takes the role of an *a priori* probability that consumer i is insensitive to price changes.

Note that in principle, we can treat π_{α_i} in (9.6) as a separate parameter or a predetermined hyper-parameter. Setting $\pi_{\alpha_i} = 0$ in the latter case clearly leads to the truncated normal specification. Alternatively, the innovative feature of the censored mixing distribution is that it explicitly relates π_{α_i} to the parameters $\boldsymbol{\psi}_\alpha, \omega_\alpha^2$ as $\pi_{\alpha_i} = \Phi(-\mathbf{z}_i' \boldsymbol{\psi}_\alpha / \omega_\alpha)$. Indeed, this a very desirable property of the censored mixing distribution for modeling price sensitivity. Whereas the general *spike and slab* distribution allows for high probabilities of price insensitivity to be associated with large values of α_i, the censored distribution heavily penalize such situations. In fact, under the censored specification, the mixing distribution approaches normal as $\mathbf{z}_i'\psi_\alpha$ increases relative to ω_α, a featured shared by the truncated normal and log-normal specifications as well.

In a similar fashion, the censored mixing distribution penalizes combinations of low price sensitivity and a low probability of price insensitivity. This is in stark contrast to the truncated normal specification, which always assigns zero-probability to a consumer being price insensitive, even when a significant amount of *a priori* weight is placed on arbitrarily low values of α_i. However, the "shrinkage" effect provided by the censored distribution may be preferable as follows.

In many applications of discrete choice models, computing quantities of interest requires division by α_i. The reason for this is obvious: since discrete choice models only identify parameters up to scale, useful inference requires some sort of normalization. The price coefficient is typically a primary choice for carrying out the normalization due to its direct interpretation as the marginal utility of income. Thus, the division of a scale-free quantity by α_i endows it with the monetary scale based on which economically meaningful inference may be drawn. For example, such an approach could be used to compute consumer i's willingness-to-pay for a unitary increase in product attribute k as $\beta_{k,i}/\alpha_i$, and an aggregated measure (over a set of consumers) might be compared with the cost of increasing (or decreasing) attribute k.

Such an approach, however, is clearly futile for a price insensitive consumer. A more subtle issue arises when α_i takes on values close to zero with a high probability. In the latter case the induced distribution of $\beta_{k,i}/\alpha_i$ may have an arbitrarily high variance, making inference difficult. Continuing with the willingness-to-pay example, even a small subset of consumers associated with low values of α_i could potentially destabilize an aggregate measure. The contribution of the censored mixing distribution in resolving this difficulty is twofold: (i) it helps to (probabilistically) separate price-sensitive from price insensitive consumers and thus apply different scales for measuring quantities of interest, and (ii) it ensures that price-sensitive consumers are not signed arbitrarily low sensitivities with a high probability such that aggregate willingness-to-pay measures are stable.

Unlike the truncated normal distribution, the positively skewed log-normal circumvents the problem of assigning significant weight to low values of price sensitivity by reallocating probability to the upper tail. Thus, it may be a more reasonable choice when computing willingness-to-pay measures are of primary interest. However, this skewness property not only eliminates the possibility of price insensitive consumers, but essentially imposes a lower-bound on price sensitivity. In many contexts, this may be not be a desirable restriction to enforce *a priori*.

Figure 9.1 further provides a visual comparison of the censored, truncated and log-normal mixing distributions. Suppose that the three distributions are derived by transforming

$$\alpha_i^* \sim \mathcal{N}(\psi_0, \omega_\alpha^2)$$

as follows:

$$\begin{aligned}
&\text{Censored:} && \alpha_i^C = \max\{0, \alpha_i^*\},\\
&\text{Truncated:} && \alpha_i^T = \psi_0 + \omega_\alpha \Phi^{-1}\left(\Phi\left(\frac{\alpha_i^* - \psi_0}{\omega_\alpha}\right)\Phi\left(\frac{\psi_0}{\omega_\alpha}\right) + \Phi\left(-\frac{\psi_0}{\omega_\alpha}\right)\right),\\
&\text{Log-normal:} && \alpha_i^L = \exp\{\alpha_i^*\}.
\end{aligned}$$

The first column demonstrates each distribution for fixed parameter values $\psi_0 = -0.5, \omega_\alpha^2 = 0.4^2$, which place a significant amount of weight on negative values in the latent process α_i^*. In this case, the censored distribution allocates most of the weight (89%) to $\alpha_i^C = 0$ thereby indicating a strong preference for price insensitivity. If this was to be interpreted with respect to a particular consumer, such information could be used to identify the consumer as being price insensitive and apply alternative techniques to draw inference about her preferences.

In contrast, the implied truncated distribution allocates most weight to very low values of α_i^L. Thus, relying on a truncated normal mixing distribution we would be forced to treat the respective consumer as price sensitive, but with sensitivity arbitrarily close to zero with high probability. As already discussed, this vagueness could be particularly obstructive in inferring willingness-to-pay and similar measures of preferences. The log-normal, under similar conditions, eliminates confusion by reallocating the weights arising from negative values of α_i^* to the upper tail of the distribution. In consequence, its mode is shifted outward (relative to the censored and truncated distributions) and values close to zero are left with arbitrarily low weights.

The comparisons change drastically, however, when α_i^* is concentrated on positive values, as depicted in the second column. For parameter values $\psi_0 = 0.5, \omega_\alpha^2 = 0.1^2$, the probability of α_i^* being negative is negligible and as a result the censored and truncated distributions are indistinguishable (i.e., both resembling the underlying normal distribution centered at $\psi_0 = 0.5$). The log-normal distribution, in this case, also assumes a "normal" shape, but centered notably above ψ_0 (i.e., its mode is $\exp\{\psi_0 - \omega_\alpha^2\} \approx 1.63$). This difference, however, is not of crucial importance since the resulting log-normal distribution can be regarded as closely aligned with another censored/truncated normal centered at 1.63, arising from a different set of parameters that also imply $\alpha_i^* > 0$ with high probability.

The final column depicts the corresponding distributions obtained by integrating the censored, truncated and log-normal specifications, respectively, over

$$\begin{aligned}
\psi_0 &\sim \mathcal{N}(0, 1),\\
\omega_\alpha^2 &\sim \mathcal{IG}(2, 0.1).
\end{aligned}$$

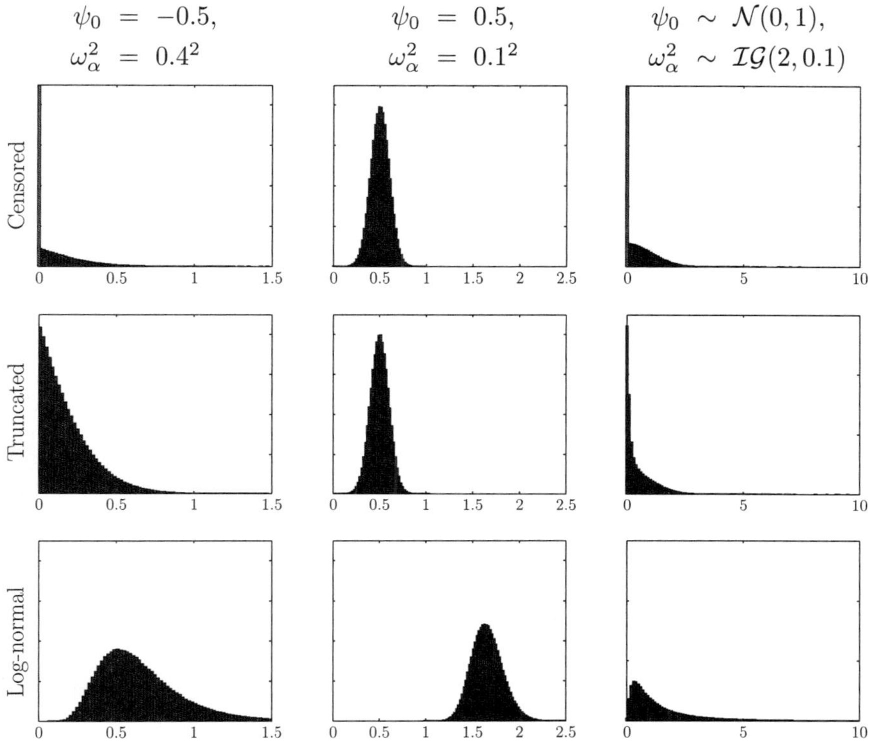

Figure 9.1: Comparison of the censored, truncated and log-normal mixing distributions for modeling price sensitivity.

This, of course, reflects a hierarchical prior on α_i (for each of the three specifications) as commonly employed in Bayesian analysis of discrete choice data. In this context, it is important to understand the properties of not only the conditional priors at the top of the hierarchies, but also the implied marginal distributions. Looking at the marginal distributions for $\alpha_i^C, \alpha_i^T, \alpha_i^L$, one particular noticeable feature is that α_i^C and α_i^L are similarly distributed over the range $(0, 10]$, and both distributions attribute fairly small weights to low (but non-zero) values of α_i [e.g., $\Pr(0 < \alpha_i^C \leq 0.1) \approx 0.04), \Pr(0 < \alpha_i^L \leq 0.1) \approx 0.01$]. The key difference once again is that α_i^T inherits a non-trivial probability of being zero [i.e., $\Pr(\alpha_i^C = 0) = 0.5, \Pr(\alpha_i^L = 0) = 0$], while the distribution of α_i^C embodies a heavier upper tail [e.g., $\Pr(\alpha_i^C > 2) \approx 0.03, \Pr(\alpha_i^L > 2) \approx 0.25$].

On the other hand, the distribution of α_i^T significantly shifted towards the zero lower-bound, with $\Pr(0 < \alpha_i^T \leq 0.1) = 0.37$ and $\Pr(\alpha_i^T > 2) \approx 0.03$. In other words, it excludes price insensitivity, yet at the same heavily discriminates against substantial sensitivity to price changes. In consequence, unless data provides strong evidence in favor of significant price sensitivity, such a prior will likely lead to large probabilities over arbitrarily small values of α_i^T as well. Therefore, proceeding with such a specification warrants a certain degree of caution with respect to the inferential goals that the underlying model is designed to achieve. In most marketing applications, however, it seems that either a mixing distribution that explicitly allows for price insensitivity (such as the censored distribution), or one that heavily discounts arbitrarily low sensitivity (i.e., the log-normal) should be always be preferred.

For reasons already discussed, in marketing applications involving segmentation, a suitable mixing distribution should allow for non-trivial price insensitivity, and the censored mixing distribution is theoretically appealing in this sense. What makes this distribution particularly attractive in practice is that it also leads to straightforward computation.[2] Train and Sonnier (2005) demonstrated that the transformation

$$\alpha_i^* \sim \mathcal{N}(\mathbf{z}_i'\boldsymbol{\psi}_\alpha, \omega_\alpha^2), \qquad \alpha_i = \max\{0, \alpha_i^*\} \tag{9.7}$$

can be easily incorporated into the (random walk) M-H algorithm that is typically implemented to sample the mixed logit posterior Train (2003). Under the probit specification, a fully tractable Gibbs sampling algorithm is readily available. Specifically, assume the random utility model given in (9.1), (9.4) and (9.5) (with $\boldsymbol{\Sigma}$ diagonal). Combining this with (9.7) implies that the algorithm for sampling

$$(\mathbf{u}_{11}, \ldots, \mathbf{u}_{nT_i}, \alpha_1^*, \ldots, \alpha_n^*, \boldsymbol{\beta}_1, \ldots, \boldsymbol{\beta}_n, \boldsymbol{\psi}_\alpha, \boldsymbol{\psi}_\beta, \omega_\alpha^2, \boldsymbol{\Omega}_\beta, \boldsymbol{\Sigma} \,|\, y_{11}, \ldots, y_{nT_i})$$

may proceed in five steps by successively sampling

1. $(\mathbf{u}_{it} \,|\, \alpha_i^*, \boldsymbol{\beta}_i, \boldsymbol{\Sigma}, y_{it})$ independently for $i = 1, \ldots, n$ and $t = 1, \ldots, T_i$,
2. $(\alpha_i^*, \boldsymbol{\beta}_i \,|\, \boldsymbol{\psi}_\alpha, \boldsymbol{\psi}_\beta, \omega_\alpha^2, \boldsymbol{\Omega}_\beta, \mathbf{u}_{i1}, \ldots, \mathbf{u}_{iT_i})$ independently for $i = 1, \ldots, n$,
3. $(\boldsymbol{\Sigma} \,|\, \alpha_1^*, \ldots, \alpha_n^*, \boldsymbol{\beta}_1, \cdots, \boldsymbol{\beta}_n, \mathbf{u}_{11}, \ldots, \mathbf{u}_{nT_n})$,
4. $(\boldsymbol{\psi}_\alpha, \boldsymbol{\psi}_\beta \,|\, \omega_\alpha^2, \boldsymbol{\Omega}_\beta, \alpha_1^*, \ldots, \alpha_n^*, \boldsymbol{\beta}_1, \cdots, \boldsymbol{\beta}_n)$,
5. $(\omega_\alpha^2, \boldsymbol{\Omega}_\beta \,|\, \boldsymbol{\psi}_\alpha, \boldsymbol{\psi}_\beta, \alpha_1^*, \ldots, \alpha_n^*, \boldsymbol{\beta}_1, \cdots, \boldsymbol{\beta}_n)$.

Conditional on $\alpha_i^*, \ldots, \alpha_n^*$ this is a standard Gibbs sampling algorithm used in marketing applications of the multinomial probit, and details regarding Step 1 and Steps 3–5 are given in Rossi et al. (1996). Implementing Step 2 is

[2] As opposed to, say, a distribution of the form

$$\alpha_i \sim \pi_{\alpha_i}\mathbf{1}(\alpha_i \leq 0) + (1 - \pi_{\alpha_i})\alpha_i^{-1}\phi(\ln \alpha_i \,|\, \mathbf{z}_i'\boldsymbol{\psi}_\alpha, \omega_\alpha^2)\mathbf{1}(\alpha_i > 0),$$

i.e., a *spike and slab* distribution involving a mixture between the degenerate point $\alpha_i = 0$ and the log-normal distribution.

also straightforward as follows. First define the posterior quantities:

$$\hat{\mathbf{v}}_{\alpha_i} = \mathbf{u}_i - \mathbf{X}_i \left(\mathrm{I}_{K_0} \otimes \mathbf{z}_i'\right) \boldsymbol{\psi}_\beta, \tag{9.8}$$

$$\hat{\omega}_{\alpha_i}^2 = \left(\omega_\alpha^{-2} + \mathbf{p}_i' \left(\mathbf{X}_i \boldsymbol{\Omega}_\beta \mathbf{X}_i' + \boldsymbol{\Sigma}\right)^{-1} \mathbf{p}_i\right)^{-1}, \tag{9.9}$$

$$\hat{\alpha}_i = \hat{\omega}_{\alpha_i}^2 \left(\mathbf{z}_i' \boldsymbol{\psi}_\alpha / \omega_\alpha^2 + \mathbf{p}_i' \left(\mathbf{X}_i \boldsymbol{\Omega}_\beta \mathbf{X}_i' + \boldsymbol{\Sigma}\right)^{-1} \hat{\mathbf{v}}_{\alpha_i}\right), \tag{9.10}$$

$$\hat{\kappa}_{\alpha_i} = \frac{\Phi(\hat{\alpha}_i / \hat{\omega}_{\alpha_i})}{\Phi(-\mathbf{z}_i' \boldsymbol{\psi}_\alpha / \omega_\alpha)} \frac{\hat{\omega}_{\alpha_i}}{\omega_\alpha} \exp\left\{\frac{1}{2}\left[\left(\frac{\hat{\alpha}_i}{\hat{\omega}_{\alpha_i}}\right)^2 - \left(\frac{\mathbf{z}_i' \boldsymbol{\psi}_\alpha}{\omega_\alpha}\right)^2\right]\right\} \tag{9.11}$$

$$\hat{\pi}_{\alpha_i} = \left(1 + \hat{\kappa}_{\alpha_i}\right)^{-1} \tag{9.12}$$

and

$$\hat{\mathbf{v}}_{\beta_i} = \mathbf{u}_i + \max\{0, \alpha_i^*\} \mathbf{p}_i,$$

$$\hat{\boldsymbol{\Omega}}_{\beta_i} = \left(\boldsymbol{\Omega}_\beta^{-1} + \mathbf{X}_i' \boldsymbol{\Sigma}^{-1} \mathbf{X}_i\right)^{-1},$$

$$\hat{\boldsymbol{\beta}}_i = \hat{\boldsymbol{\Omega}}_{\beta_i} \left(\boldsymbol{\Omega}_\beta^{-1} \left(\mathrm{I}_{K_0} \otimes \mathbf{z}_i'\right) \boldsymbol{\psi}_\beta + \mathbf{X}_i' \boldsymbol{\Sigma}^{-1} \hat{\mathbf{v}}_{\beta_i}\right),$$

where

$$\mathbf{u}_i = \begin{pmatrix} \mathbf{u}_{i1} \\ \vdots \\ \mathbf{u}_{iT_i} \end{pmatrix}, \qquad \mathbf{X}_i = \begin{pmatrix} \mathbf{X}_{i1} \\ \vdots \\ \mathbf{X}_{iT_i} \end{pmatrix}, \qquad \mathbf{p}_i = \begin{pmatrix} \mathbf{p}_{i1} \\ \vdots \\ \mathbf{p}_{iT_i} \end{pmatrix}.$$

Note that in computing $\hat{\omega}_{\alpha_i}^2$ and $\hat{\alpha}_i$, the potentially high-dimensional matrix $\mathbf{X}_i \boldsymbol{\Omega}_\beta \mathbf{X}_i' + \boldsymbol{\Sigma}$ does not require direct inversion in practice because a quantity of the form $\boldsymbol{\Upsilon}^{-1}\boldsymbol{\upsilon}$, for a $q \times 1$ vector $\boldsymbol{\upsilon}$, is computed in $\mathcal{O}(q)$ operations by solving the system of equations $\boldsymbol{\upsilon} = \boldsymbol{\Upsilon}\boldsymbol{\varpi}$. Next, proceed by sampling

$$(\alpha_i^* \mid \cdot) \sim \begin{cases} \mathcal{N}_{(-\infty,0]}(\mathbf{z}_i' \boldsymbol{\psi}_\alpha, \omega_\alpha^2) & \text{w.p.} \quad \hat{\pi}_{\alpha_i} \\ \mathcal{N}_{(0,\infty)}(\hat{\alpha}_i, \hat{\omega}_{\alpha_i}^2) & \text{w.p.} \quad 1 - \hat{\pi}_{\alpha_i} \end{cases}, \tag{9.13}$$

marginally of $\boldsymbol{\beta}_i$, followed by

$$(\boldsymbol{\beta}_i \mid \alpha_i^*, \cdot) \sim \mathcal{N}(\hat{\boldsymbol{\beta}}_i, \hat{\boldsymbol{\Omega}}_{\beta_i}). \tag{9.14}$$

In particular, to sample from (9.13), first draw an indicator $\delta \in \{0, 1\}$ as a Bernoulli trial with $\Pr(\delta = 1) = \hat{\pi}_{\alpha_i}$. If $\delta = 1$ is obtained, then α_i^* is sampled form the normal distribution truncated below zero, otherwise it is sampled from the normal distribution truncated above zero. Finally, conditional on the draw of α_i^*, sample $\boldsymbol{\beta}_i$ from the unrestricted normal distribution given in (9.14).

Implementing this algorithm yields an estimate of $\Pr(\alpha_i = 0 \,|\, \text{Data}) = \Pr(\alpha_i^* \leq 0 \,|\, \text{Data})$ for each individual i in the sample. The estimate can be obtained by simply counting the number of samples with $\alpha_i^* \leq 0$ relative to the total draws; alternatively, an entire posterior distribution can be constructed for the price insensitivity probability by retaining a sample of $\hat{\pi}_{\alpha_i}$, which requires no additional computation since it is already computed at each iteration in sampling α_i^*. In turn, this probability may be used to designate i as either price sensitive or price insensitive for target marketing purposes by defining an appropriate probability threshold. If the sample of consumers is representative for a particular population, then the estimated probabilities can also be used in combination with the observed personal characteristics to construct a price insensitive consumer profile.

Moreover, working with $\hat{\pi}_{\alpha_i}$ it is possible to construct further quantities of interest, such as elasticities of the form

$$\epsilon_i = \frac{\partial \ln \Pr(\alpha_i = 0 \,|\, w_i, \cdot)}{\partial \ln w_i},$$

where w_i could be a product (i.e., an element in $\mathbf{x}_{ijt}$) or personal (i.e., an element in $\mathbf{z}_i$) attribute. For example, if w_i measures the exposure of i to a particular advertising campaign, then the derivative may be interpreted as the effect on price insensitivity from increasing consumer i's level of exposure. Since such a derivative may be computed analytically using (9.8)-(9.12), its posterior distribution is straightforward to construct as well.

9.4 Reference Price Models with Price Thresholds

The simple algorithm presented in the previous section provides a crude way of probabilistically identifying price insensitive consumers in the sample. To the extent that prices would not exceed the limits of those observed in the sample, this in itself could be a useful tool to marketers in designing price-based and non-price promotions. However, the specification in (9.7) is uninformative about the nature the potential price insensitivity attributed to consumer i; thus, all that is known is that consumer i is "locally" insensitive, but nothing regarding when this insensitivity would break down is inferrable. In academic marketing, on the other hand, price insensitivity is typically endowed with a more profound theoretical framework. Specifically, the marketing literature regards insensitivity to prices as arising from the existence of price thresholds within the reference price model (Helson, 1964; Kalyanaram and Winer, 1995). In this section, we demonstrate how the ideas developed so far regarding the censored distribution can be easily extended derive an effective econometric specification in this context.

The random utility specification based on the reference price model may be generally presented as

$$u_{ijt} = \mathbf{x}_{ijt}'\boldsymbol{\beta}_i - \lambda(p_{ijt})p_{ijt} + \epsilon_{ijt}, \tag{9.15}$$

where price-sensitivity $\lambda(p_{ijt})$ is now a discontinuous function of the *relative* price p_{ijt}. In this context, relative price refers to the difference between the *transaction* price and a *reference* price—i.e., $p^*_{ijt} - \bar{p}_{ijt}$. The fact that this relative price could be positive or negative plays a crucial role in the analysis that follows. Also, it is important to emphasize that the reference price $\bar{p}_{ijt}$ is a mental construct, specific to each consumer, product and timing of the purchase. We follow the literature in assuming that $\bar{p}_{ijt}$ is calculated from data.

This gives rise to the following formulation in terms of price thresholds:

$$\lambda(p_{ijt}) = \alpha_i^{(0)}\mathbf{1}\left(p_{ijt} \geq \zeta_{ijt}^{(0)}\right) + \alpha_i^{(1)}\mathbf{1}\left(p \leq -\zeta_{ijt}^{(1)}\right). \tag{9.16}$$

By imposing

$$\alpha_i^{(l)} \geq 0, \qquad 0 \leq \zeta_{ijt}^{(1)} \leq \bar{\zeta}_{ijt}^{(l)} \tag{9.17}$$

for $l = 0, 1$ and some predetermined (possibly infinite) bound $\bar{\zeta}_{ijt}^{(l)}$, (9.16) embodies a number of interpretations in terms of consumer behavior. Existing literature on price thresholds has generally focused on the following:

- *utility loss*—when $\alpha_i^{(0)} > 0$ and $p_{ijt} \geq \zeta_{ijt}^{(0)}$, an increase in the transaction price above the reference price reduces utility at constant rate of $\alpha_i^{(0)}$ per monetary unit;
- *utility gain*—when $\alpha_i^{(1)} > 0$ and $p_{ijt} \leq -\zeta_{ijt}^{(1)}$, a decrease in the transaction price below the reference price increases utility at constant rate of $\alpha_i^{(1)}$ per monetary unit;
- *insensitivity regions*—utility is unaffected by "small" changes in the relative price, i.e., over the region $-\zeta_{ijt}^{(1)} < p_{ijt} < \zeta_{ijt}^{(0)}$;
- *asymmetric effects*—sensitivity to price may differ depending on whether prices are relatively positive or negative (i.e., $\alpha_i^{(0)} \neq \alpha_i^{(1)}$), and in particular a consumer may be more sensitive to increased losses than to increased gains (i.e., $\alpha_i^{(0)} > \alpha_i^{(1)}$), as predicted by prospect theory Kahneman and Tversky (1979). In the same spirit, the thresholds themselves need not be symmetric (i.e., $\zeta_{ijt}^{(0)} \neq \zeta_{ijt}^{(1)}$), and a significant interest in marketing centers on understanding what drives these differences (e.g., Han et al., 2001; Terui and Dahana, 2006).

However, existing empirical investigations of the price threshold model have yet to employ an empirical specification that fully corresponds to (9.16) and (9.17). For example, most models that investigate price thresholds do not allow for heterogeneity in taste parameters, including price sensitivity (e.g., Gupta and Cooper, 1992; Kalwani and Yim, 1992; Kalyanaram and Little, 1994; Han et al., 2001). Alternatively, Terui and Dahana (2006) formulate a discontinuous utility regimes design and propose an MCMC algorithm (to our knowledge, the only existing Bayesian implementation of the price threshold model) to infer thresholds in a fully heterogeneous specification. However, their specification does not allow for a true insensitivity region; instead they impose that when the relative price is within the thresholds (e.g., $-\zeta_{ijt}^{(1)} < p_{ijt} < \zeta_{ijt}^{(0)}$), sensitivity is *a priori* centered at zero with a "low variance." Moreover, no existing econometric specification imposes that $\alpha_i^{(l)} \geq 0$, despite the crucial role that these restrictions play in interpreting asymmetric effects—i.e., without them, it would be possible for a consumer to, say, *prefer* price *increases* above the "loss" threshold while preferring price decreases below the "gain" threshold.

A simple alternative to implementing (9.16)–(9.17) empirically is through the censored distribution as follows:

$$\alpha_i^{*(l)} \sim \mathcal{N}\left(\mathbf{z}_i'\boldsymbol{\psi}_\alpha^{(l)}, \omega_\alpha^{2(l)}\right), \qquad \alpha_i^{(l)} = \max\left\{0, \alpha_i^{*(l)}\right\}, \tag{9.18}$$

$$\zeta_{ijt}^{*(l)} \sim \mathcal{N}\left(\mathbf{w}_{ijt}'\gamma_i^{(l)}, \tau^{2(l)}\right), \qquad \zeta_{ijt}^{(l)} = \min\left\{\bar{\zeta}_{ijt}^{(l)}, \max\left\{0, \zeta_{ijt}^{*(l)}\right\}\right\}, \tag{9.19}$$

for $l = 0, 1$. The resulting empirical model captures all the salient features of a random utility model with heterogeneous price thresholds and price sensitivities, and allows for a straightforward investigation of the behavioral phenomena discussed above. In addition, employing censored distributions to model price threshold and price sensitivities facilitates a number of other interesting interpretations. For example, $\zeta_{ijt}^{(0)} > 0, \zeta_{ijt}^{(1)} = 0$ and $\alpha_i^{(1)} > 0$ implies that only an "loss" threshold exists and the consumer only cares about price changes when relative price exceeds a certain level. Likewise, $\zeta_{ijt}^{(0)} = \zeta_{ijt}^{(1)} = 0$ while $\alpha_i^{(0)} > 0, \alpha_i^{(1)} > 0$ implies asymmetric price effects *without* thresholds—i.e., the consumer is always sensitive to price changes, but reacts differently to "loses" than to "gains" (as in Bell and Lattin, 2002, for example).

In terms of price insensitivity, the proposed specification allows this phenomenon to be explained by price thresholds (e.g., when $\zeta_{ijt}^{(0)} > 0, \zeta_{ijt}^{(1)} > 0$ and $\alpha_i^{(0)} > 0, \alpha_i^{(1)} > 0$) or as being driven by an entirely different force (e.g., whenever $\alpha_i^{(0)} = \alpha_i^{(1)} = 0$). To distinguish between the two types of price insensitivities, it is reasonable to impose a finite upper bound on $\zeta_{ijt}^{(l)}$ of the form

$$\bar{\zeta}_{ijt}^{(l)} = \max\left\{(-1)^l \mathbf{p}_i\right\}.$$

Doing so imposes that for each consumer i, at least one binding "high" price, and at least one binding "low" price, are observed for at least one chosen product over the choice occasions. Moreover, it is worth emphasizing that price thresholds in (9.16)-(9.19) are not only heterogeneous but also vary across products and time. This is inline with the view of Han et al. (2001) who argue that thresholds may change with, among other things, product-specific factors such as the number of times it has been previously discounted (e.g., reference price p_{ijt} has been negative in the past). Accordingly, the $K_2 \times 1$

data vector $\mathbf{w}_{ijt}$ is assumed to contain observable explanatory variables to this effect. We complete the specification with a mixing distribution on $\gamma_i^{(l)}$ in the standard way:

$$\gamma_i^{(l)} \sim \mathcal{N}\left(\mathbf{z}_i'\boldsymbol{\psi}_\gamma^{(l)}, \boldsymbol{\Omega}_\gamma^{(l)}\right)$$

Finally, implementing the censored distribution to model price thresholds still retains the computation simplicity discussed in Section 9.3 in terms of the simple price insensitivity specification. In particular, while the price threshold model contains a richer set of parameters, the basic framework of the Gibbs sampling algorithm outlined in Section 9.3 still applies: all conditional distributions are fully tractable, and therefore, the sampler is straightforward to implement. The algorithm involves iteratively sampling the following parameter blocks:

1. $\left(\mathbf{u}_{it} \,|\, \alpha_i^{*(0)}, \alpha_i^{*(1)}, \boldsymbol{\zeta}_i^{*(0)}, \boldsymbol{\zeta}_i^{*(1)}, \boldsymbol{\beta}_i, \boldsymbol{\Sigma}, y_{it}\right)$ independently for $i = 1, \ldots, n$ and $t = 1, \ldots, T_i$,
2. $\left(\zeta_{ijt}^{*(l)} \,|\, \zeta_{ijt}^{*(1-l)}, \boldsymbol{\gamma}_i^{(l)}, \tau^{2(l)}, \alpha_i^{*(0)}, \alpha_i^{*(1)}, \sigma_j, u_{ijt}\right)$ independently for $i = 1, \ldots, n$ $j = 1, \ldots, m$, $t = 1, \ldots, T_i$ and for each $l = 0, 1$,
3. $\left(\boldsymbol{\gamma}_i^{(l)} \,|\, \boldsymbol{\zeta}_i^{(l)}, \tau^{2(l)}\right)$ independently for $i = 1, \ldots, n$ and $l = 0, 1$,
4. $\left(\tau^{2(l)} \,|\, \boldsymbol{\zeta}_i^{*(l)}, \ldots, \boldsymbol{\zeta}_n^{*(l)}, \boldsymbol{\gamma}_1^{(l)}, \ldots, \boldsymbol{\gamma}_n^{(l)}\right)$ independently for $l = 0, 1$,
5. $\left(\alpha_i^{*(l)} \,|\, \alpha_i^{*(1-l)}, \boldsymbol{\zeta}_i^{*(0)}, \boldsymbol{\zeta}_i^{*(1)}, \boldsymbol{\psi}_\alpha^{(l)}, \boldsymbol{\psi}_\beta, \omega_\alpha^{2(l)}, \boldsymbol{\Omega}_\beta, \mathbf{u}_i\right)$ independently for $i = 1, \ldots, n$ and for each $l = 0, 1$,
6. $\left(\boldsymbol{\beta}_i \,|\, \alpha_i^{*(0)}, \alpha_i^{*(1)}, \boldsymbol{\zeta}_i^{*(0)}, \boldsymbol{\zeta}_i^{*(1)}, \boldsymbol{\psi}_\beta, \boldsymbol{\Omega}_\beta, \mathbf{u}_i\right)$ independently for $i = 1, \ldots, n$,
7. $\left(\boldsymbol{\Sigma} \,|\, \left\{\alpha_i^{*(0)}, \alpha_i^{*(1)}, \boldsymbol{\beta}_i, \boldsymbol{\zeta}_i^{*(0)}, \boldsymbol{\zeta}_i^{*(1)}\right\}_{i=1}^n, \mathbf{u}_1, \ldots, \mathbf{u}_n\right)$,
8. $\left(\boldsymbol{\psi}_\alpha^{(0)}, \boldsymbol{\psi}_\alpha^{(1)}, \boldsymbol{\psi}_\beta, \boldsymbol{\psi}_\gamma^{(0)}, \boldsymbol{\psi}_\gamma^{(1)} \,|\, \omega_\alpha^{2(0)}, \omega_\alpha^{2(1)}, \boldsymbol{\Omega}_\beta, \boldsymbol{\Omega}_\gamma^{(0)}, \boldsymbol{\Omega}_\gamma^{(1)}, \mathbb{B}\right)$, where $\mathbb{B} = \left\{\alpha_i^{*(0)}, \alpha_i^{*(1)}, \boldsymbol{\beta}_i, \boldsymbol{\gamma}_i^{(0)}, \boldsymbol{\gamma}_i^{(1)}\right\}_{i=1}^n$,
9. $\left(\omega_\alpha^{2(0)}, \omega_\alpha^{2(1)}, \boldsymbol{\Omega}_\beta, \boldsymbol{\Omega}_\gamma^{(0)}, \boldsymbol{\Omega}_\gamma^{(1)} \,|\, \boldsymbol{\psi}_\alpha^{(0)}, \boldsymbol{\psi}_\alpha^{(1)}, \boldsymbol{\psi}_\beta, \boldsymbol{\psi}_\gamma^{(0)}, \boldsymbol{\psi}_\gamma^{(1)}, \mathbb{B}\right)$, where $\mathbb{B} = \left\{\alpha_i^{*(0)}, \alpha_i^{*(1)}, \boldsymbol{\beta}_i, \boldsymbol{\gamma}_i^{(0)}, \boldsymbol{\gamma}_i^{(1)}\right\}_{i=1}^n$.

Again, Step 1 and Steps 7–9 are standard, as given in Rossi et al. (1996). Steps 5–6 likewise require only minor modifications to the algorithm outlined in Section 9.3. Specifically, using (9.18)-(9.19) define $\tilde{p}_{ijt}^{(l)} = (-1)^l p_{ijt}$ along with $\tilde{\mathbf{p}}_i^{(l)} = (-1)^l \mathbf{p}_i$ and

$$q_{ijt}^{(l)} = \mathbf{1}\left(\tilde{p}_{ijt}^{(l)} \geq \zeta_{ijt}^{(l)}\right) p_{ijt}.$$

Stacking $\mathbf{q}_i^{(l)} = (q_{i11}^{(l)}, \ldots, q_{imT_i})'$, let

$$\hat{\mathbf{v}}_{\alpha_i}^{(l)} = \mathbf{u}_i - \mathbf{X}_i \left(\mathrm{I}_{K_0} \otimes \mathbf{z}_i'\right) \boldsymbol{\psi}_\beta + \alpha_i^{(l-1)} \mathbf{q}_i^{(1-l)}.$$

To sample $\alpha_i^{*(l)}$, just replace $(\boldsymbol{\psi}_\alpha, \omega_\alpha^2)$ with $(\boldsymbol{\psi}_\alpha^{(l)}, \omega_\alpha^{2(l)})$ as well as $\hat{\mathbf{v}}_{\alpha_i}$ with $\hat{\mathbf{v}}_{\alpha_i}^{(l)}$ and $\mathbf{p}_i$ with $\mathbf{q}_i^{(l)}$ in (9.13). Subsequently, sample $\boldsymbol{\beta}_i$ by making the analogous adjustments to (9.14).

To implement Steps 2–4, assume the prior $\tau^{2(l)} \sim \mathcal{IG}(\nu_0, \eta_0)$ and let

$$\begin{aligned}
v_{\zeta_{ijt}}^{(l)} &= u_{ijt} - \mathbf{x}_{ijt}'\boldsymbol{\beta}_i + \alpha_i^{(l-1)} q_{ijt}^{(1-l)}, \\
\hat{\kappa}_{\zeta_{ijt}}^{(l)} &= \frac{1 - \Phi\left(\left(p_{ijt} - \mathbf{w}_{ijt}'\gamma_i^{(l)}\right)/\tau^{(l)}\right)}{\Phi\left(\left(p_{ijt} - \mathbf{w}_{ijt}'\gamma_i^{(l)}\right)/\tau^{(l)}\right)} \\
&\quad \times \exp\left\{\alpha_i^{(l)} p_{ijt}\left(v_{\zeta_{ijt}}^{(l)} + 0.5\alpha_i^{(l)} p_{ijt}\right)/\sigma_j^2\right\}, \\
\hat{\pi}_{\zeta_{ijt}}^{(l)} &= \left(1 + \hat{\kappa}_{\zeta_{ijt}}^{(l)}\right)^{-1},
\end{aligned}$$

and

$$\begin{aligned}
\hat{\boldsymbol{\Omega}}_{\gamma_i}^{(l)} &= \left(\boldsymbol{\Omega}_\gamma^{-1(l)} + \mathbf{W}_i'\mathbf{W}_i/\tau^{2(l)}\right)^{-1}, \\
\hat{\gamma}_i^{(l)} &= \hat{\boldsymbol{\Omega}}_{\gamma_i}^{(l)} \left(\boldsymbol{\Omega}_\gamma^{-1(l)} \left(\mathrm{I}_{K_2} \otimes \mathbf{z}_i'\right) \boldsymbol{\psi}_\gamma^{(l)} + \mathbf{W}_i'\boldsymbol{\zeta}_i^{*(l)}/\tau^{2(l)}\right)^{-1}, \\
\hat{\nu}^{(l)} &= \nu_0 + 0.5m \sum_{i=1}^{n} T_i, \\
\hat{\eta}^{(l)} &= \eta_0 + 0.5 \sum_{i=1}^{n} \sum_{j=1}^{m} \sum_{t=1}^{T_i} \left(\zeta_{ijt}^{*(l)} - \mathbf{w}_{ijt}'\gamma_i^{(l)}\right)^2.
\end{aligned}$$

Sample $\zeta_{ijt}^{*(l)}$ from the two-point mixture of truncated normals

$$\left(\zeta_{ijt}^{*(l)} \mid \cdot\right) \sim \begin{cases} \mathcal{N}_{(-\infty, p_{ijt}]}\left(\mathbf{w}_{ijt}'\boldsymbol{\psi}_\gamma^{(l)}, \tau^{2(l)}\right) & \text{w.p.} \quad \hat{\pi}_{\zeta_{ijt}}^{(l)} \\ \mathcal{N}_{(p_{ijt}, \infty)}\left(\mathbf{w}_{ijt}'\boldsymbol{\psi}_\gamma^{(l)}, \tau^{2(l)}\right) & \text{w.p.} \quad 1 - \hat{\pi}_{\zeta_{ijt}}^{(l)} \end{cases}.$$

Conditional on these draws, sample

$$\begin{aligned}
\left(\boldsymbol{\gamma}_i^{(l)} \mid \boldsymbol{\zeta}_i^{*(l)}, \cdot\right) &\sim \mathcal{N}\left(\hat{\gamma}_i^{(l)}, \hat{\boldsymbol{\Omega}}_{\gamma_i}^{(l)}\right), \\
\left(\tau^{2(l)} \mid \left\{\boldsymbol{\zeta}_i^{*(l)}, \boldsymbol{\gamma}_i^{(l)}\right\}_{i=1}^{n}\right) &\sim \mathcal{IG}\left(\hat{\nu}^{(l)}, \hat{\eta}^{(l)}\right).
\end{aligned}$$

9.5 Conclusion

This chapter takes a more in-depth look at the way that the price variable is modeled in random utility applications, particularly within the marketing context. We argue that allowing for a non-trivial probability of price insensitivity is desirable form a number of theoretical and practical perspectives, and that the censored distribution is especially suitable for that purpose. Indeed, assigning a censored mixing distribution to heterogenous price responses in the simple linear utility model leads to not only probabilistic price insensitivity, but also desirable shrinkage effects and straightforward computation. To illustrate the latter, we outlined a fully tractable Gibbs sampler under the probit specification that can be readily used to implement a stochastic search algorithm for segmentation purposes. Moreover, it was shown that the censored distribution can be further applied in a more elaborate setting—i.e., it can be used to model both asymmetric responses as well as heterogeneous price thresholds in empirical implementations of the reference price model. The latter still leads to a fully tractable Gibbs sampling algorithm.

Of course, the theoretical work presented here only provides an orientation for empirical analysis. To establish the algorithms presented in this chapter as viable empirical instruments, it will be necessary to implement them in practice with real-world data. On that note, we point out that an algorithm similar to that presented in Section 9.3 has already been successfully implemented in a time-series context (e.g., Eisenstat et al., 2013). Therefore, we would except little computational difficulties in applying similar procedures to discrete choice data. However, the sampler of Section 9.4 is based on additional levels of hierarchy involving $\zeta_{ijt}^{*(l)}$ and added interdependence between $\alpha_i^{*(0)}, \alpha_i^{*(1)}, \boldsymbol{\zeta}_i^{*(0)}$, and $\boldsymbol{\zeta}_i^{*(1)}$. This, in turn, may potentially produce adverse effects in terms of sampling efficiency. The extent to which such complications are important in practice remains to be tested.

Bibliography

Allenby, G. M. and P. E. Rossi (1998). Marketing models of consumer heterogeneity. *Journal of Econometrics 89*(1-2), 57–78.

Bell, D. R. and J. M. Lattin (2002). Looking for loss aversion in scanner panel data: The confounding effect of price response heterogeneity. *Marketing Science 19*(2), 185–200.

Berry, S. T. (1994). Estimating discrete-choice models of product differentiation. *The RAND Journal of Economics 25*(2), 242–262.

Berry, S. T., J. Levinsohn, and A. Pakes (1995). Automobile prices in market equilibrium. *Econometrica 63*(4), 841–890.

Boatwright, P., R. McCulloch, and P. Rossi (1999). Account-level modeling for trade promotion: An application of a constrained parameter hierarchical model. *Journal of the American Statistical Association 94*(448), 1063–1073.

Brownstone, D., A. Ghosh, T. F. Golob, C. Kazimi, and D. V. Amelsfort (2003). Drivers' willingness-to-pay to reduce travel time: Evidence from the San Diego I-15 Congestion Pricing Project. *Transportation Research A 37*, 373–387.

Eisenstat, E., J. C. C. Chan, and R. W. Strachan (2013). Stochastic model specification search for time-varying parameter VARs. Working Paper.

Frühwirth-Schnatter, S. and H. Wagner (2011). Stochastic model specification search for Gaussian and partial non-Gaussian state space models. *Journal of Econometrics 154*, 85–100.

Gázquez-Abad, J. C. and M. Sánchez-Pérez (2009). Characterising the deal-proneness of consumers by analysis of price sensitivity and brand loyalty: An analysis in the retail environment. *The International Review of Retail, Distribution and Consumer Research 19*(1), 1–28.

George, E. I. and R. McCulloch (1993). Variable selection via Gibbs sampling. *Journal of the American Statistical Association 88*, 881–889.

George, E. I. and R. McCulloch (1997). Approaches for Bayesian variable selection. *Statistica Sinica 7*, 339–373.

Gupta, S. and L. G. Cooper (1992). The discounting of discounts and promotion thresholds. *Journal of Consumer Research 19*(3), 401–11.

Han, S., S. Gupta, and D. R. Lehmann (2001). Consumer price sensitivity and price thresholds. *Journal of Retailing 77*(4), 435 – 456.

Helson, H. (1964). *Adaptation-Level Theory.* New York: Harper and Row Inc.

Jedidi, K. and J. Z. Zhang (2002). Augmenting conjoint analysis to estimate consumer reservation price. *Management Science 48*(10), 1350–1368.

Kahneman, D. and A. Tversky (1979). Prospect theory: An analysis of decision under risk. *Econometrica 47*, 263–291.

Kalwani, M. U. and C. K. Yim (1992). Consumer price and promotion expectations: An experimental study. *Journal of Marketing Research 29*, 90–100.

Kalyanaram, G. and J. D. C. Little (1994). An empirical analysis of latitude of price acceptance in consumer package goods. *Journal of Consumer Research 21*, 408–418.

Kalyanaram, G. and R. S. Winer (1995). Empirical generalizations from reference price search. *Marketing Science 14*(3), G161–G169.

Kim, J., G. M. Allenby, and P. E. Rossi (2002). Modeling consumer demand for variety. *Marketing Science 21*(3), 229–250.

Marschak, J. (1960). Binary choice constraints on random utility indications. In K. J. Arrow (Ed.), *Stanford Symposium on Mathematical Methods in the Social Sciences*, pp. 312–329. Stanford: Stanford University Press.

McFadden, D. (1974). Conditional logit analysis of qualitative choice behavior. In P. Zarembka (Ed.), *Frontiers in Econometrics*, pp. 105–142. New York: Academic Press.

McFadden, D. (1986). The choice theory approach to market research. *Marketing Science 5*, 275–279.

Mitchell, T. and J. J. Beauchamp (1988). Bayesian variable selection in linear regression. *Journal of the American Statistical Association 404*, 1023–1032.

Nevo, A. (2000). Mergers with differentiated products: The case of the ready-to-eat cereal industry. *The RAND Journal of Economics 31*(3), 395.

Ofek, E. and V. Srinivasan (2002). How much does the market value an improvement in a product attribute? *Marketing Science 21*(4), 398–411.

Pinkse, J. and M. E. Slade (2004). Mergers, brand competition, and the price of a pint. *European Economic Review 48*(3), 617–643.

Reichl, J. and S. Frühwirth-Schnatter (2012). A censored random coefficients model for the detection of zero willingness to pay. *Quantitative Marketing and Economics 10*(2), 259–281.

Rossi, P. E., R. E. McCulloch, and G. M. Allenby (1995). Hierarchical modelling of consumer heterogeneity: An application to target marketing. In C. Gatsonis, J. S. Hodges, R. E. Kass, and N. D. Singpurwalla (Eds.), *Case Studies in Bayesian Statistics*, Chapter 10, pp. 323–350. New York: Springer-Verlag.

Rossi, P. E., R. E. McCulloch, and G. M. Allenby (1996). The value of purchase history data in target marketing. *Marketing Science 15*(4), 321–340.

Sherif, C. W. (1963). Social categorization as a function of latitude of acceptance and series range. *Journal of Abnormal and Social Psychology 67*, 148–156.

Small, K. A. and H. S. Rosen (1981). Applied welfare economics with discrete choice models. *Econometrica 49*(1), 105–130.

Swait, J., T. Erdem, J. Louviere, and C. Dubelaar (1993). The equalization price: A measure of consumer-perceived brand equity. *International Journal of Research in Marketing 10*(1), 23–45.

Terui, N. and W. D. Dahana (2006). Research note—estimating heterogeneous price thresholds. *Marketing Science 25*(4), 384–391.

Train, K. (2003). *Discrete Choice Methods with Simulation.* Cambridge: Cambridge University Press.

Train, K. and G. Sonnier (2005). Mixed logit with bounded distributions of correlated partworths. In R. Scarpa and A. Alberini (Eds.), *Applications of Simulation Methods in Environmental and Resource Economics*, Volume 6 of *The Economics of Non-Market Goods and Resources*, Chapter 7, pp. 117–134. Berlin/Heidelberg: Springer-Verlag.

CHAPTER 10

HIERARCHICAL MODELING OF CHOICE CONCENTRATION OF U.S. HOUSEHOLDS

KARSTEN T. HANSEN[1], ROMANA KHAN[2] AND VISHAL SINGH[3]

[1]University of California, San Diego, USA.
[2]Ozyegin University, Istanbul, Turkey.
[3]New York University, New York, USA.

In this chapter we provide an empirical analysis of the purchase decisions made by over 100,000 US households observed over multiple years in hundreds of product categories. In particular, we study the degree of *choice concentration*, i.e., how concentrated are the repeated choices made by a household in a category over time? We use hierarchical modeling to decompose the variation in observed choice concentration measures across households and product categories. The context is grocery retail with households making choices in a large number of grocery categories. In terms of measuring choice concentration we follow the methodology in Hansen et al. (2013) and present a new empirical analysis.

10.1 Introduction

Suppose we recorded our shopping history for several years for a variety of frequently purchased grocery products such as cereal, detergent, toothpaste, or soup. What would that product portfolio look like? Would purchases in a category be concentrated on a small set or even a single product as illustrated on the left side of Figure 10.1? Or would they be diversified across a large variety that varies by attributes such as brand name, price, flavor, or size as shown on the right side of the plot?

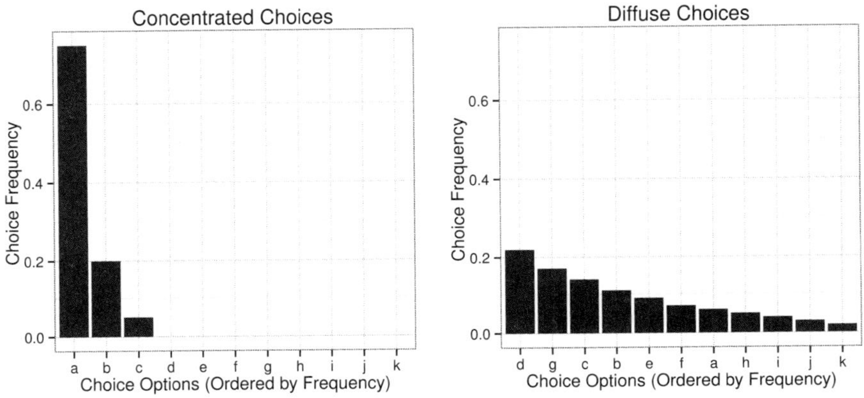

Figure 10.1: Hypothetical choice concentration distributions.

Given the high frequency and low ticket value of most grocery products, consumers are unlikely to spend significant cognitive resources in evaluating options on every purchase occasion. This in turn suggests a routinized, almost habitual shopping behavior characterized by high levels of repeat buying. An extensive literature in psychology and cognitive science documents that repetitive behaviors (and thoughts) in a stable context can take the form of automated reflexive responses to external or internal triggers, i.e., take the form of habits (Verplanken and Aarts, 1999; Wood and Neal, 2007; Danner et al., 2008; Graybiel, 2008; Neal et al., 2012).

A counter argument to habitual buying behavior is an extensive literature documenting consumer motivation to seek variety even in the presence of their most favorite option (McAlister and Pessemier, 1982; Read and Loewenstein, 1995; Ratner et al., 1999). The desire for variety could be driven by a number of intrinsic factors such as uncertainty of preferences (Pessemier, 1978; Kahn, 1995), signaling uniqueness and open mindedness (Ariely and Levav, 2000), or simply satiation with existing options and a desire for novelty (McAlister and Pessemier, 1982; Inman, 2001). At the same time, deviation from the default or favorite option could also be triggered by external factors. Although grocery shopping shares many features of mundane habitual activities considered in experimental work, it occurs in a stimulus-rich environment of advertising and frequent point-of-purchase promotions. Furthermore, the CPG industry is characterized by high level of new product introductions that include new brands as well as modifications (e.g., new flavors) of existing products. Such intrinsic and environmental factors could in turn lead to a product portfolio characterized by a wide array of product choices over time.

A related literature in Marketing has sought to analyze the degree of state dependence in purchase behavior: Conditional on a household having bought brand A on the previous purchase occasion, what is the probability of buying brand A again? The objective in this line of research is to econometrically identify the influence of "preference heterogeneity" versus so-called "pure state dependence" (Keane, 1997; Dubé et al., 2010). Given the primary objectives, as well as data limitations, applications in this literature tend to be with small sample sizes (around 200-800 households), a handful (often single) product categories, and short time periods (usually 2 years). This literature models the purchase-to-purchase state transitions using discrete choice models with an emphasis on trying to explain the causes behind brand loyalty/variety seeking.

In the current work we don't seek to understand what drives individual households' decisions from purchase to purchase. Rather, we are interested in the overall result of a household's decision process when observed over an extended period of time. How many different products does a household buy in a certain category and how concentrated are these purchases? For example, how large a fraction of total category expenditure is spent on the household's most popular choice? How much does this metric vary across households within a category and across categories? What are the characteristics of categories where choices are highly concentrated? How fast do choices diffuse across options – if at all – as the number of choices increase?

We answer these questions by explaining the overall variation in choice concentration by variation in four components: A category effect, an effect due to number of purchases, a household level "preference for concentration" and a choice horizon effect. The category effect captures variation in average choice concentration across categories, while the effect of number of purchases measures how fast or slow choice concentration diffuses as purchases increase. The household concentration preference represents a household general choice concentration across categories. Finally, the choice horizon effect measures the effect of time on concentration, e.g., are choices made over six years more or less concentrated than choices made over one year (holding other factors constant)? A hierarchical model is used to explain variation in these effects across categories using category characteristics and variation in the concentration preference using household demographics.

10.2 Data Description

The cornerstone of our empirical strategy is an extensive consumer panel data provided by A.C. Nielsen. Nielsen provides households in its panel with in-home optical scanners that are used to record details of grocery purchases from all retail outlets. The data used in this chapter contain purchase histories of approximately 100,000 demographically balanced households spanning 50 US states. The data are observed from 2004-09 and households stay in the panel for an average of 3 years. For approximately 18,000 households, we observe purchase histories for the entire six year period. The database contains information on each store visit including date of purchase, identification of the retail outlet, and the total amount spent on the shopping occasion. Furthermore, the database records detailed information on every product in the shopping basket such as the product description (brand name, size, flavor, etc.), number of units purchased, price paid, and indicators for any promotion or coupon usage. We also observe a large number of household demographics that are discussed in detail below. Over the past several decades, these data have become one of the most important tools for understanding consumer buying behavior in the CPG industry and has sprawned a vast academic literature in Marketing.

10.3 Measures of Choice Concentration

The primary objective of the chapter is to analyze the degree to which product purchases are concentrated on a few options. The first natural question that arises is how to define a product. A typical supermarket in the US contains approximately 50,000 unique items that are organized into several hundred product categories (e.g., Yogurt, Breakfast Cereal, Paper Towels). Each category in turn contains several brands and variants within brands based on packaging, size, flavor, fat content, and so forth. For example, the Yogurt category contains store brand (or private label) as well as several national brands such as Dannon, Yoplait, Stonyfield, and Chobani. Within Dannon, there are a total of 314 unique UPCs that vary by packaging, size, flavor, and fat content. Naturally the degree to which consumers display concentration in purchase behavior will crucially depend on the level of aggregation. For example, a brand loyal customer of Dannon may have high product concentration when measured at the brand level, but may display high level of purchase diversity in other attributes such as flavor. Our approach in the chapter is to create concentration measures at the UPC level – the most granular unit of analysis possible. If high concentration is found at this level, then – by definition – it will be found at more aggregate levels. Having defined the unit of analysis, we next need to create measures of purchase concentration. Several metrics are possible: Concentration ratios, Herfindahl-Hirschman Indices or various Entropy measures. We found that these measures tend to be very highly correlated and that the results of the final analysis are robust to the choice of concentration measure. In the following we will use the top expenditure concentration ratio (CR1), i.e., the fraction of overall category dollar expenditure spend on the most popular product. For example, a concentration ratio of 40% means that out of the household's total spending in the category, 40% went to the top choice.

For the analysis, we selected the top 150 product categories in the data based on dollar volume. One of these product categories had missing values, and another 3 had highly irregular data. We retained the remaining 146 categories and use two of these for out-of-sample predictions. The product categories used include a wide variety of food (e.g., Cereal, Soup), non-food (e.g., Detergent, Toilet Paper), condiments (e.g., Ketchup, Salad Dressing), health and beauty (e.g., Shampoo, Toothpaste), and over the counter medicines (e.g., Headache Remedies, Multi-vitamins).

Not all households buy in all categories and some are only occasional users of a category. We defined a household as a user of a category if the household made at least 5 purchases in the category over the observation period (anywhere from one to six years). Figure 10.2 shows the number of users per category as well as the average concentration ratio by category. The number of observations range from just under 20,000 to over 70,000 for the high penetration categories. The most concentrated categories have average concentration ratios above 50% with the least concentrated having ratios around 25%. Not surprisingly, the length of time over which choices are observed tends to reduce concentration ratios – but not by much: the one year sample has average concentration ratios about 10 percentage point higher than the six year sample. However, this is raw data and we are not controlling for the longer sample having more purchases than shorter sample.

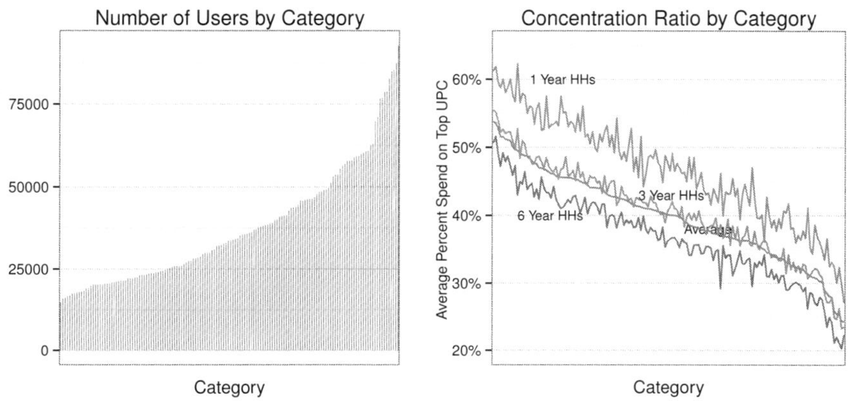

Figure 10.2: Category summary statistics: Number of users per category (left) and average concentration ratio of top UPC by category (right).

The average concentration ratios in Figure 10.2 show extensive variation across categories. They also mask substantial heterogeneity at the household level: These are means over thousands of households and the raw data show that all categories have segments of households with either very small or very large concentration ratios. What explains these differences both at the category and household level? This is the question we turn to in the next section.

10.4 Methodology

Our methodology decomposes variation in observed choice concentration measures into category effects, household effects and time effects. The building block for our final specification is a simple linear model of the form

$$y_{ic} = \beta_c + \gamma_c x_{ic} + \alpha_c \theta_i + \varepsilon_{ic}, \tag{10.1}$$

where y_{ic} is the choice concentration measure for household i in category c, x_{ic} is the number of choices made by household i in context c (i.e., number of purchases in category c) and θ_i is household i's unobserved general preference for concentration. The specification in (10.1) lacks one important feature: It doesn't allow for an effect of time, i.e., the time over which y_{ic} is measured. For example, holding the concentration preference θ_i constant, (10.1) would predict that a household making 20 choices over 1 year would have the same choice concentration as a household making 20 choices over 5 years. While it is possible that this is true, we don't want to impose this on the specification directly. Therefore, we allow for a general interaction with time:

$$\begin{aligned} y_{ic} &= \beta_c(t_i) + \gamma_c(t_i) x_{ic} + \alpha_c(t_i)\theta_i + \varepsilon_{ic}, \\ \varepsilon_{ic} &\sim \mathrm{N}\big(0,\ \tau_c(t_i)^{-1}\big), \end{aligned} \tag{10.2}$$

where $t_i \in [1, 2, \ldots, 6]$ is the length of time over which i's behavior is measured. Based on this model we can quantify the effects of varying t, x and θ on the choice concentration measure. If we normalize both x and θ to have mean zero, then we can think of β as a baseline concentration measure – the average concentration measure for households with an average number of purchases and diversity preference.

Model (10.2) is the first layer in a hierarchical structure. In the second layer we specify a model for the variation of $\lambda_c(t) \equiv \big(\beta_c(t), \gamma_c(t), \alpha_c(t)\big)$ and of θ_i. Specifically, we link the distribution of these parameters to a set of category characteristics while allowing for a time-specific intercept:

$$\begin{aligned} \beta_c(t) &= \beta_0(t) + \pi'_\beta Z_c + u_{\beta,c}(t), \\ \gamma_c(t) &= \gamma_0(t) + \pi'_\gamma Z_c + u_{\gamma,c}(t), \\ \alpha_c(t) &= \alpha_0(t) + \pi'_\alpha Z_c + u_{\alpha,c}(t), \end{aligned} \tag{10.3}$$

where Z_c is a vector of observable category characteristics. If Z_c is centered to have mean zero, $\beta_0(t)$ is the average of the baseline concentrations across categories for sample length t. Similarly, $\gamma_0(t)$ and $\alpha_0(t)$ is the average sensitivity to the number of purchases and diversity preference for sample length t. We assume that the errors are jointly normal distributed:

$$u_c(t) \equiv \big(u_{\beta,c}(t), u_{\gamma,c}(t), u_{\alpha,c}(t)\big) \sim \mathrm{N}\big(0, \Omega(t)\big), \tag{10.4}$$

where $\Omega(t)$ is cohort specific covariance matrix.

For θ we assume

$$\theta_i = \pi_\theta' W_i + \epsilon_i, \tag{10.5}$$

where W_i is a vector of observable household characteristics. We assume that $\epsilon_i \sim \mathrm{N}(0, \sigma_\varepsilon)$. Due to the factor structure used in (10.2), we cannot separately identify the variance of $\alpha_c(t)$ and the variance of θ_i. We use the restriction $\sigma_\varepsilon = 1$ to identify the variance $\alpha_c(t)$. From (10.5) we can test hypotheses about which household characteristics can explain variation in the latent choice concentration preference.

10.4.1 Specification

As a dependent variable we use

$$y_{ic} = \log\left(\frac{cr_{ic}}{1 - cr_{ic}}\right),$$

where cr_{ic} is the concentration ratio for household i in category c.[1] Number of purchases, x_{ic} is entered in logged form, centered to have mean zero. As category characteristics, Z_c, we include the percent of transactions made on promotion (a measure of the category's promotional activity), the number of UPCs in the category relative to the average across categories (a measure of assortment size or choice dimension of the category), the penetration of the category (a measure of the necessity of the category) and the average monthly purchase frequency of the category (a measure of how often a purchase decision is required). We include standard household level demographics in W_i, i.e., income, age, education and household size. These were also entered using dummy categories. We also include measures of religiosity and conservatism measured at the US census level.[2]

[1] In some categories a small number of households have concentration ratios equal to 1. We recode these as 0.99. We also estimated a version of the model where we separately modeled the probability of observing a concentration ratio of 1. This two-equation model generated very similar results to the ones shown here.

[2] To measure religiosity we use the number of churches per capita. This was entered as a dummy variable with three categories: "very high religiosity" if a household lived in a county with number of churches per capita in the highest decile, "high religiosity" for the next highest decile with the baseline being decile 1 to 8. A similar construction was used to measure conservatism using vote percentages for the republican candidate for president in the 2000 election.

We use independent normal priors on the intercept and Z_c parameters of (10.3):

$$\{\beta_0(t), \gamma_0(t), \alpha_0(t)\}_{t=1}^6 \sim \mathrm{N}(0, k{*}I),\ (\pi_\beta, \pi_\gamma, \pi_\alpha) \sim \mathrm{N}(0, k{*}I),\ \pi_\theta \sim \mathrm{N}(0, k{*}I),$$

where $k = 1000$. We use a Wishart prior for each of the $\Omega(t)^{-1}$ parameters with 5 degrees of freedom and prior expectation equal to the identity matrix and Gamma priors for the precision parameters in (10.2):

$$\Omega(t)^{-1} \sim \text{Wishart}(5, \tfrac{1}{5}I);\quad \tau_c(t) \sim \text{Gamma}(0.1, 0.1), \qquad t = 1, \ldots, 6.$$

This is a linear hierarchical model with easily derived conditional posteriors. We use a standard Gibbs sampler to simulate draws from the posterior distribution. Due to the relatively large sample size we coded the algorithm using the Intel Visual Fortran XE Compiler 12.1 taking advantage of parallel processing capabilities. We ran the sampler for 6000 iterations, discarding the first 1000 and used the remaining 5000 for summaries.

10.5 Results

We start by documenting the estimates of the parameters in (10.2). For each of 144 categories and for each sample length $t = 1$ year, $\ldots, 6$ years, we have an estimate of $\lambda = (\beta, \gamma, \alpha)$. In Figure 10.3 we plot estimates of λ and $\sigma(t) \equiv \sqrt{\tau(t)^{-1}}$ for the one year and six year sample.

Recall from above that the intercepts can be thought of as a baseline concentration measure. As expected we see large amounts of heterogeneity in baseline concentration across categories. It is also evident that – for most but not all categories – baseline concentration is reduced in the six year sample compared to the one year sample. Note that the model controls for the larger number of purchases in the six year sample versus the one year sample. Therefore, the lower baseline in the six year sample suggests that time in itself makes choices more diffuse. We investigate this in more detail below.

An increased number of choices (i.e., purchases) leads to a lower choice concentration in most categories. In a small number of categories households' choice concentration display no sensitivity to the number of choices made and a few categories even have positive effects (although the effects sizes are tiny). For most categories the effects sizes are slightly smaller (in absolute value) for the six year sample versus the one year sample.

The third plot in Figure 10.3 shows the sensitivity of choice concentration to the latent factor θ. The impact of θ is positive for all categories so we may think of this as a *latent preference for choice concentration*: Households with large θ's will on average have more concentrated choices than households with small θ's. The sensitivity to this concentration preference is either similar or smaller for the six year sample versus the one year sample, suggesting a lower importance of this factor for the long sample. However, the error term

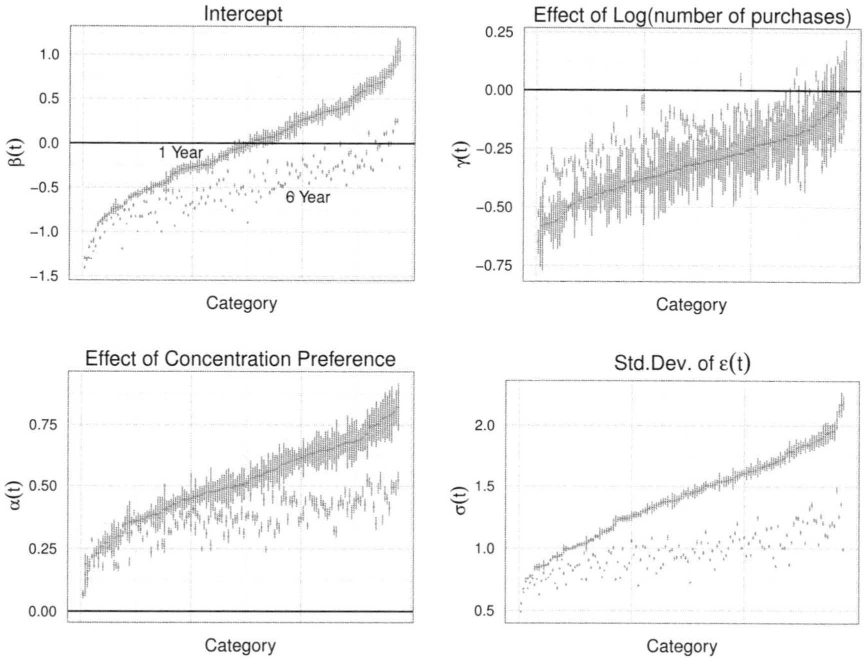

Figure 10.3: Estimates of the model parameters for 1 and 6 year sample. (Note: Plot shows posterior mean and 95% quantile range for each parameter; for each sub-plot categories are ordered based on estimates for 1 year sample.)

variance is also smaller for the six year sample (shown in the fourth plot) so in a relative sense θ might be as important for the six year sample versus the one year sample. In fact, if we calculate the fraction of the variance in $\alpha_c\theta+\varepsilon$ that is due to variation in θ, we get an average of 11.9 percent across all categories for the one year sample and 12.4 percent for the six year sample. 12.4 percent may not sound like much but remember that the model is trying to explain variation in choice concentration in a large number of categories using a single factor θ. In light of this it is actually remarkable that the explained variation is as high as it is. Furthermore, we will document below that θ can predict out-of-sample behavior.

10.5.1 Explaining Variation in Category Effects

As shown in the previous section, there is considerable heterogeneity in the model parameters across categories and across sample duration. The effects of sample duration, i.e., the length of time over which decisions are made, are captured by the intercepts $(\beta_0(t), \gamma_0(t), \alpha_0(t))$ in (10.3). Figure 10.4 shows the estimated intercepts. There are large effects of sample duration: The average

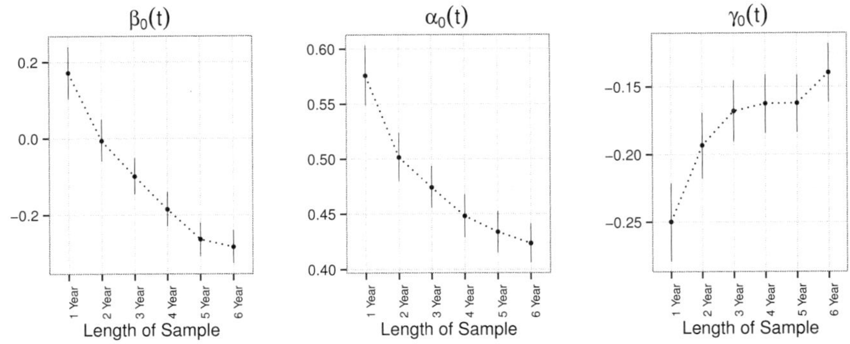

Figure 10.4: Estimates of cohort specific intercepts.

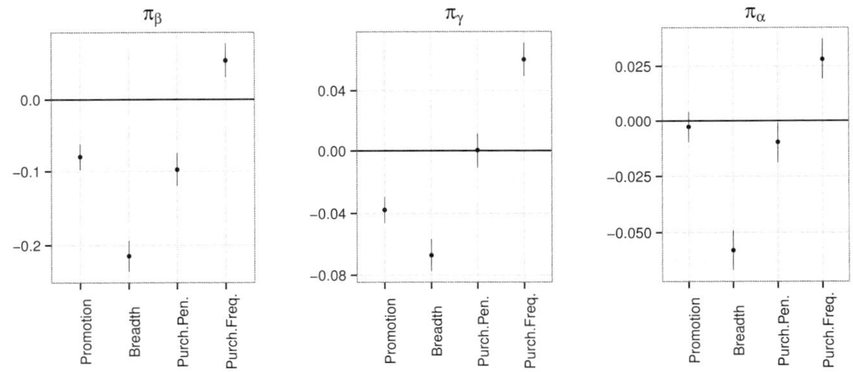

Figure 10.5: Effect of category characteristics. The plot shows posterior mean and 95 percent quantile range for $\pi_\beta(t), \pi_\gamma(t), \pi_\alpha(t)$ scaled to correspond to a one standard deviation increase in each Z variable.

baseline concentration across categories declines with sample duration. For the average category, the baseline concentration declines from 0.17 in the one year sample to -0.28 in the six year sample.[3] However, as is evident in the top left plot in Figure 10.3 this is not true for all categories. It is also clear that on average across categories the effect of an increase in number of decisions decreases (in absolute value). This decrease is most dramatic from the one year to the three year sample. For the longer samples the effects size levels off at around -0.15. The same is true for the effect of the latent concentration

[3]Note that this is the effects size on the log-ratio transformed concentration scale. Below we will calculate the impact of sample duration on the original concentration measure.

preference: On average the effect is stronger in the one year sample compared to the longer samples.

We next address the question of whether variation in category characteristics can explain the variation in the $\beta_c, \gamma_c, \alpha_c$ parameters. Figure 10.5 shows estimates of π in (10.3). It is apparent that an increase in a category's promotional activity decreases both baseline choice concentration and the sensitivity to number of choices. In other words, choices are less concentrated in categories with high promotional activity and in these categories choices diffuse faster with number of choices. Not surprisingly, an increase in the size of the choice set ("breadth" of the category) leads to a large decrease in baseline concentration. It also increases the sensitivity of a category's concentration to the number of purchases made. Finally, an increase in the breadth of a category on average decreases the sensitivity to the latent concentration preference. So we can conclude that in categories with large choice sets, choices tend to diffuse quickly as the number of decisions increase and a household's general preference for concentration has less of an impact.

An increase in the purchase penetration of a category decreases the baseline choice concentration of a category. This implies that essential categories, e.g., Paper Towels or Toothpaste, which have high penetration, tend to have lower choice concentration. On the other hand, categories which are used at a household's discretion and appeal to a narrow segment, e.g., Frozen Entrees or Apple Juice, have on average high choice concentration. Purchase penetration appears to have little or no impact on the sensitivity to number of purchases and concentration preference. Finally, the figure shows that the higher the average purchase frequency of a category is, the higher is choice concentration in that category, choices diffuse slower with number of purchases and the sensitivity to the latent concentration preference increases. Categories where choices are made infrequently, e.g., Vinegar or Toothbrushes, tend to have more diffuse choices than categories with frequent choices such as Milk or Carbonated Soft Drinks. This is consistent with a theory suggesting that frequently made decisions are more likely to be "automatic" than rare decisions.

10.5.2 Explaining Variation in θ

Figure 10.6 reports the estimates of the effects of observable decision maker characteristics on the choice concentration preference θ [the π_θ parameter in equation (10.5)]. Recall from above that larger θ scores are consistent with higher choice concentration. We see that higher income leads to a higher concentration preference. The effects size from the smallest to the biggest income category is about 10% of the standard deviation of θ. Similarly, older households are seen to have a higher concentration preference. The effects from youngest to oldest age category is over 20%. Not surprisingly, we see a substantial effect of household size: The largest households have a concentration preference that on average is about half a θ-standard deviation smaller than

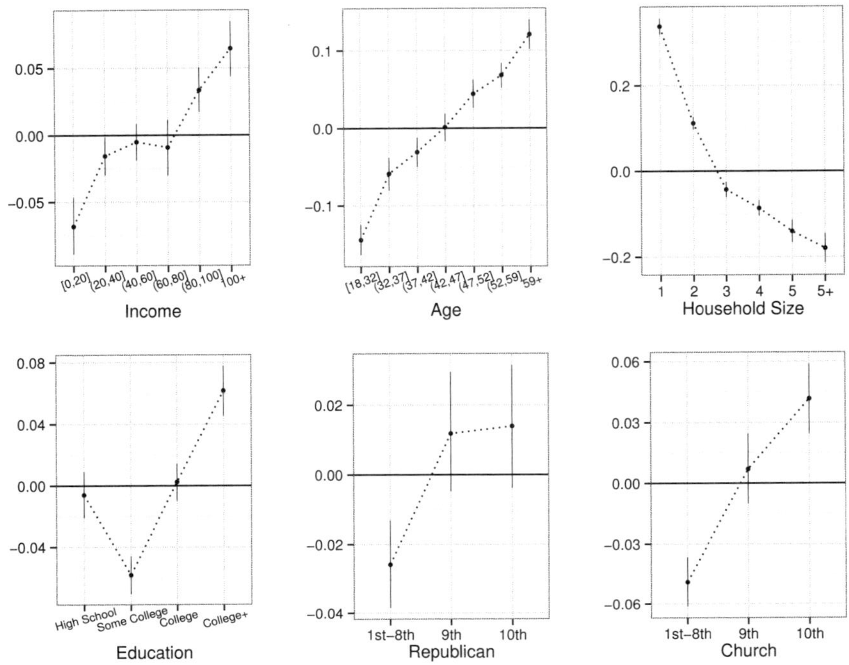

Figure 10.6: Estimates of π_θ. (Note: Plot shows posterior mean and 95% quantile range.)

single households. Length of education has a U-shaped effect on θ with high school graduates having a similar effects size as college graduates. Post-college graduates have a slightly higher concentration preference compared to lower education households. Households in counties that are heavily Republican tends to have higher concentration preferences than non-Republican counties. However, the effects size is very small. Households in counties with higher religiosity (measured as the concentration of churches) have higher choice concentration preferences: Compared to less religious counties, high religion counties have on average about a 10% higher concentration preference.

10.6 Interpreting θ

Above we gave θ the interpretation as a choice concentration preference. To demonstrate that θ indeed is highly correlated with households' observed choice concentration, we divided the 100,443 estimated θ's into deciles. For each decile we computed the average number of purchased brands and UPCs and the average concentration ratio for both brands and UPCs. We also computed the average Herfindahl-Hirschman Index for UPCs and brands. This is an alternative measure of concentration.[4] These averages were computed for both the 144 categories used for estimation and for 2 hold-out categories. The results (shown in Figure 10.7) shows that θ is strongly predictive of average choice concentration behavior within each decile. For example, households in the top θ-decile buys on average about 50 percent fewer unique UPCs and brands within a category than households in the first decile. They also have a concentration ratio almost twice that of the first decile. These patterns hold true for both the estimation and holdout categories.

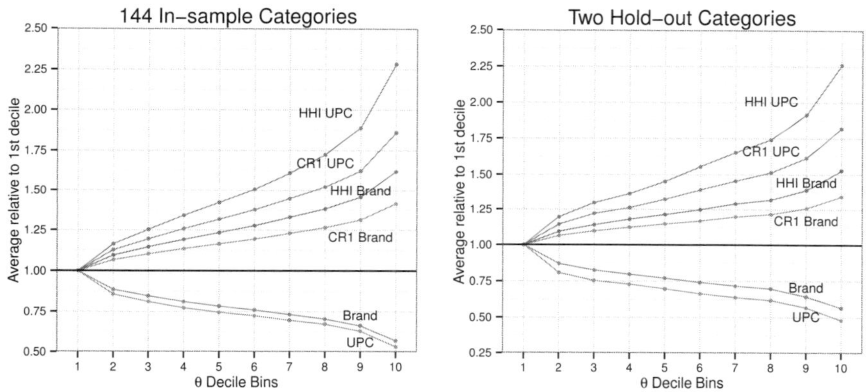

Figure 10.7: Predictive validity of θ.

If θ measures a decision makers general preference towards "consistency" in choice, one might suspect that θ can predict choice concentration in a wider range of contexts than the ones we used to extract it. By definition, θ measures a household's general desire for concentrated choices in 144 product categories and Figure 10.7 documented that it does a good job at this. Can θ predict other types of concentration behavior? We tested this using three concentration metrics:

- *The total number of UPCs purchased by a household.* This is using all purchases in the data – more than 800 product categories or "modules." Note that we only used 146 of these for estimation and holdout predictions. As mentioned above, most modules don't fit the usual notion of a category which is why we dropped them for estimation. Above we saw

[4] The Herfindahl-Hirschman Index is typically used to measure firm concentration of an industry. It is defined as

$$HHI = \sum_{j=1}^{J} ms_j^2,$$

where ms_j is the market share of firm j. The closer the index is to one, the more concentrated and less competitive is the industry. In our case we use choice shares for each household for each category.

that θ is predictive of the total number of UPCs purchased in the 146 modules that are well defined categories. Can θ also predict the total number of UPCs for all modules?

- *The total number of purchased modules.* In how many of the more than 800 modules did each household make a purchase? This is a fairly strong out-of-context test of θ: As documented above θ can predict choice concentration within a category. Can θ also predict how diffuse your choices are in terms of *whether or not to buy at all* in each module?
- *Retailer Concentration Measure.* We computed a concentration measure for retail visits: How large a fraction of overall expenditure was spent at the most preferred retailer? Is is possible that θ can explain how concentrated a household's expenditure is across retail outlets?

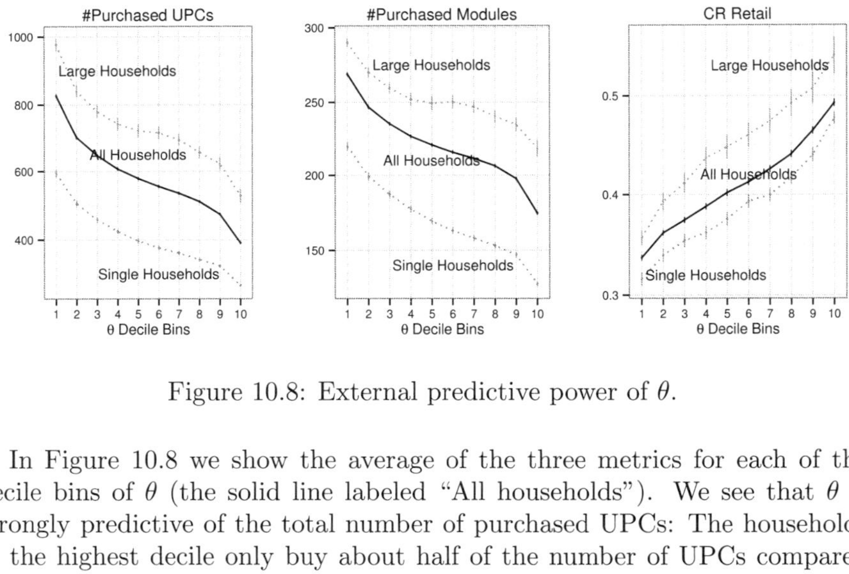

Figure 10.8: External predictive power of θ.

In Figure 10.8 we show the average of the three metrics for each of the decile bins of θ (the solid line labeled "All households"). We see that θ is strongly predictive of the total number of purchased UPCs: The households in the highest decile only buy about half of the number of UPCs compared to the the lowest decile. This is interesting and provides further hold-out evidence for the interpretation of θ. What is even more interesting is the second plot in Figure 10.8. This shows that θ is strongly predictive of the total number of modules purchased by a household. That this should be true is not obvious: by design θ captures choice concentration within a category with large θ household having highly concentrated decisions. What we see in Figure 10.8 is that high θ households also are much more likely to be concentrated in their decisions of *which* modules to buy: Households in the 10th decile on average makes purchases in 175 modules, while households the first decile on average makes purchases in 269 modules. So θ is a trait general enough that it can explain concentration in choosing among UPCs in a category *and* choosing among modules in a retail outlet. Can θ also predict concentration in choices among retail outlets? Yes. The last panel in Figure 10.8 shows that the retail CR1 measure is monotonically increasing in θ: Household's with the highest concentration preference spend about half of their total expenditure at their most preferred retailer. This is close to 30 percent for the households with the lowest concentration preference.[5]

We documented above that the most important driver of θ was household size. Large households had on average much lower θ values. It is possible that what we see in Figure 10.8 is simply a size effect. For example, large households tend to have low θ's, i.e., low concentration preference, and are also more likely to buy a larger number of UPCs, buy in a larger number of modules and a wider variety of retail outlets. To analyze this question, we calculate the average value of the three metrics for each θ decile within households with a single member and within households with 5 or more members. These are shown as "Single" and "Large" households in Figure 10.8. We see that the same monotonic relationship between θ and the three concentration metrics holds true even when controlling for household size. Therefore, household size cannot explain the finding that θ can predict a wide range of observed concentration measures.

10.7 Decomposing the Effects of Time, Number of Decisions and Concentration Preference

Above we have documented that the time over which purchases are made (one to six years), the total number of decisions and a latent choice concentration preference all impact the observed choice concentration measure. Which of these factors has the biggest effect? Since the model in (10.2) has interactions, we cannot just inspect the individual coefficients to gauge this. For example, the effect of making purchase decisions over six years versus only one year will depend on the concentration preference θ. Similarly, the effect of increasing number of purchases depends on the time length over which purchases are made.

In order to analyze the effects of time, number of decisions and concentration preference, we predict concentration ratios implied by the model in (10.2) and compute comparative statics by changing one factor at a time holding all

[5]We have documented a positive correlation between θ and concentration in choice of retail outlet. However, it is not clear what the causal direction is. Do households who prefer low concentration in choices within a category choose to visit a more diverse set of retailers? Or are households, who visit a diverse set of retailers with different layouts and store environments, induced to make more diverse choices within a category? We do not attempt to answer this question in the current chapter.

other factors fixed. We can predict concentration ratios by computing

$$\mathrm{E}[cr|x,\theta,t,\lambda_c,\tau_c] = \int \frac{\exp\{\beta_c(t)+\gamma_c(t)x+\alpha_c(t)\theta+\tau_c^{-1/2}\varepsilon\}}{1+\exp\{\beta_c(t)+\gamma_c(t)x+\alpha_c(t)\theta+\tau_c^{-1/2}\varepsilon\}}\phi(\varepsilon)d\varepsilon, \quad (10.6)$$

where $\phi(\varepsilon)$ is a standard normal density, for different values of x, θ and t.[6] We can then quantify the effects of, for example, changing the number of purchases from some x_0 to x_1, for different values of t. Figure 10.9 shows the results of this calculation. The top-left of the figure shows the effect of increasing sample length from one to six years at three different values of θ (the 5th percentile, mean and 95th percentile), holding number of purchases fixed at the mean level for each category. Increasing the length of the sample period from one to six years reduces choice concentration for almost all categories. At least two explanations can be given for this phenomenon. First, increasing sample length also raises the probability of variability in choice environment (new products, changes in marketing activities, new stores, etc.). Second, observing choices over a longer time period may also increase the probability of a change in households tastes/preferences. The size of the effect depends on θ: At the mean θ level the effect varies from around 20 percentage points down to zero. The effect is bigger for higher θ values and smaller for lower values of θ. In other words, households with strong preferences for concentrated choices, reduce their choice concentration over time by a bigger amount than households with weaker concentration preferences. The top-right plot shows the effect of increasing θ from the 5th to 95th percentile. The size of the effect varies highly across categories: for some categories choice concentration increases by less than 5 percentage points. These are categories where the latent concentration preference has little impact on observed choices. However, in other categories the effect is over 30 percentage points for the 1 year sample and between 20 and 30 percentage points in the six year sample. Finally, the bottom-left plot shows the effects of increasing the number of purchases on choice concentration for the one year and six year sample – holding θ fixed at the mean. For almost all categories an increase in number of purchases, i.e., decisions, reduce choice concentration. The effects are smaller for the six year sample with a small number of categories actually having positive effects. Overall, these calculations reveal that time, number of decisions and latent concentration preferences all can have substantial impact on observed choice concentrations.

Further refinement of this analysis can be obtained by inspecting the types of categories that are sensitive to each of the three factors. We do not undertake a detailed classification analysis here. To give a brief sense of this type

[6]We use simple Monte Carlo integration to compute the integral. We evaluate the integral at the mean posterior estimates of λ_c and τ_c. Alternatively, one could integrate over these parameters using the simulated posterior to get a posterior predictive expectation.

of analysis we provide in Figure 10.10 a scatter plot of the θ and number-of-purchase effects for two types of categories: Dairy and Health & Beauty Care. An interesting pattern emerges from the plot: The Dairy categories tend to be highly sensitive to the latent concentration preference and not impacted by number of purchases. The choice concentration in Health & Beauty Care categories are more sensitive to number of purchases and less sensitive to households' concentration preferences.

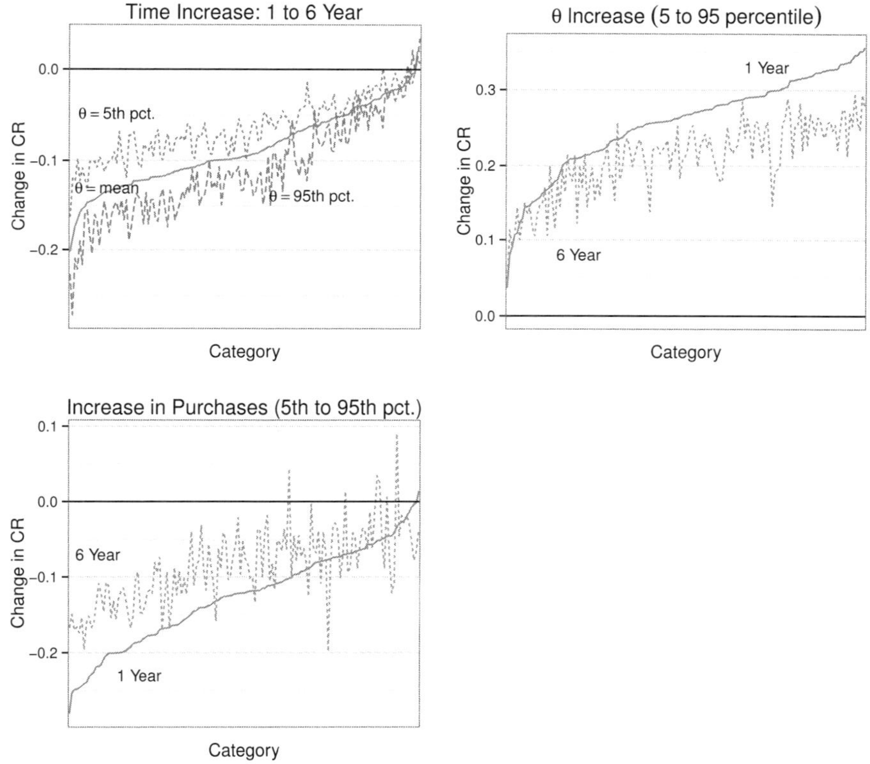

Figure 10.9: Effects of time, concentration preference and number of purchases on choice concentration.

10.8 Conclusion

In this chapter we developed an empirical model of choice concentration of US households using a multilevel model structure. We uncovered substantial variation in baseline concentration of choices across categories and showed that this could be explained by category characteristics. Controlling for the number of purchases made by a household and an unobservable choice con-

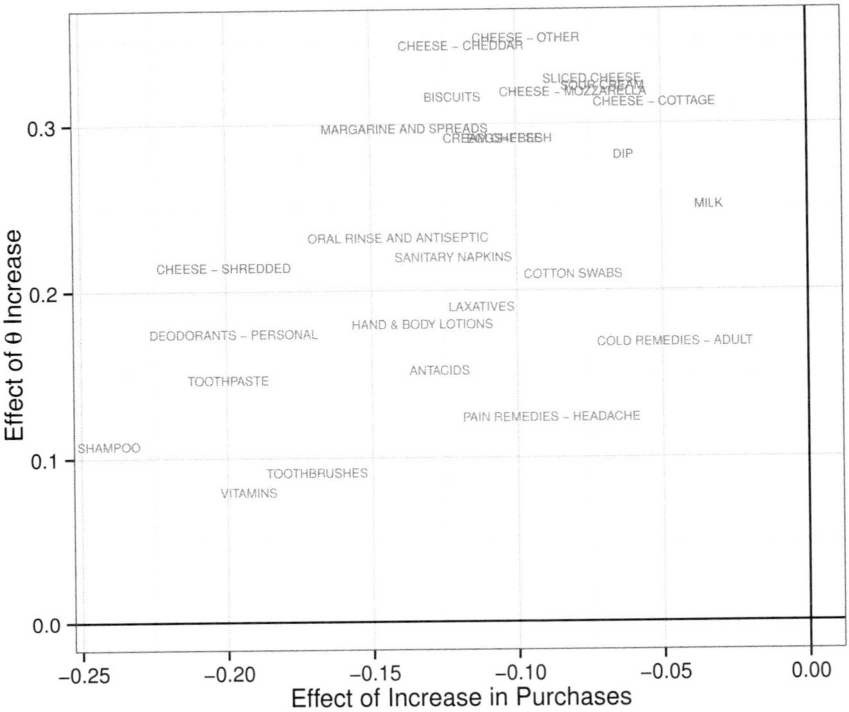

Figure 10.10: Effects for Dairy and Health & Beauty Care categories.

centration preference, we found that a category's average choice concentration can be explained by the category's promotional activity, breadth, penetration and purchase frequency. We also found that for most categories choice concentration is reduced by an increase number of purchases. In categories with large assortment, choice concentration diffuses more rapidly as the number of purchases increase. We showed that choice concentration preferences had predictive power both in terms of hold-out product categories but also in terms of different types of concentration behavior (the number of product categories where a household is a user and the number of retail outlets frequented by the household). We found that the time over which decisions were made reduces choice concentration and households with inherent strong preferences for concentration reduce their concentration by more than households with weaker concentration preferences.

Acknowledgements: The authors would like to acknowledge the support of the James M. Kilts Center for Marketing at The University of Chicago Booth

School of Business and the Nielsen Company for providing the data used in this chapter.

Bibliography

Ariely, D. and J. Levav (2000). Sequential choice in group settings: Taking the road less traveled and less enjoyed. *Journal of Consumer Research 27*(3), 279–290.

Danner, U. N., H. Aarts, and N. K. Vries (2008). Habit vs. intention in the prediction of future behaviour: The role of frequency, context stability and mental accessibility of past behaviour. *British Journal of Social Psychology 47*(2), 245–265.

Dubé, J.-P., G. J. Hitsch, and P. E. Rossi (2010). State dependence and alternative explanations for consumer inertia. *The RAND Journal of Economics 41*(3), 417–445.

Graybiel, A. M. (2008). Habits, rituals, and the evaluative brain. *Annual Review of Neuroscience 31*, 359–387.

Hansen, K. T., R. Khan, and V. Singh (2013). Habitual buying behavior: A large-scale empirical analysis. Working Paper.

Inman, J. J. (2001). The role of sensory-specific satiety in attribute-level variety seeking. *Journal of Consumer Research 28*(1), 105–120.

Kahn, B. E. (1995). Consumer variety-seeking among goods and services: An integrative review. *Journal of Retailing and Consumer Services 2*(3), 139–148.

Keane, M. P. (1997). Modeling heterogeneity and state dependence in consumer choice behavior. *Journal of Business & Economic Statistics 15*(3), 310–327.

McAlister, L. and E. Pessemier (1982). Variety seeking behavior: An interdisciplinary review. *Journal of Consumer Research*, 311–322.

Neal, D. T., W. Wood, J. S. Labrecque, and P. Lally (2012). How do habits guide behavior? Perceived and actual triggers of habits in daily life. *Journal of Experimental Social Psychology 48*(2), 492–498.

Pessemier, E. A. (1978). Stochastic properties of changing preferences. *The American Economic Review 68*(2), 380–385.

Ratner, R. K., B. E. Kahn, and D. Kahneman (1999). Choosing less-preferred experiences for the sake of variety. *Journal of Consumer Research 26*(1), 1–15.

Read, D. and G. Loewenstein (1995). Diversification bias: Explaining the discrepancy in variety seeking between combined and separated choices. *Journal of Experimental Psychology: Applied 1*(1), 34.

Verplanken, B. and H. Aarts (1999). Habit, attitude, and planned behaviour: Is habit an empty construct or an interesting case of goal-directed automaticity? *European Review of Social Psychology 10*(1), 101–134.

Wood, W. and D. T. Neal (2007). A new look at habits and the habit-goal interface. *Psychological Review 114*(4), 843.

CHAPTER 11

APPROXIMATE BAYESIAN INFERENCE IN MODELS DEFINED THROUGH ESTIMATING EQUATIONS

GIUSEPPE RAGUSA

Department of Economics and Finance, Luiss University, Rome, Italy.

This chapter reviews recent approaches to Bayesian inference when the only information linking parameters and data is in the form of estimating equations. Bayesian inference in these class of models is difficult to implement since the likelihood function is left unspecified. The approaches reviewed in the chapter are those that rely on posterior constructed using frequentist objective functions, such as the Generalized Method of Moments and the Generalized Empirical Likelihood, as approximate likelihoods functions. We explore the Bayesian and frequentist properties of inference based on these posteriors.

11.1 Introduction

Over the last three decades, econometric models characterized in terms of restrictions on the moments of underlying random vectors have gained a prominent role in economics and in the social sciences in general. The class of models we have in mind here are those that can be summarized by the following population equation:

$$E[g(X, \theta)] = 0, \tag{11.1}$$

where the random vector X has dimension D and support $\mathcal{X} \subseteq \Re^D$, θ is a parameter vector, an element of $\Theta \subset \Re^K$, and the moment function $g(\cdot,\cdot) : \mathcal{X} \times \Theta \to \Re^M$ is a known vector-valued function. In the next section we will provide some examples of models that satisfy (11.1), but it should already be clear that many conventional estimation methods such as maximum likelihood, ordinary least squares, panel data models, and instrumental variables can be casted into this framework. Keep in mind though that (11.1) is flexible enough to also accommodate specification of more complex statistical models.

While the Generalized Method of Moments (GMM) as formalized by Hansen (1982) and the Generalized Empirical Likelihood (GEL) of Newey and Smith (2004) are the de-facto standard for conducting inference about the parameter θ of (11.1), there is not an accepted way of doing so in a manner consistent with the Bayesian framework. This deficiency is problematic because, as Sims (2002) rightly puts it, models defined in terms of estimating equations are used in real decision making problems and, thus, classical confidence bands for parameters are going to be interpreted as posterior credible regions.

The key difficulty in carrying out proper Bayesian inference in this setting has to do with the fact that usually the conditional distribution of the data given the parameter—the likelihood function—is not available and thus the Bayes' theorem cannot be readily applied. There have been many attempts at overcoming such problem. For specific instances of the class of models that we are concerned with here, it is feasible to carry out a full fledged semi-parametric Bayesian analysis. For example, Hirano (2002) and Conley et al. (2008) propose semiparametric Bayesian methods for inference in dynamic panel data models with fixed effects and in instrumental variables problems, respectively. It is however difficult to directly broaden the scope of these methods to other models defined through the moment condition equations in (11.1). Always for specific models, in particular linear ones, Zellner (1994) introduced the Bayesian Method of Moments (BMOM); see, for applications and further discussions, Zellner (1991, 1997) and Zellner and Tobias (2001).

A more general approach to carry out Bayesian inference in the estimating equation framework has been proposed by Chamberlain and Imbens (2003). They extend the Bayesian bootstrap of Rubin (1981) to obtain a posterior distribution for θ that is justified by assuming an improper Dirichlet prior on the space of probability distributions. There are however two issues with this approach. The main problem is that it only applies directly to cases in which the dimension of θ is equal to the dimension of the moment function, the exactly identified case. When instead the model is over-identified, that is there are more equations than parameters, the method is only applicable indirectly. Furthermore, since the computation of the posterior relies on solving potentially non-linear equations for each iteration of the bootstrap, the method may become infeasible, especially when the dimension of θ is large.

A recent trend has seen the attempt of replacing the likelihood function in the Bayes' theorem with approximate likelihoods that are transformations of frequentist objective functions. Kim (2002) proposes using as a likelihood a transformation of the efficient GMM objective function. Yin (2009) shows that Kim's likelihood can potentially provide a good starting point for approximate Bayesian inference for certain instances of the class of models in (11.1). Yu and Stander (2007) develop a Bayesian framework for Tobit quantile regressions that is organized around a likelihood function based on the objective function of Powell (1986).

Lazar (2003) investigates the properties of Bayesian posteriors based on the empirical likelihood function. Empirical likelihood (EL) (Owen, 1988, 1990, 2001) acts as a nonparametric analog of the usual likelihood function. Lazar (2003) conducts simulations to asses the coverage validity of EL based posteriors and the frequentist coverage of credible intervals. Although she only considers very simple models, the simulation results allow her to conclude that it is reasonable to use empirical likelihood within the Bayesian paradigm. Schennach (2005) introduces another approximate likelihood which is similar in spirit to that based on empirical likelihood. This likelihood, aptly named Bayesian exponentially tilted empirical likelihood (BETEL), is obtained from an exponential tilted multinomial distribution (DiCiccio and Romano, 1990). Ragusa (2011) considers various other likelihoods that can be thought as extension of the BETEL.

The main objective of the chapter is to explore Bayesian and frequentist properties of inference based on posterior distributions resulting from using these likelihood functions. To be able to do so, we need to be more precise about the formulation of the statistical model under consideration.

11.2 Examples

As said, most models considered in econometrics can be expressed in terms of estimating equations. We provide same examples for which Bayesian inference is not yet developed.

Generalized Linear Model

The generalized linear model provides a unifying framework for the analysis of discrete and continuous outcomes. For the $i = 1, \ldots, n$ subject the outcome y_i is observed together with the corresponding covariate vector X_i. The density function of y_i given X_i takes the form

$$\varphi(Y_i|X_i) = \exp\left\{\frac{Y_i\zeta_i - b(\zeta_i)}{a_i(\phi)} + c(Y_i, \phi)\right\},$$

where ζ_i and ϕ are location and scale parameters, respectively. The linear predictor $\theta^\top X_i$ is linked to the location parameter by a monotone differential function $h(\cdot)$, such that $\zeta_i = h(X_i^\top \theta)$. The conditional mean of Y_i is then given by $\mu_i = E[Y_i|X_i] = b'(\zeta_i)$, where $\mu_i = h^{-1}(X_i^\top \theta)$, and the conditional variance is $v_i = \text{Var}(Y_i|X_i) = b''(\zeta_i)a_i(\phi)$. The population score equation satisfies (11.1) with

$$g(Y_i, X_i, \theta) = \frac{\partial \mu_i}{\partial \theta} v_i^{-1}(Y_i - \mu_i).$$

Linear Instrumental Variables

Consider the linear model

$$Y = X^\top \theta + u,$$

where the conditional mean of u given X is not constant. The parameter θ can be estimated consistently if there is a vector of instruments Z that is orthogonal to u and correlated with X. In this case, the model boils down to the moment function

$$g(Y, X, Z, \theta) = Z^\top (Y - X^\top \theta).$$

The linear instrumental variable model is overidentified if the dimension of Z is larger than the dimension of θ.

Quantile Instrumental Variables

Let D be a binary treatment indicator. There is a potential outcome, Y_D, associated with the potential treatment status $d \in \{0, 1\}$, $Y_{d=1}$ and $Y_{d=0}$. The outcome is latent because for each individual unit we only observe either $Y_{d=1}$ or $Y_{d=0}$. The quantile treatment response (QTR) function is

$$q(d, x, \tau) = d\,\theta_1(\tau) + x^\top \theta_2(\tau),$$

which represents the τth quantile of potential outcomes under various treatments d, conditional on observed characteristics $X = x$. The quantity of interest is the quantile treatment effect (QTE)

$$q(1, x, \tau) - q(0, x, \tau) = \theta_1(\tau), \quad \tau \in (0, 1),$$

which is the differential impact of treatment on different quantiles of Y. If the treatment is endogenous, Chernozhukov and Hansen (2005) show that the QTR can be identified by a conditional moment restriction

$$E\left[\mathbf{1}_{[Y_d \leqslant D\theta_1(\tau) + X^\top \theta_2(\tau)]} - \tau | X, Z\right] = 0,$$

where Z is an instrument that affects D but is independent of potential outcomes. Thus, the model can be casted in the form of (11.1) with the following moment function

$$g(Y_D, X, Z, \theta_1(\tau), \theta_2(\tau)) = \begin{pmatrix} X \\ Z \end{pmatrix} \mathbf{1}_{[Y_d \leqslant D\theta_1(\tau) + X^\top \theta_2(\tau)]}.$$

Side Information

Estimating equations are key when marginal population information about some variables involved in a regression is available. As shown by Imbens and Lancaster (1994), this information can be easily incorporated by imposing moment restrictions. Consider the following linear model:

$$Y = X^\top \theta + u,$$

where $E[u|X] = 0$. Suppose now that information is available about certain moments of the distribution of X. For instance, the researcher knows that

$$h_1^* = E[X|X > c], \quad h_2^* = E[X|X \leqslant c].$$

Knowledge of h_1^* and h_2^* restricts the marginal distribution of X in a non trivial way. The equations above lead to the following set of restrictions:

$$E[\mathbf{1}_{X>c}(h_1^* - X^\top \theta)] = 0, \quad E[\mathbf{1}_{X^\top \leqslant c}(h_2^* - X^\top \theta)] = 0.$$

The moment function corresponding to this model is

$$g(Y, X, \theta) = \begin{bmatrix} X_i(Y_i - X_i^\top \beta) \\ \mathbf{1}_{X_i > c}(h_1^* - X_i^\top \beta) \\ \mathbf{1}_{X_i \leqslant c}(h_2^* - X_i^\top \beta) \end{bmatrix}.$$

11.3 Frequentist Estimation

For a frequentist point of view, the parameter θ is unknown but "fixed." To reflect this view, the model in (11.1) is often complemented by two additional assumptions. First, there exists a "true" parameter $\theta^* \in \Theta$ such that

$$E[g(X, \theta^*)] = 0. \tag{11.2}$$

Second, the model is identified in the sense that the moment conditions only hold at θ^*, or more formally,

$$\|E[g(X, \theta)]\| \neq 0, \quad \text{for all } \theta^* \neq \theta.$$

Given an i.i.d. random sample $\mathbb{X} = \{x_1, x_2, \ldots, x_n\}$ let $g_i(\theta) = g(x_i, \beta)$, $g_n(\theta) = \sum_{i=1}^n g_i(\theta)/n$ and $\widehat{\Omega}(\theta) = \sum_{i=1}^n g_i(\theta) g_i(\theta)'/n$. The objective is to carry out inference about the finite dimensional parameter β_0.

The most common frequentist procedure to estimate θ^* is the Generalized Method of Moments (GMM) of Hansen (1982). The procedure consists in minimizing a weighted quadratic expression in the empirical estimating equations, where the weighting matrix is a consistent estimator of the covariance of $\sum_{i=1}^n g(x_i, \theta)/n$. Formally, the GMM estimator is given by

$$\hat{\theta}^{GMM} = \arg\min_{\theta \in \Theta} J_n(\theta), \quad J_n(\theta) := n \cdot g_n(\theta)' \left[\frac{1}{n} \sum_{i=1}^n g_i(\bar{\theta}) g_i(\bar{\theta})^\top \right]^{-1} g_n(\theta),$$

where $\bar{\theta}$ is a preliminary consistent estimate of θ^*. In a wide array of settings (Hansen, 1982; Gallant and White, 1988; Newey and McFadden, 1994) the GMM estimator is consistent and asymptotically normal.

A number of alternative estimators to the GMM have been proposed in recent years. The motivation for these estimators is often the bias of the two-step GMM estimator and the lack of accuracy of the confidence intervals based on the normal approximation to the distribution of $\hat{\theta}^{GMM}$.

Qin and Lawless (1994) and Imbens (1997) have proposed estimators that embed estimating equations into the empirical likelihood framework (Owen, 1988, 1990). The intuition behind the empirical likelihood is important for understanding the logic of some of the approximate posteriors we will be discussing in the next sections. Suppose that the moment function does not depend on any unknown parameter and takes the form $g(x, \theta) = x$. The condition $E[g(X, \theta)] = 0$ implies that the expected value of X is 0. The empirical distribution function does not satisfy the moment condition, since with probability 1, $\sum_{i=1}^n x_i/n \neq 0$. The idea behind empirical likelihood is to modify the empirical distribution in such a way that it satisfies the moment conditions. As there exist many modifications of the empirical distribution satisfying the moment condition, the idea is to pick the one that is the closest to the empirical one. In other words, among all systems of weights $\{p_i\}_{i=1}^n$ such that $\sum_{i=1}^n p_i x_i = 0$, we pick the one that maximizes $\prod_{i=1}^n p_i$. We can think of this optimal weighting as an estimator of the distribution function of X.

Consider now the general case where the moment condition depends on the unknown parameter θ. In this case, for a given θ, the empirical likelihood estimator of the probability distribution is given by

$$\max_p \prod_{i=1}^n p_i, \text{ subject to } p_i > 0, \quad \sum_{i=1}^n p_i = 1, \quad \sum_{i=1}^n p_i g(x_i, \theta) = 0.$$

The empirical distribution for this given value of θ has a simple solution, namely

$$p_i^{EL} = \frac{1}{n} \frac{1}{1 + \lambda_\theta^\top g(x_i, \theta)}, \tag{11.3}$$

where λ_θ solves the following set of nonlinear equations:

$$\sum_{i=1}^{n} \frac{1}{1+\lambda_\theta^\top g(x_i,\theta)} g(x_i,\theta) = 0. \tag{11.4}$$

Substituting (11.3) into the likelihood, we obtain the "empirical likelihood" at θ:

$$f_{EL}(\mathbb{X}|\theta) = \prod_{i=1}^{n} \frac{1}{n[1+\lambda_\theta^\top g(x_i,\theta)]}. \tag{11.5}$$

The empirical likelihood estimator is obtained by maximizing $f_{EL}(\mathbb{X}|\theta)$ over $\theta \in \Theta$.

The empirical likelihood idea can be generalized by noting that maximizing $\prod_{i=1}^n p_i$ is equivalent to minimizing $-\sum_{i=1}^n \log p_i$. From this point of view, the empirical likelihood operationalizes the concept of closeness between empirical distribution and the system of weights compatible with the moment conditions. As such, other measures of distance can be employed to estimate the distribution of the underlying random variables. Kitamura and Stutzer (1997) propose to use the Kullback-Leibler distance measure. In this case, the estimator of the probability distribution is

$$\min_p \sum_{i=1}^{n} p_i \log p_i, \text{ subject to } p_i \geq 0, \quad \sum_{i=1}^{n} p_i = 1, \quad \sum_{i=1}^{n} p_i g(x_i,\theta) = 0.$$

The solution is then

$$p_i^{ET} = \frac{e^{\lambda_\theta^\top g(x_i,\theta)}}{\sum_{i=1}^n e^{\lambda_\theta^\top g(x_i,\theta)}}, \tag{11.6}$$

and, as for the empirical likelihood case, λ_θ is chosen in such a way to guarantee that

$$\sum_{i=1}^{n} e^{\lambda_\theta^\top g(x_i,\theta)} g(x_i,\theta) = 0. \tag{11.7}$$

Kitamura and Stutzer's exponential tilting estimator of θ^* is then obtained by substituting this expression into the objective function and minimizing over $\theta \in \Theta$.

A generalization of these estimators is obtained by considering, as measure of distance between the empirical distribution and the system of weights, the Cressie-Read divergence, which is given by

$$\frac{1}{n[\alpha(1+\alpha)]} \sum_{i=1}^{n} \left[\left(\frac{1}{np_i} \right)^\alpha - 1 \right],$$

where α is a parameter. In this case, the estimator of the probability distribution is

$$p_i^{CR} = \frac{(1+\alpha\lambda_\theta^\top g(x_i,\theta))^{1/\alpha}}{\sum_{i=1}^n (1+\alpha\lambda_\theta^\top g(x_i,\theta))^{1/\alpha}}$$

and, once again, the estimator of θ^* is obtained by substituting this expression in the objective function and maximizing over $\theta \in \Theta$. The empirical likelihood and the exponential tilting estimate of the probability distribution are special cases which are obtained for $\alpha \to 0$ and $\alpha \to -1$, respectively. As shown by Newey and Smith (2004), when $\alpha = -2$, the Cressie-Read estimator coincides with the Continuous Updating (CUE) estimator of Hansen et al. (1996), which is obtained by minimizing over $\theta \in \Theta$ the following objective function

$$Q_n(\theta) = n \cdot g_n(\theta)^\top S_n^{-1}(\theta) g_n(\theta), \tag{11.8}$$

where

$$S_n(\theta) = \frac{1}{n}\sum_{i=1}^{n} g_i(\theta) g_i(\theta)^\top - \frac{1}{n^2}\sum_{i=1}^{n} g(x_i,\theta)\sum_{i=1}^{n} g(x_i,\theta)^\top.$$

The main difference between the traditional GMM and the CUE is that the weighting matrix is not held fixed, but it is seen as a function of the parameter vector.

The Cressie-Read estimators are all asymptotically equivalent to the two-step GMM estimator. However, as shown by Newey and Smith (2004) and Ragusa (2011), they possess better second order properties, such as smaller bias with respect to the GMM. Furthermore, the empirical likelihood estimator is third order efficient.

11.4 Bayesian Estimation

While the frequentist estimation of models specified by estimating equations is well developed, Bayesian inference is more problematic. Bayesian inference is run by the posterior distribution, which by Bayes' theorem is given by

$$\pi(\theta|\mathbb{X}) = \frac{f(\mathbb{X}|\theta)\pi(\theta)}{\int f(\mathbb{X}|\theta)\pi(\theta)d\theta},$$

where $f(\mathbb{X}|\theta)$ is the likelihood function for the observed datum $\mathbb{X} = \{x_{1,\ldots,}x_n\}$ and $\pi(\theta)$ is the prior density on the parameter θ. In parametric models, the likelihood is the conditional density of the data given the parameter θ and it arises from distributional assumptions. When the model is specified through estimating equations, it is semi-parametric and thus the likelihood is not readily available without further assumptions.

Before digging further into the issue of conducting inference, we should carefully highlight the difference of interpretation of the estimating equations in the Bayesian framework. For a frequentist, the parameter θ is unknown but fixed and it takes a true value, θ^*. Thus, the moment conditions (11.1) hold only at θ^*. From a Bayesian perspective, the parameter θ is random and the expectation should be considered with respect to the conditional distribution of X given θ, P_θ, that is,

$$E[g(X,\theta)|\theta] \equiv \int g(X,\theta)dP_\theta = 0. \tag{11.9}$$

Writing the moment condition this way clarifies that in this class of models the likelihood function is left unspecified and Bayesian inference is problematic. The recent literature has proposed different approaches to obviate the lack of a likelihood function. In what follows, we review these approaches.

11.4.1 Bayesian Bootstrap

Chamberlain and Imbens (2003) extend the Bayesian bootstrap (BB) of Rubin (1981) to semiparametric models by embedding the model in a multinomial framework.[1] To describe the setup, let $a = (a_1, \ldots, a_J)$ be the vector of all possible distinct values of X, and let $\xi = (\xi_1, \ldots, \xi_J)$ be the associated vector of probabilities, $P(X = a_j) = \xi_j$, $\sum_j \xi_j = 1$. Let n_j denote the number of observations equal to a_j. The likelihood of the data, $\mathbb{X} = \{x_1, \ldots, x_n\}$, is proportional to

$$\prod_{j=1}^{J} \xi_j^{n_j}.$$

The prior on ξ is assumed to belong to the Dirichlet family, that is, the prior has density proportional to $\prod_{j=1} \xi_j^{l_j - 1}$, $l_j > 0$. The posterior density of ξ is proportional to

$$\pi(\xi|\mathbb{X}) \propto \prod_{j=1}^{J} \xi_j^{n_j + l_j - 1}, \quad l_j > 0,$$

which is as well a Dirichlet with parameters $n_j + l_j$, $j = 1, \ldots, J$. So far we have not considered the estimating equation in (11.9). The idea of Chamberlain and Imbens (2003) is to turn the posterior for ξ into a posterior distribution for the parameter θ by drawing θ's which are consistent with $\pi(\xi|\mathbb{X})$. To get the intuition, suppose ξ^s is a draw from $\pi(\xi|\mathbb{X})$; associated with this draw of ξ, there is a value of the parameter θ, say θ^s, such that

$$\sum_{j=1}^{J} g(a_j, \theta^s)\xi_j^s = 0.$$

All the θ^s consistent with all the possible draws from $\pi(\xi|\mathbb{X})$ are draws from the posterior of θ. Chamberlain and Imbens (2003) consider drawing from this

[1] Bayesian bootstrap in nonparametric settings has also been considered by Ferguson (1973, 1974) and Gasparini (1995). Hahn (1997) studies the frequentist properties of Bayesian Bootstrap for the quantile regression case and Lancaster (1994) applies Bayesian bootstrap to the analysis of choice-based sampling. Poirier (2011) utilizes the Bayesian bootstrap to provide a Bayesian interpretation of the heteroskedastic consistent covariance estimator of White (1980).

posterior when $l_j \to 0$ for all j, in which case support points not observed in the sample receive posterior probability 0. In this case, to simulate from the posterior distribution of θ we can repeatedly draw exponential i.i.d. random variables, V_i^s, $i = 1, \ldots, n$, and solve

$$\sum_{i=1}^{n} V_i^s g(x_i, \theta^s) = 0.$$

Repeating this for $s = 1, \ldots, S$, gives us S independent draws from the posterior distribution of θ.

As pointed out by Lancaster (2004), there are few issues with this approach. First, it is not clear how to incorporate prior knowledge about θ, as the prior on θ is implicitly elicited. Second, drawings are obtained by solving a potentially high-dimensional nonlinear set of equations which can result in computational problems. Third, when the model is overidentified, that is when the dimension of the parameter space is smaller than the dimension of the moment function, the method can be applied only after projecting out the redundant equations. The projection takes place outside the model and hence the resulting inference might be arbitrary.

11.4.2 GMM-Based Likelihoods

Kim (2002) proposes conducting Bayesian inference by using a transformation of the GMM objective function. The procedure is, at least conceptually, extremely simple as it involves using the Bayes' formula replacing the likelihood with a transformation of the objective function $J_n(\theta)$. More specifically, for

$$f_{GMM}(\mathbb{X}|\theta) = e^{-\frac{1}{2}J_n(\theta)},$$

inference is based on

$$\pi_{GMM}(\theta|\mathbb{X}) \propto f_{GMM}(\mathbb{X}|\theta)\pi(\theta).$$

Kim (2002) motivates this choice of a likelihood by embedding the moment condition in an information theoretical framework. However, the justifications for using $\pi_{GMM}(\theta|\mathbb{X})$ are only asymptotic and no "small sample" justification is provided.

An approach similar in spirit to the one just presented is the one by Yin (2009). He considers using as posterior

$$\pi_{CU}(\theta|\mathbb{X}) \propto f_{CU}(\mathbb{X}|\theta)\pi(\theta),$$

where the likelihood function, $f_{CU}(\mathbb{X}|\theta)$, is the exponential transformation of the CUE objective function in (11.8):

$$f_{CU}(\mathbb{X}|\theta) = e^{-\frac{1}{2}Q_n(\theta)}.$$

The same approach was suggested by Chernozhukov and Hong (2003) who considered quasi posterior inference from a frequentist point of view, mainly as a computational device to construct consistent and asymptotically normal estimators of parameters defined through moment conditions.

These likelihoods can be motivated asymptotically by noticing that under fairly general conditions, as $n \to \infty$,

$$\frac{1}{n}\sum_{i=1}^{n} g(X_i, \theta) \xrightarrow{d} N\left(0, S(\theta)/n\right), \quad S(\theta) = E\left[g(X_i, \theta)g(X_i, \theta)^\top\right].$$

The CU likelihood can be thought as being based on the limit distribution of the empirical estimating equation when the normalizing constant $(2\pi)^{-k/2}|S^{-1}(\theta)|^{1/2}$ is not taken into account. This suggests a modified version of the CU likelihood which also uses the empirical normalizing constant,

$$f_{NCU}(\mathbb{X}|\theta) = (2\pi)^{-m/2}|S_n^{-1}(\theta)|^{1/2}e^{-\frac{1}{2}Q_n(\theta)},$$

where $S(\theta)$ is replaced by its sample counterpart $S_n(\theta)$.

While we will carefully discuss the issue of simulating from $\pi_{GMM}(\theta|\mathbb{X})$ or $\pi_{CU}(\theta|\mathbb{X})$ in a later section, it is instructive to compare these two posterior distributions with the "true" posterior for one simple example. To do so, we consider data generated from the following model $Y|X = x \sim N(x\theta, 1)$, $X \sim N(0, 1)$, and $\theta = 0$. The parameter θ follows a normal distribution centered at 0 and variance $\tau^2 = \sqrt{.2}$. In this case, the exact posterior distribution of θ is normal with mean $\left(\sum_{i=1}^n x_i^2 + \tau^{-2}\right)^{-1}\sum_{i=1}^n x_i y_i$ and variance $\left(\sum_{i=1}^n x_i^2 + \tau^{-2}\right)^{-1}$. The posterior distributions $\pi_{GMM}(\theta|\mathbb{X})$, $\pi_{CU}(\theta|\mathbb{X})$, and $\pi_{NCU}(\theta|\mathbb{X})$ are calculated using the following estimating equations:

$$g(Y, X, \theta) = X^\top(Y - X^\top\theta).$$

Figure 11.1 plots the posterior distributions for different draws from the model and for different sample sizes. As it can be seen, when the sample size is large all the three posterior distributions based on the estimating equations are close to the exact posterior, especially near the tails. At smaller sample sizes, the three posteriors put more mass toward the center, with a marked tendency of the posterior based on the GMM to have the thinnest tails. Whether this issue is specific to the linear model or it is a standard feature of the posteriors will be briefly discussed in a later section.

11.4.3 Empirical Likelihood-Type Posteriors

As discussed in Section 11.3, the empirical likelihood behaves for some purposes as a parametric likelihood. For instance, the maximum empirical likelihood is consistent, asymptotically normal, and like the parametric MLE is third order efficient in the sense of Pfanzagl and Wefelmeyer (1978) (see Newey

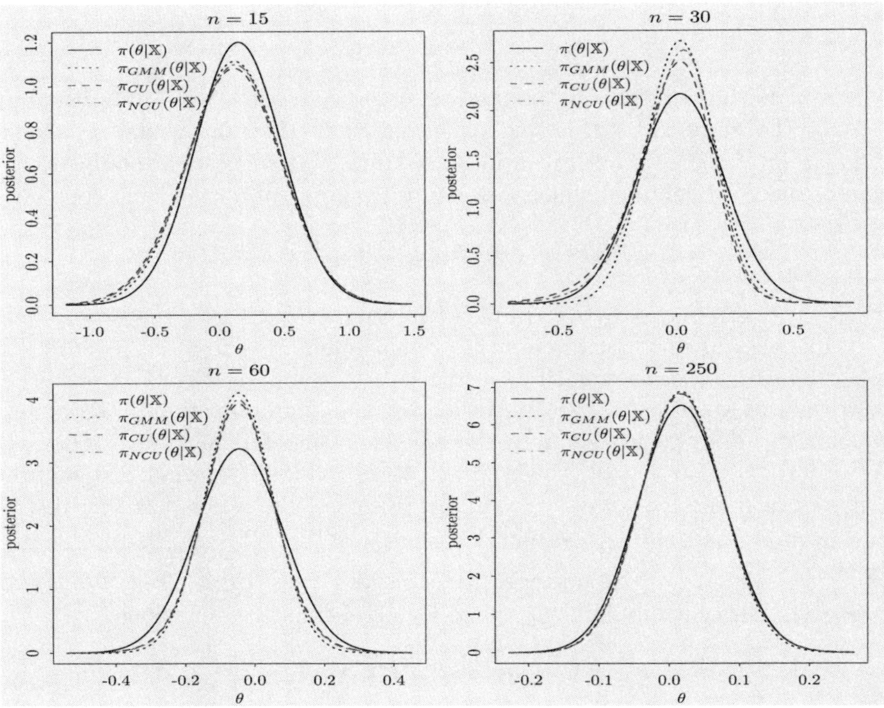

Figure 11.1: GMM, CU, and normalized CU based posteriors. The posterior distributions are based on data generated from the following model: $Y|X = x \sim N(x\theta, 1)$, $X \sim N(0,1)$, $\theta = 0$. The prior is $\theta \sim N(0, \tau^2)$, $\tau = \sqrt{.2}$. The posterior distributions $\pi_{GMM}(\theta|\mathbb{X})$, $\pi_{CU}(\theta|\mathbb{X})$, and $\pi_{NCU}(\theta|\mathbb{X})$ are calculated from the estimating equation $g(Y, X, \theta) = X^\top(Y - X^\top\theta)$. The normalizing constants are calculated by numerical integration.

and Smith, 2004; Ragusa, 2011). The empirical likelihood is also Bartlett correctable (DiCiccio et al., 1991; Chen and Cui, 2007). It is natural then to try to go a step further and see whether the empirical likelihood can be used in the Bayesian framework.

Lazar (2003) follows up on this simple idea and proposes using the empirical likelihood instead of the conditional distribution of the data in the Bayes' theorem. This amounts to consider the following posterior distribution:

$$\pi_{EL}(\theta|\mathbb{X}) \propto f_{EL}(\mathbb{X}|\theta)\pi(\theta),$$

where the expression for $f_{EL}(\mathbb{X}|\theta)$ is given in equation (11.5). Instead of using the empirical likelihood, Schennach (2005) suggests a posterior distribution based on the exponentially tilted empirical distribution function. More formally, Schennach's posterior is given by

$$\pi_{ET}(\theta|\mathbb{X}) \propto f_{ET}(\mathbb{X}|\theta)\pi(\theta), \quad f_{ET}(\mathbb{X}|\theta) = \prod_{i=1}^{n} p_i^{ET}, \tag{11.10}$$

where p_i^{ET}, $i = 1, \ldots, n$, are defined in (11.6). Following the same idea, other posteriors could potentially be constructed by using other Cressie-Read empirical distributions.

Calculating the empirical likelihood and the exponentially tilted posteriors for a given value of θ amounts to calculating the weights and solving (11.4) and (11.7) for λ, respectively. At certain values of the parameter, the nonlinear equations might not have a solution. In this case, the value of the likelihood is 0, since there exists no system of weights that sets the empirical moment condition to zero.

Figure 11.2 shows $\pi_{EL}(\theta|\mathbb{X})$ and $\pi_{ET}(\theta|\mathbb{X})$ for the same model and the same data for which we calculated the posteriors presented in Section 11.4.2. A comparison with the distributions depicted in Figure 11.1 reveals that, at least for the regression model considered here, empirical likelihood and exponential tilting posteriors behave similarly to those based on GMM, CU, and the normalized CU.

11.5 Simulating from the Posteriors

While the posterior distributions described in the previous section can be calculated quite easily point-wise over the space Θ, to calculate their moments one needs to resort to Monte Carlo simulation techniques. In principle any algorithm that allows simulating from a distribution that can be computed up to a proportionality constant can be employed. In this section, we restrict the discussion to the Metropolis-Hastings (MH) algorithm. Keep in mind that a detailed discussion of the MH algorithm and its properties is beyond the scope of the present chapter; for specific issues and implementation details see, among others, Robert and Casella (2004).

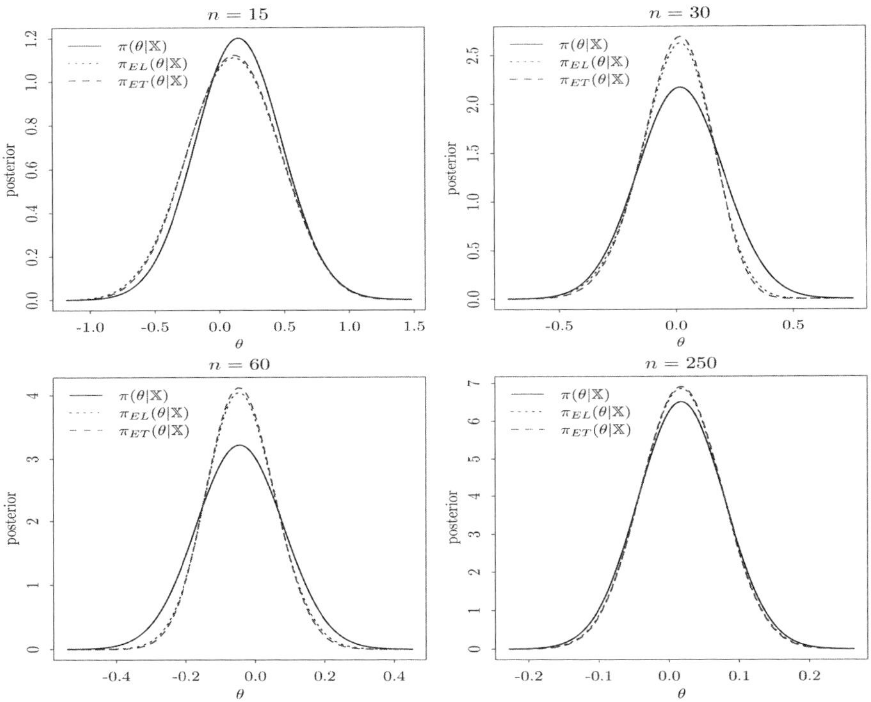

Figure 11.2: Empirical likelihood based posteriors. The posterior distributions are based on data generated from the following model: $Y|X = x \sim N(x\theta, 1)$, $X \sim N(0, 1)$, $\theta = 0$. The prior is $\theta \sim N(0, \tau^2)$, $\tau = \sqrt{.2}$. The posterior distributions $\pi_{EL}(\theta|\mathbb{X})$, $\pi_{ET}(\theta|\mathbb{X})$ are calculated from the estimating equation $g(Y, X, \theta) = X^\top(Y - X^\top\theta)$.

The MH algorithm is a method to produce Markov chains with a given ergodic distribution. The output is a dependent sample $(\theta^{(1)}, \ldots, \theta^{(S)})$ whose empirical distribution approaches, as $S \to \infty$, $\pi_j(\theta|\mathbb{X})$, and the index j ranges over the different posteriors $j = \{GMM, CU, NCU, EL, ET\}$.[2] The ergodicity of the chain implies that

$$\frac{1}{S}\sum_{s=1}^{S} h(\theta^s) \xrightarrow[s\to\infty]{p} \int h(\theta)\pi_j(\theta|\mathbb{X})d\theta.$$

Given the posterior $\pi_j(\theta|\mathbb{X})$ and a pre-specified density $q(\theta|\theta')$, a Markov chain with limiting distribution $\pi_j(\theta|\mathbb{X})$ can be constructed iteratively as follows:

- Choose a starting value $\theta^{(0)}$
- Generate $\tilde{\theta} \sim q(\theta|\theta^{(s)})$ and calculate

$$\alpha = \inf\left\{\frac{\pi_j(\tilde{\theta}|\mathbb{X})}{\pi_j(\theta^{(s)}|\mathbb{X})}\frac{q(\theta^{(s)}|\tilde{\theta})}{q(\tilde{\theta}|\theta^{(s)})}, 1\right\}$$

- Set

$$\theta^{(s+1)} = \begin{cases} \tilde{\theta} & \text{with probability } \alpha \\ \theta^{(s)} & \text{with probability } 1-\alpha \end{cases}.$$

A common choice for the proposal density is

$$q(\theta|\theta') = \xi(|\theta - \theta'|),$$

where $\xi(\cdot)$ is a density symmetric about the origin, such as the Gaussian, the Student-t, or the Cauchy density. In this case, the Markov chain is a random walk and the MH algorithm is called the random walk Metropolis-Hastings (RWMH). The advantage of RWMH over the standard HM algorithm is that it allows a local exploration of the neighborhood of the current value of the Markov chain, since the candidate distribution depends on the current state of the chain.

11.6 Asymptotic Theory

Under well-known regularity conditions, standard Bayesian inference has a well defined asymptotic behavior. Ibragimov and Has'minskii (1981) have shown that in regular and correctly specified models the posterior distribution approaches a normal density. From these results it follows that posterior

[2]Since most of the discussion in this section is valid for all the posterior distributions presented in Section 11.4 with the exception of the Bayesian bootstrap, we will refer to them generically with $\pi_j(\theta|\mathbb{X})$.

mean, posterior median and other symmetric moments of the posterior are consistent for the "true" parameter. Chernozhukov and Hong (2003) have extended these results to the setting considered in this chapter. In particular, they have developed an asymptotic theory for quasi-posterior distributions constructed from models whose parameters are specified through estimating equations.

Their framework can be easily adapted to study the asymptotic behavior of the following generalized Bayesian estimator

$$\hat{\theta}^{\rho} = \text{arginf}_{\xi} \sqrt{n} \int \rho(\theta - \xi)\pi_j(\theta|\mathbb{X})\, d\theta, \tag{11.11}$$

where $\rho(u)$ is a non-negative symmetric convex criterion function normalized in such a way that $\rho(0) = 0$ only when $u = 0$, $\rho(u) \leq 1 + |u|^p$, for some $p \geq 1$, and $\int \rho(u - \xi) \exp(-u^\top au)du$ is uniquely minimized at some point $\xi \in \mathbb{R}^k$. For instance, $\hat{\theta}^{\rho}$ is posterior mean for $\rho(u) = u^2$; $\rho(u) = \sum_{j=1}^{k} |u_j|$ gives instead the posterior median.

Chernozhukov and Hong (2003) show that, as $n \to \infty$, $\pi_{CU}(\theta|\mathbb{X})$ concentrates on H_n, a $1/\sqrt{n}$ neighborhood of θ^*; on this neighborhood the CU posteriors is approximately normal. Formally, the result can be stated as

$$\int_{H_n} |\pi_{CU}(u|\mathbb{X}) - \pi_{\infty}(u)|\, du \xrightarrow{p} 0,$$

where

$$\pi_{\infty}(u|\mathbb{X}) = \sqrt{\frac{|V|}{(2\pi)^k}} \exp(-u^\top Ju/2).$$

A consequence of the normality of the posterior is that the estimators defined in (11.11) are asymptotically normal, that is,

$$V^{-1}\sqrt{n}(\hat{\theta}^{\rho} - \theta^*) \xrightarrow{d} N(0, I_k),$$

where

$$V = \left(\frac{\partial Eg(X, \theta^*)}{\partial \theta^\top}\right) E\left[g(X, \theta^*)g(X, \theta^*)^\top\right] \left(\frac{\partial Eg(X, \theta^*)}{\partial \theta}\right).$$

These results are proved under very general conditions that do not require differentiability of $g(\cdot, \theta)$ by using empirical process theory and Law of Large Numbers and Central Limit Theorem for Donsker classes. When g is differentiable, V coincides with the semiparametric efficiency bound of Chamberlain (1987).

The results of Chernozhukov and Hong (2003) can be extended to the other posteriors discussed in Section (11.4) by showing that on the same shrinking neighborhood of θ^* all the distributions are close to $\pi_{CU}(\theta|\mathbb{X})$. Indeed, it can be shown that

$$\lim_{n\to\infty} \int_{H_n} |\pi_j(\theta|\mathbb{X}) - \pi_{CU}(\theta|\mathbb{X})| = o_p(1), \quad j = \{GMM, NCU, EL, ET\},$$

which implies that the asymptotic analysis of Chernozhukov and Hong (2003) extend to the other posterior distributions.

11.7 Bayesian Validity

If the approximate posteriors are to be used from a Bayesian point of view, it is important to asses whether they provide valid posterior inference. Strictly speaking, a posterior is valid if the likelihood function is based on the conditional density of the data given the parameters. If one takes such a narrow view, the posterior distributions of Section (11.4) are clearly all invalid and so is the resulting inference.

Monahan and Boos (1992) introduce a milder concept of validity for posteriors based on likelihood functions that are different from the conditional distribution of the data given the parameter. Under this definition, posterior inference is valid if the likelihood function satisfies a simple coverage condition.

Definition 1 (Validity by Coverage) *A likelihood function $f(\mathbb{X}|\theta)$ is valid by coverage if, for every continuous proper prior distribution $\pi(\theta)$, the statistic*

$$a(\mathbb{X},\theta) = \int_{-\infty}^{\theta} \frac{f(\mathbb{X}|u)\pi(u)}{f(\mathbb{X})} du,$$

is unconditionally distributed as $\mathrm{Unif}(0,1)$.

The rationale behind this definition of validity is that if the posterior is based on the conditional density of the data, then $a(\mathbb{X},\theta)$ is unconditionally $\mathrm{Unif}(0,1)$ for every continuous prior distribution. To see this, let $f^*(\mathbb{X}|\theta)$ denote the "true" conditional density. Then, the joint density of $(\mathbb{X},\beta)$ is $f^*(\mathbb{X},\beta) = f^*(\mathbb{X}|\theta)\pi(\theta)$, and thus

$$\begin{aligned}
\Pr\big(a(x,\theta) \leqslant z\big) &= \int\int 1(a(\mathrm{x},u) \leq z) f^*(x,u)dud\mathrm{x} \\
&= \int\int 1(a(\mathrm{x},u) \leq z)\pi(u|\mathrm{x})duf^*(x)dx \\
&= \int z f^*(x)dx = z.
\end{aligned}$$

Examining the deviations of $a(\mathbb{X},\theta)$ from $\mathrm{Unif}(0,1)$ can be thought as a *goodness of fit* test and the distance between the two distributions of $a(\mathrm{x},\beta)$ provides a metric to asses the quality of resulting the Bayesian inference. The validity by coverage concept leads to a simulation method which consists in generating θ^s, $s = 1,\ldots,S$, independently from the pre-specified prior $\pi(\theta)$ and then generate data $x^{(s)}$ according to $f(\cdot|\theta^s)$ and compute the statistic $a(\mathrm{x}^{(s)},\theta^{(s)})$. The quality of the nonparametric likelihoods can be gauged either graphically—by means of quantile-quantile plots—or formally—by calculating Kolmogorov-Smirnov distances from uniformity of the distribution of a_k.

We study the performance of the posteriors derived in Section 11.4 for a two moment, one parameter problem. The moment vector is

$$g(X,\theta) = \begin{pmatrix} X-\theta \\ X^2-\theta^2-2\theta \end{pmatrix}. \tag{11.12}$$

The distribution of X is $N(\theta,1)$. Given n realizations of X, parametric Bayesian inference about θ is based on a posterior proportional to

$$e^{-\sum_{i=1}^{n}(\theta-x_i)^2/2}\pi(\theta).$$

Instead of the true normal likelihood, we consider using the nonparametric likelihoods based on EL, ET and CUE implied probabilities.

As a summary, Table 11.1 gives the Kolmogorov-Smirnov distances from uniformity of the distribution of $a(\mathbb{X},\theta)$ for the case of a uniform prior with different variances. For all of cases considered, the distributions of $a(\mathbb{X},\beta)$ for $f_{EL}(\mathbb{X},\theta)$, $f_{ET}(\mathbb{X},\theta)$ and $f_{CU}(\mathbb{X},\theta)$ are statistically indistinguishable from Unif$(0,1)$ when n is larger than 50. For $f_{NCU}(\mathbb{X},\theta)$, the Kolmogorov-Smirnov null hypothesis of uniformity is rejected at $n=\{20,50\}$ for the more disperse priors.

The results of this simple simulation exercise suggest that it is feasible to consider a Bayesian inferential procedure based on the posteriors discussed in this chapter. Of course, some caveats should be kept in mind. First, the set of simulations is limited: only uniform priors with different variances have been considered. The concept of validity by coverage requires that the statistics $a(\mathbb{X},\theta)$ be uniform for all possible priors. We used a uniform prior because it simplifies the numerical integrations required to calculate the quantities involved in the calculation of $a(\mathbb{X},\theta)$. Second, we are considering a very simple model and it is possible that for more involved models the posteriors could be easily invalidated.

11.8 Application

In this section we conduct an empirical investigation using the approximate posteriors discussed so far. We will focus on a reanalysis of data from the Pennsylvania Reemployment Bonus Demonstration described in detail in Corson et al. (1992). The data set collects information about randomized controlled experiment conducted in Pennsylvania by the U.S. Department of Labor between July 1988 and October 1989. The objective of the experiment was to test the effectiveness of cash bonuses paid for early reemployment in shortening the length of insured unemployment spells. During the enrollment period, claimants who became unemployed and registered for unemployment benefits in one of the 12 selected local offices throughout the state were randomly assigned either to a control group or an experimental treatment groups.

Table 11.1: (Overidentified location model) Kolmogorov-Smirnov distances of the empirical distributions of $a(\mathrm{x}, \beta)$ for $\mathscr{L}^{el}$, $\mathscr{L}^{et}$, and $\mathscr{L}^{cue}$ with uniform prior centered at 0 and variances .33, 1.33, and 3, sample sizes of $n = \{20, 50, 100\}$, and simulation size of $S = 1,500$. *p-values* are reported in parentheses.

			$\sup_t \|F_n(t) - (t)\|$		
			n = 20	n = 50	n = 100
Uniform Prior	$Var = .33$	$f_{EL}(\mathbb{X}\|\theta)$	0.040	0.024	0.023
			(0.018)	*(0.333)*	*(0.423)*
		$f_{ET}(\mathbb{X}\|\theta)$	0.060	0.023	0.018
			(0.000)	*(0.3854)*	*(0.710)*
		$f_{CU}(\mathbb{X}\|\theta)$	0.047	0.036	0.026
			(0.0000)	*(0.042)*	*(0.256)*
		$f_{NCU}(\mathbb{X}\|\theta)$	0.078	0.030	0.025
			(0.000)	*(0.148)*	*(0.311)*
	$Var = 1.33$	$f_{EL}(\mathbb{X}\|\theta)$	0.050	0.025	0.015
			(0.001)	*(0.312)*	*(0.895)*
		$f_{ET}(\mathbb{X}\|\theta)$	0.050	0.03	0.026
			(0.001)	*(0.147)*	*(0.264)*
		$f_{CU}(\mathbb{X}\|\theta)$	0.072	0.035	0.024
			(0.000)	*(0.044)*	*(0.333)*
		$f_{NCU}(\mathbb{X}\|\theta)$	0.070	0.047	0.029
			(0.0002)	*(0.002)*	*(0.180)*
	$Var = 3$	$f_{EL}(\mathbb{X}\|\theta)$	0.055	0.0290	0.0288
			(0.000)	*(0.159)*	*(0.165)*
		$f_{CU}(\mathbb{X}\|\theta)$	0.0661	0.0230	0.029
			(0.000)	*(0.405)*	*(0.139)*
		$f_{CU}(\mathbb{X}\|\theta)$	0.070	0.021	0.019
			(0.000)	*(0.470)*	*(0.640)*
		$f_{NCU}(\mathbb{X}\|\theta)$	0.057	0.050	0.029
			(0.000)	*(0.001)*	*(0.150)*

In the control group the existing rules of the unemployment insurance system applied. Individuals in the treatment groups were offered a cash bonus if they became reemployed in a full-time, working more than 32 hours per week. A detailed description of the characteristics of claimants under study is presented in Koenker and Bilias (2001) which has information on age, race, gender, number of dependents, location in the state, existence of recall expectations, and type of occupation.

We follow Koenker and Xiao (2002) and we model the log-duration of the individual UI benefits as a linear conditional quantile function

$$Q_{\log(T)}(\tau|D,X) = D\beta(\tau) + X^\top\theta(\tau),$$

where the vector X contains demographic characteristics of the subject (sex, race, age indicators and number of dependents) and other variables that are relevant to the experiment and D is the treatment indicator. The parameter of interest is $\beta(\tau)$, that can be interpreted as the treatment effect at quantile τ. For brevity, we will concentrate on the median case, that is $\tau = 0.5$.

Frequentist inference is easily obtained by applying the methodology of Koenker and Bassett Jr (1978). In particular, the point estimate of $\beta(.5)$ is -0.1279, with a standard error of 0.043, numbers that indicates a statistically significant effect of the cash bonus on the UI benefits duration of about 14%. Approximate Bayesian inference can be obtained by using the posteriors based on the following estimating equations

$$E\left[\begin{pmatrix} D \\ X \end{pmatrix} \mathbf{1}_{\log(T)-D\beta(\tau)-X^\top\theta(\tau)} - \tau\right] = 0,$$

and we put a uniform independent prior $U(-10,10)$ on all the parameters of the model. We simulate from $\pi_{EL}(\beta|\mathbb{X})$, $\pi_{ET}(\beta|\mathbb{X})$, $\pi_{CU}(\beta|\mathbb{X})$, and $\pi_{NCU}(\beta|\mathbb{X})$ using the Metropolis-Hastings algorithm described in Section 11.5.

In Figure 11.3 we present a concise visual representation of the results of the model. Each of the panels of the figure illustrate the histogram of ten thousands draws from the posterior draws of $\beta(.5)$ for each posterior considered. Posterior medians, means, and the highest probability credible intervals are summarized in Table 11.2. All posteriors give basically very similar inference, the main difference being the credible regions that are wider for π_{CU} and π_{NCU}. Notice also, that these credible intervals are also shorter than the confidence interval based on the frequentist point estimator and the normal calibration of the asymptotic distribution of the quantile regression estimator.

11.9 Conclusions

This chapter has surveyed different approximate posteriors to conduct Bayesian inference. These posteriors are useful when the only information about the

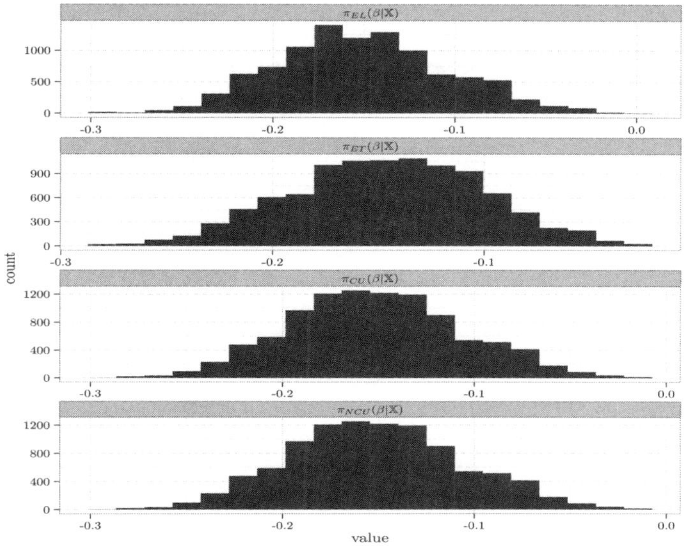

Figure 11.3: Pennsylvania reemployment bonus demonstration. Posterior inference for the median treatment effect of cash bonus on unemployment duration.

Table 11.2: Pennsylvania reemployment bonus demonstration. Posterior inference for the median treatment effect of cash bonus on unemployment duration.

	Mean	Median	lower	upper
$\pi_{EL}(\beta\|\mathbb{X})$	-0.143	-0.145	-0.227	-0.049
$\pi_{ET}(\beta\|\mathbb{X})$	-0.149	-0.138	-0.225	-0.049
$\pi_{CU}(\beta\|\mathbb{X})$	-0.144	-0.146	-0.223	-0.057
$\pi_{NCU}(\beta\|\mathbb{X})$	-0.144	-0.146	-0.223	-0.056

parameters of interest is expressed in terms of estimating equations. These posteriors are based on likelihood functions that use ideas behind frequentist estimations. These likelihoods have, at least in a simple model, good Bayesian and frequentist properties. We have illustrated their usefulness in a simple application using data from Pennsylvania reemployment bonus demonstration using data from the U.S. Department of Labor as in Koenker and Bilias (2001).

Bibliography

Chamberlain, G. (1987). Asymptotic efficiency in estimation with conditional moment restrictions. *Journal of Econometrics 34*(3), 305–334.

Chamberlain, G. and G. W. Imbens (2003, January). Nonparametric applications of Bayesian inference. *Journal of Business and Economic Statistics 21*(1), 12–18.

Chen, S. X. and H. Cui (2007). On the second-order properties of empirical likelihood with moment restrictions. *Journal of Econometrics 141*(2), 492–516.

Chernozhukov, V. and C. Hansen (2005). An IV model of quantile treatment effects. *Econometrica 73*(1), 245–261.

Chernozhukov, V. and H. Hong (2003, August). An MCMC approach to classical estimation. *Journal of Econometrics 115*(2), 293–346.

Conley, T. G., C. B. Hansen, R. E. McCulloch, and P. E. Rossi (2008). A semi-parametric Bayesian approach to the instrumental variable problem. *Journal of Econometrics 144*(1), 276–305.

Corson, W., P. Decker, S. Dunstan, and S. Keransky (1992). Pennsylvania 'reemployment bonus' demonstration: Final report. Technical report, Unemployment Insurance Occasional Paper, 92-1, Dept of Labor, Employment and Training Administration, Washington, DC.

DiCiccio, T., P. Hall, and J. Romano (1991). Empirical likelihood is Bartlett-correctable. *The Annals of Statistics 19*(2), 1053–1061.

DiCiccio, T. and J. Romano (1990). Nonparametric confidence limits by resampling methods and least favorable families. *International Statistical Review 58*(1), 59–76.

Ferguson, T. (1973). A Bayesian analysis of some nonparametric problems. *The Annals of Statistics 1*, 209–230.

Ferguson, T. (1974). Prior of distributions on spaces of probability measures. *The Annals of Statistics 2*, 615–629.

Gallant, R. A. and H. White (1988). *A Unified Theory of Estimation and Inference for Nonlinear Dynamic Models.* New York: Basil Blackwell.

Gasparini, M. (1995). Exact multivariate Bayesian bootstrap distributions of moments. *The Annals of Statistics 23*(3), 762–768.

Hahn, J. (1997). Bayesian bootstrap of the quantile regression estimator: A large sample study. *International Economic Review 38*(4), 206–213.

Hansen, L. P. (1982, July). Large sample properties of generalized method of moments estimators. *Econometrica 50*(4), 1029–54.

Hansen, L. P., J. Heaton, and A. Yaron (1996, July). Finite-sample properties of some alternative GMM estimators. *Journal of Business and Economic Statistics 14*(3), 262–80.

Hirano, K. (2002). Semiparametric Bayesian inference in autoregressive panel data models. *Econometrica 70*(2), 781–799.

Ibragimov, I. A. and R. Z. Has'minskii (1981). *Statistical Estimation: Asymptotic Theory.* New York: Springer-Verlag.

Imbens, G. W. (1997, July). One-step estimators for over-identified generalized method of moments models. *Review of Economic Studies 64*(3), 359–83.

Imbens, G. W. and T. Lancaster (1994). Combining micro and macro data in microeconometric models. *The Review of Economic Studies 61*(4), 655–680.

Kim, J. Y. (2002, March). Limited information likelihood and Bayesian analysis. *Journal of Econometrics 107*(1-2), 175–93.

Kitamura, Y. and M. Stutzer (1997, July). An information-theoretic alternative to generalized method of moments estimation. *Econometrica 65*(4), 861–74.

Koenker, R. and G. Bassett, Jr. (1978). Regression quantiles. *Econometrica 46*(1), 33–50.

Koenker, R. and Y. Bilias (2001). Quantile regression for duration data: A reappraisal of the Pennsylvania reemployment bonus experiments. *Empirical Economics 26*(1), 199–220.

Koenker, R. and Z. Xiao (2002). Inference on the quantile regression process. *Econometrica 70*(4), 1583–1612.

Lancaster, T. (1994). Bayes WESML: Posterior inference from choice-based samples. unpublished manuscript, Brown University, Providence, RI.

Lancaster, T. (2004). *An Introduction to Modern Bayesian Econometrics*. Wiley-Blackwell.

Lazar, N. A. (2003). Bayesian empirical likelihood. *Biometrika 90*(2), 319–326.

Monahan, J. F. and D. Boos (1992). Proper likelihoods for Bayesian analysis. *Biometrika 79*, 271–278.

Newey, W. K. and D. McFadden (1994). Estimation and inference in large samples. In R. Engle and D. McFadden (Eds.), *Handbook of Econometrics*, Amsterdam, pp. 2113–2245. North-Holland.

Newey, W. K. and R. J. Smith (2004). Higher order properties of GMM and generalized empirical likelihood estimators. *Econometrica 72*(1), 219–55.

Owen, A. (1990). Empirical likelihood ratio confidence regions. *The Annals of Statistics 18*(1), 90–120.

Owen, A. (2001). *Empirical Likelihood*, Volume 92. Chapman & Hall/CRC.

Owen, A. B. (1988). Empirical likelihood ratio confidence intervals for a single functional. *Biometrika 75*, 237–49.

Pfanzagl, J. and W. Wefelmeyer (1978). A third-order optimum property of the maximum likelihood estimator. *Journal of Multivariate Analysis 8*(1), 1–29.

Poirier, D. J. (2011). Bayesian interpretations of heteroskedastic consistent covariance estimators using the informed Bayesian bootstrap. *Econometric Reviews 30*(4), 457–468.

Powell, J. (1986). Censored regression quantiles. *Journal of Econometrics 32*(1), 143–155.

Qin, J. and J. Lawless (1994). Empirical likelihood and general estimating equations. *Annals of Statistics 22*, 300–325.

Ragusa, G. (2011). Minimum divergence, generalized empirical likelihoods, and higher order expansions. *Econometric Reviews 30*(4), 406–456.

Robert, C. P. and G. Casella (2004). *Monte Carlo Statistical Methods* (Second ed.). Springer.

Rubin, D. (1981). Bayesian bootstrap. *Annals of Statistics 9*, 130–34.

Schennach, S. C. (2005). Bayesian exponentially tilted empirical likelihood. *Biometrika 92*(1), 31–46.

Sims, C. A. (2002). Interview with Christopher A. Sims. *Journal of Business and Economic Statistics 20*(4), 448–49.

White, H. (1980). A heteroskedasticity-consistent covariance matrix estimator and a direct test for heteroskedasticity. *Econometrica 48*(4), 817–838.

Yin, G. (2009). Bayesian generalized method of moments. *Bayesian Analysis 4*(2), 191–207.

Yu, K. and J. Stander (2007). Bayesian analysis of a Tobit quantile regression model. *Journal of Econometrics 137*(1), 260–276.

Zellner, A. (1991). Bayesian methods and entropy in economics and econometrics. *Maximum Entropy and Bayesian Methods*, 17–31.

Zellner, A. (1994). Bayesian method of moments/instrumental variables analysis of mean and regression models. Manuscript, Graduate School of Business, University of Chicago.

Zellner, A. (1997). The Bayesian method of moments (BMOM): Theory and applications. *Advances in Econometrics 12*, 85–105.

Zellner, A. and J. Tobias (2001). Further results on Bayesian method of moments analysis of the multiple regression model. *International Economic Review 42*(1), 121–139.

CHAPTER 12

REACTING TO SURPRISING SEEMINGLY INAPPROPRIATE RESULTS

Dale J. Poirier

Department of Economics, University of California, Irvine

This chapter draws on two well-known tools in Bayesian econometrics to provide a simple framework to view research that adds "seemingly inappropriate" covariates to a familiar regression and finds them surprisingly statistically and substantively significant. An application to variation in per-capita income growth rates across countries is included.

12.1 Introduction

"Seemingly inappropriate" covariates are covariates for which the reader cannot imagine why anyone would be interested in conditioning on them in the first place, and furthermore, cannot fathom why they would have any effect. The prior for the coefficients of these covariates is centered over zero to capture the "seemingly inappropriateness." The tightness of the prior depends on a single hyperparameter g. The primary goal here is to provide the reader (Bayesian or non-Bayesian) a simple yes/no question concerning the magnitude of g as a way to determine how concerned they should feel about the statistical significance of the surprising seemingly inappropriate regressor. The prior that is the basis for this question is flexible and can be used in any regression situation. Section 12.2 provides the econometric details and Section 12.3 provides an empirical application. Finally, Section 12.4 draws some conclusions.

12.2 Statistical Framework

Consider the standard normal linear regression

$$y = X_1\beta_1 + X_2\beta_2 + u, \tag{12.1}$$

where y and u are $N \times 1$, X_1 is $N \times K_i$, $\beta_i = [\beta_{i1}, ..., \beta_{iK_i}]'$ $(i = 1, 2)$, and $\beta = [\beta_1', \beta_2']'$. All analysis is conditional on $X = [X_1, X_2]$, where $\text{rank}(X) = K = K_1 + K_2$. The assumption that $u|X \sim N(0_N, \sigma^2 I_N)$ implies the OLS estimator $b = [b_1', b_2']' = (X'X)^{-1}X'y$ of $\beta = [\beta_1', \beta_2']'$ has a multivariate normal sampling distribution with mean β and covariance matrix $\sigma^2(X'X)^{-1}$, conditional on X, β, and σ^2. For further reference, note that a .95 confidence region for β_2 is the ellipse

$$\frac{(\beta_2 - b_2)'X_2'M_1X_2(\beta_2 - b_2)}{K_2 s^2} \leq F(.95; K_2, N - K), \tag{12.2}$$

where $s^2 = y'My/(N - K)$ with $M = I_N - X(X'X)^{-1}X$, $M_1 = I_N - X_1(X_1'X_1)^{-1}X_1$, and $F(.95; K_2, N - K)$ denotes the .95 critical value of an F-distribution with K_2 and $N - K$ degrees of freedom. It is assumed that X_2 is "seemingly inappropriate." A "surprise" occurs if equation (12.2) does not include $\beta_2 = 0_{K_2}$. What should a reader of such results think?

The corresponding Bayesian results under the "noninformative" prior distribution $p(\beta, \sigma^{-2}) \propto \sigma^2$ are numerically identical to these OLS results. The difference is the Bayesian conditions on the observed y, and equation (12.2) becomes a highest posterior density (HPD) region. Because the "surprising results" reflect the existence of conflicting information that is plausible, an informed Bayesian approach seems natural.

The specification here combines Leamer (1982) and Zellner (1986). Consider the prior $\beta_2|\sigma^{-2} \sim N(0_{K_2}, \sigma^2\underline{A}^{-1})$, where the precision matrix $\underline{A}$ is nonsingular. The centering of the prior over $\beta_2 = 0_{K_2}$ is part of the definition of the problem and should not be controversial. The precision matrix $\underline{A}$ (more below) indicates how tightly the reader's prior is around this mean. Assuming the prior for β_1 is independent of β_2 given σ^{-2} and that $p(\beta_2, \sigma^{-2}) \propto \sigma^2$, the posterior density is

$$\begin{aligned} p(\beta, \sigma^{-2}|y) &\propto \sigma^2|\sigma^2\underline{A}|^{-\frac{1}{2}}e^{-\frac{1}{2\sigma^2}\beta_2'\underline{A}\beta_2}|\sigma^2 I_N|^{-\frac{1}{2}}e^{-\frac{1}{2\sigma^2}(y-X\beta)'(y-X\beta)} \\ &\propto \left(\sigma^{-2}\right)^{\frac{N+K_2-2}{2}} e^{-\frac{1}{2\sigma^2}\left[\beta_2'\underline{A}\beta_2+\nu s^2+(\beta-b)'X'X(\beta-b)\right]} \\ &\propto \left(\sigma^{-2}\right)^{\frac{N-K_1+K-2}{2}} e^{-\frac{1}{2\sigma^2}\left[\nu s^2+\left(\beta-\bar{b}\right)'\left(A+X'X\right)\left(\beta-\bar{b}\right)\right]}, \end{aligned} \tag{12.3}$$

where $\nu = N - K$ and

$$\begin{aligned} \bar{b} = \begin{bmatrix} \bar{b}_1 \\ \bar{b}_2 \end{bmatrix} &= \bar{Q}X'Xb \\ &= \left[I_K - \left(A + X'X\right)^{-1} A\right] b \qquad (12.4) \\ &= \left(I_K - \begin{bmatrix} 0_{K_1\times K_2} & \bar{Q}^{12}\underline{A} \\ 0_{K_2\times K_1} & \bar{Q}^{22}\underline{A} \end{bmatrix}\right) \begin{bmatrix} \bar{b}_1 \\ \bar{b}_2 \end{bmatrix}, \end{aligned}$$

$$\bar{Q} = \begin{bmatrix} \left(X_1'X_1\right)^{-1} - \left(X_1'X_1\right)^{-1} X_1'X_2\bar{Q}^{22}X_2'X_1\left(X_1'X_1\right)^{-1} & -\left(X_1'X_1\right)^{-1} X_1'X_2\bar{Q}^{22} \\ -\bar{Q}^{22}X_2'X_1\left(X_1'X_1\right)^{-1} & \left(\underline{A} + X_2'M_1X_2\right)^{-1} \end{bmatrix},$$

and

$$A = \begin{bmatrix} 0_{K_1\times K_1} & 0_{K_1\times K_2} \\ 0_{K_2\times K_1} & \underline{A} \end{bmatrix}.$$

The posterior in equation (13.1) is recognized as a normal-gamma density. The marginal posterior distribution of β is a multivariate-t with mean $\bar{b}$, covariance matrix proportional to $\bar{s}^2\bar{Q}$, and degrees of freedom $\bar{\nu} = N - K_1$, where

$$\bar{\nu}\bar{s}^2 = \nu s^2 + \left(\bar{b} - b\right)' A \left(\bar{b} - b\right). \tag{12.5}$$

The marginal posterior distribution of σ^{-2} is gamma with mean $\bar{s}^{-2}$ and variance $\frac{2}{\nu s^2}$.

A convenient parsimonious formulation is to choose $\underline{A}$ proportional to the sampling precision of the OLS estimator b_2, i.e.,

$$\underline{A} = gX_2'M_1X_2, \tag{12.6}$$

where $g > 0$ is a hyperparameter to be chosen. For example, $g = .25$ indicates the prior precision is one-fourth of the precision in the sample. Given this choice of $\underline{A}$, $\bar{Q}^{22} = (g+1)^{-1}\left(X_2'M_1X_2\right)^{-1}$, $\bar{Q}^{22}\underline{A} = \left(\frac{g}{g+1}\right) I_{K_2}$, and $\bar{Q}^{12}\underline{A} = -\left(\frac{g}{g+1}\right)\left(X_1'X_1\right)^{-1} X_1'X_2$. Plugging (12.6) into (12.4) and (12.5) yields

$$\bar{b} = \begin{bmatrix} b_1 + \left(\frac{g}{g+1}\right)\left(X_1'X_1\right)^{-1} X_1'X_2b_2 \\ \left(\frac{1}{g+1}\right) b_2 \end{bmatrix} = \begin{bmatrix} \left(\frac{g}{g+1}\right) b_1^* + \left(\frac{1}{g+1}\right) b_1 \\ \left(\frac{1}{g+1}\right) b_2 \end{bmatrix},$$

$$\bar{\nu}\bar{s}^2 = \nu s^2 + \frac{g}{g+1} b_2'X_2'M_1X_2b_2, \tag{12.7}$$

where $b_1^* = (X_1'X_1)^{-1} X_1'y$ is the restricted OLS estimator using only X_1. The posterior mean $\bar{b}_1$ is a convex combination of the restricted and unrestricted OLS estimators. As g increases, $\bar{b}_1$ approaches the restricted estimator. Posterior mean $\bar{b}_2$ shrinks the OLS estimate b_2 toward zero.

A .95 HPD region for β_2 is the set of all β_2 that satisfy

$$\frac{\left[\beta_2 - \left(\frac{1}{g+1}\right) b_2\right]' X_2 M_1 X_2 \left[\beta_2 - \left(\frac{1}{g+1}\right) b_2\right]}{K_2 (g+1)\, \bar{s}^2(g)} \leq F(.95; K_2, \bar{v}). \tag{12.8}$$

Since (12.6) implies the posterior precision of β_2 is proportional to the sampling precision of b_2, the HPD region in (12.8) is the same shape as the confidence region in (12.2), but with a different location and scale. Depending on g, (12.8) may or may not contain the origin. Note that g also enters the right-hand side of (12.7) through $\bar{s}^2$ given in (12.5).

It is instructive to first consider the regions corresponding to (12.2) and (12.8) given σ^{-2}:

$$(\beta_2 - b_2)' \left(\frac{X_2' M_1 X_2}{\sigma^2}\right) (\beta_2 - b_2) \leq \chi^2(.95; K_2), \tag{12.9}$$

$$\begin{aligned} &\left[\beta_2 - \left(\frac{1}{g+1}\right) b_2\right]' \left(\frac{X_2 M_1 X_2}{\sigma^2}\right) \left[\beta_2 - \left(\frac{1}{g+1}\right) b_2\right] \\ &\qquad \leq \left(\frac{1}{g+1}\right) \chi^2(.95; K_2) \end{aligned}, \tag{12.10}$$

where $\chi^2(.95; K_2)$ denotes the .95 critical value of a chi-square distribution with K_2 degrees of freedom. The case of interest is when $\beta_2 = 0_2$ does not lie in (12.9), i.e.,

$$\frac{b_2' X_2' M_1 X_2 b_2}{\sigma^2} > \chi^2(.95; K_2). \tag{12.11}$$

Does $\beta_2 = 0_2$ lie in (12.10)? It does if and only if

$$\frac{b_2' X_2' M_1 X_2 b_2}{\sigma^2} \leq (g+1)\, \chi^2(.95; K_2),$$

or equivalently, if and only if

$$g \geq \frac{b_2' X_2' M_1 X_2 b_2}{\sigma^2 \chi^2(.95; K_2)} - 1. \tag{12.12}$$

Note that (12.11) guarantees the right hand side of (12.12) is positive. In other words, if g is greater than the ratio of the classical test statistic to its

.05 critical value minus one, then HPD region (12.10) contains $\beta_2 = 0_2$. For example, suppose the test statistic exceeds the critical value by 10%. Then, provided the prior precision is at least 10% of the sample precision, the HPD region in (12.10) contains $\beta_2 = 0_2$. If the test statistic is twice its critical value, then the prior must be at least as precise as the sample.

Now return to the case where σ^{-2} is unknown. The case in which $\beta_2 = 0_2$ does not fall in (12.2), but does fall in (12.8), is not as easy to interpret because $\bar{s}^2$ also depends on g. $\beta_2 = 0_2$ does not lie in (12.2) if and only if

$$\frac{b_2' X_2' M_1 X_2 b_2}{k_2 s^2} > F\left(.95; K_2, N - K\right), \tag{12.13}$$

and $\beta_2 = 0_2$ lies in (12.8) if and only if

$$\frac{b_2' X_2' M_1 X_2 b_2}{k_2\left(g+1\right)\bar{s}^2\left(g\right)} \leq F\left(.95; K_2, N - K_1\right). \tag{12.14}$$

Using equation (12.7), inequality (12.14) can be written as

$$\frac{b_2' X_2' M_1 X_2 b_2}{k_2\left(\frac{\upsilon s^2 + g\left[\nu s^2 + b_2' X_2' M_1 X_2 b_2\right]}{N - K_1}\right)} \leq F\left(.95; K_2, N - K_1\right),$$

or equivalently, as

$$g \geq \frac{\left(\frac{N-K_1}{N-K_1-K_2}\right)\left(\frac{f}{F(.95; K_2, N-K_1)}\right) - 1}{1 + \left(\frac{K_2}{N-K_1-K_2}\right) f}, \tag{12.15}$$

where f is the usual F-statistic appearing on the left-hand side of inequality (12.13). While not as clean as (12.12), (12.15) has some similarities. Like (12.12), the numerator of (12.15) contains the ratio of the test statistic to the critical value, multiplied by a number slightly greater than one, minus one. The denominator in (12.15) is one plus a small number times the test statistic. Most importantly, the only data required in (12.15) is the test statistic f.

While the usual Bayesian analysis starts with a choice of g and then proceeds to the posterior analysis, I suggest another approach aimed at non-Bayesian researchers who encounter the "surprising seemingly inappropriate" result that $\beta_2 = 0_2$ does not lie in confidence region (12.2). Put on your Bayesian hat and adopt the prior described above. The right-hand side of (12.15) gives the minimal prior precision relative to the sample precision that is required for $\beta_2 = 0_2$ to fall inside the HPD region in equation (12.8). gN converts this relative precision into the minimal size of a fictitious data that would correspond to the same information. If the researcher is not embarrassed by this minimal amount of prior information, then the "surprising seemingly inappropriate" results are not so striking.

Inequality (12.15) is the basis of a simple binary self inquisition for a reader. Is the precision of my prior beliefs, as captured by g, greater than the right hand side of (12.15)? The right-hand side is a simple function of the test statistic and should be readily available.

The next section provides an empirical example that most readers will likely agree involves "surprising seemingly inappropriate" results.

12.3 Empirical Illustration

There is a vast literature on differences across countries in terms of their levels and growth of per capita GDP over time. Connoisseurs of this literature need to be flexible in accepting a wide variety of potential covariates - the proverbial kitchen sink. In this regard (Westling, 2011) breaks new ground in imagination.

Westling (2011) drew on the 98 non-oil producing countries used in a well-known study (Mankiw et al., 1992) and focused on a sub-sample of $N = 75$ countries which had data on one additional variable. Westling (like most authors in the field and here as well) ignores numerous potential complications (e.g., missing data issues, errors-in-variables, endogeneity), and focuses on what the new variable adds to a slightly modified specification of (Mankiw et al., 1992).

I consider two of his regressions. In the first (Westling, 2011, Table 3, specification 4), y = ln(GDP per working age person) in 1985. Included as regressors are an intercept, x_2 = ln(investment/GDP per capita) [averaged over 1960-85], x_3 = ln(human capital accumulation) (HCA), x_4 = ln(school), and x_5 = Polity [1980 score: -10 (authoritarian) $\ldots +10$ (democracy)]. x_5 was not used in Mankiw et al. (1992). In addition Westling added his new variable, w_n in quadratic form, so $K_1 = 5$, $K_2 = 2$, $x_{n2} = [w_n, w_n^2]'$, and $\beta_2 = [\beta_{21}, \beta_{22}]'$. All data are included in Westling (2011). At this point I am intentionally not telling you what is w_n, but its sample mean is 14.5 and sample standard deviation is 1.9.

The .95 HPD region for β_2 does not include the origin. The F-statistic for testing $\beta_{21} = \beta_{22} = 0$ is 10.3 and the corresponding p-value is 0.0001. An already high R^2 goes up noticeably from .81 to .84 when w_n and w_n^2 are added. The estimated coefficients (with estimated standard errors in parentheses) are 1.54 (.34) and $-.05$ (.01) on w_n and w_n^2, respectively. Together the coefficients are jointly significant at the 0.001 level. The marginal effect of a one-unit change in w_n on ln(GDP85 per capita) evaluated at $\bar{w}_n = 14.5$ is a substantively significant $1.54 - (.05)(14.5) = .81$ with a posterior standard deviation of .33.

Impressed? I'm not, but I know what is w_n. Let's perform the analysis of Section 12.2. Under the prior $\beta_2|\sigma^{-2}\sim\mathbb{N}\left(0_{K_2}, \sigma^2\left[gX_2'M_1X_2\right]^{-1}\right)$, the .95 HPD (12.14) includes $\beta_{21} = \beta_{22} = 0$ for all g satisfying (12.15) which in this case is

$$g \succeq \frac{\left(\frac{75-5}{75-7}\right)\left(\frac{10.3}{3.13}\right) - 1}{1 + \left(\frac{2}{75-7}\right) 10.3} = \frac{1.03\,(3.29) - 1}{1.30} = 1.84. \tag{12.16}$$

Therefore, if the precision of my prior beliefs is 84% larger than the sample precision, then $\beta_{21} = \beta_{22} = 0$ is in HPD region (12.9). In other words, if my prior beliefs are as if they were based on a sample of size at least $(1.826)(75) = 138$ from specification (12.1), then the "surprising" result ceases to be surprising. The posterior mean for β_2 when $g = 1.84$ is $\bar{b}_2 = [.54, .02]'$.

In the second regression (Westling, 2011, Table 3, specification 4), y is the log difference in GDP per working age person over 1960-1985. The same regressors are used plus log GDP per working age person in 1960. In this case Westling only reports a regression adding $w_n$with an estimated coefficient $-.04$ (.01). Here $K_1 = 6$ and $K_2 = 2$. In this case a .95 HPD interval includes zero provided the prior precision satisfies

$$g \geq \frac{\left(\frac{75-6}{75-7}\right)\left(\frac{4}{2}\right) - 1}{1 + \left(\frac{1}{75-7}\right) 4} = \frac{1.01\,(2) - 1}{1.06} = .96. \tag{12.17}$$

Or in other words, if the prior precision is equivalent to a fictitious sample of at least $(.96)(75) = 72$ observations, then the surprising seemingly inappropriate result vanishes. In this case $\bar{b}_2 = .02$.

12.4 Discussion

This chapter has considered the question of how to react to surprising seemingly inappropriate results in the standard normal linear model. The key result is (12.15) which provides a minimal prior precision for which the "surprising seemingly inappropriate" results disappear. In terms of the two examples in Section 12.3, my priors are easily sufficiently precise. In Westling (2011), w_n purports to be an accurate measurement of the erect length (in cms.) of the average male organ in country n. After learning the specification of w_n, does your implicit g satisfy (12.16) and (12.17)? Mine does. I think (12.15) serves as a useful metric for how surprising are "seemingly inappropriate results."

Acknowledgement: I would like to thank John Geweke for helpful comments on an earlier draft.

Bibliography

Leamer, E. (1982). Sets of posterior means with bounded variance priors. *Econometrica 50*, 725–736.

Mankiw, N. G., D. Romer, and D. N. Weil (1992). A contribution to the empirics of economic growth. *Quarterly Journal of Economics 107*, 407–437.

Westling, T. (2011). Male organ and economic growth: Does size matter? *Munich Personal RePEc Archive Paper No. 32302*, available online at http://mpra.ub.uni–muenchen.de/32302/.

Zellner, A. (1986). On assessing prior distributions and Bayesian regression analysis with g-prior distributions. In P. K. Goel and A. Zellner (Eds.), *Bayesian Inference and Decision Techniques: Essays in Honor of Bruno de Finetti*, pp. 233–243. Amsterdam: North-Holland.

CHAPTER 13

IDENTIFICATION AND MCMC ESTIMATION OF BIVARIATE PROBIT MODELS WITH PARTIAL OBSERVABILITY

ASHISH RAJBHANDARI

StataCorp LP, College Station, TX, USA.

13.1 Introduction

Partial observability in a bivariate probit model refers to the case where the response variable is the outcome of a paired decision. In models as such, an outsider only observes the joint decision taken by the pair of agents without observing the individual decisions. For example, in a bilateral agreement to decide on loan arrangements between the IMF and a nation, the IMF decides whether or not to approve a loan and the participating nation decides whether or not to accept the loan. Nonetheless, to an outsider, only the final outcome of whether the agreement was a success or not is observed. In this case, a successful loan agreement implies that both the IMF and the participating nation were in favor of the loan conditions. Whereas failure to secure a loan arrangement could result from any one of the following conditions; both IMF and the nation reject the loan agreement or either one party rejects due to unsatisfactory loan conditions.

Identification in a partially observable model is weak as discussed in later sections as the outsider has to make inference on individual decisions based on the joint outcome. Specifically in the case when 'failure' is observed, one cannot distinguish between the three possible scenarios that might have generated this outcome. In Poirier (1980), the author discusses identification issues in a partially observable model and concludes that the parameters are weakly identified as long as they do not lead to observationally equivalent equations. More importantly, the author suggests that "the exogenous variable must take on at least as many distinct data configurations as there are unknown parameters."

The issue of identification in bivariate probit models with various levels of observability is undertaken in a study by Meng and Schmidt (1985). The authors quantify the loss of asymptotic efficiency of the parameter estimates in a partially observable model compared to different levels of observability such as a censored probit model and so on. A fully observable model is considered as a benchmark and by computing ratio of asymptotic variance of a partially observable to fully observable they compare efficiency across various models. In their study, the magnitude of efficiency loss of a partial observable model was found to be the highest compared to all other levels of observability.

Following Poirier (1980), this model has widespread application in the realm of business finance, macroeconomics, political science, etc. Dwyer and Hasan (2007) apply this model to study the effect of suspension of payments on bank failures. The dependent variable is whether a bank remains open or closes which depend on two characteristics namely the bank's present value of earnings and the value of bank's notes. Boyes et al. (1989) study bank's credit scoring decision based on cardholder status and default status as a model of partial observability. In political science, this model has been applied in evaluating the role of IMF. Przeworski and Vreeland (2002) study the effect of the IMF on economic growth of participating nations. They also develop a dynamic version of the partial observability model to study loan agreements between the IMF and national governments. Agostino (2008) uses this model as a robustness check to investigate the effect of World Bank loans and private investments in member countries. Ghosh et al. (2008) study aggregate use of IMF resources and applies this model for robustness checks. Stone (2008) studies the bargaining process between the IMF and the borrowing country where the decision to participate is made by both and an outsider only observes the joint outcome. Knight and Santaella (1997) also study the IMF's decision of loan approval to a borrowing country. The joint decision involves IMF's approval of a loan and whether a borrowing country meets the IMF's criteria for approval. We only observe the presence or absence of loan approval for a particular country. However, unlike Przeworski and Vreeland (2002) they allow the decisions of agents to be correlated. In corporate finance, Firth et al. (2011) study the costs of falsified financial statements in China.

In all of the aforementioned works, the model is usually estimated assuming independence between the two parties involved in the decision making process. In some applications where the independence assumption is dropped, the estimated values of correlation are unreliable or not reported. In this chapter, we show through Monte Carlo simulations the difficulty of recovering the correlation parameter. We also provide a Bayesian framework to estimate a partially observable bivariate probit model using data augmentation technique for multivariate probit models as in Chib and Greenberg (1998). Finally as an application, we estimate the partially observable bivariate probit model using maximum likelihood and Bayesian MCMC methods to a data set used in Przeworski and Vreeland (2002). We find that estimating correlation in this model highly depends on the initial value which more than often leads to unreliable estimates.

In Section 13.2 we layout a bivariate probit model and discuss two levels of observability associated with the model. In Section 13.3 we discuss identification issues related to the partial observability model and Section 13.4 reports results of various Monte Carlo simulation experiments. In Section 13.5 we present a Bayesian method of estimating this model and discuss results based on the simulated experiments. Section 13.6 provides an application and Section 13.7 concludes.

13.2 Bivariate Probit Model

In a bivariate probit model, the binary response variable Y_j for $j = 1, 2$ is the outcome of a joint decision taken by a pair of agents. The decisions depend on individual characteristics and are related to the binary outcome through a set of latent variables given by

$$y_{i1}^* = x_{i1}'\beta_1 + \epsilon_{i1},$$

$$y_{i2}^* = x_{i2}'\beta_2 + \epsilon_{i2},$$

where $i = 1, \ldots, n$ is the number of observations. y_{ij}^* represents latent variables that describe a threshold for choosing one alternative to the other. x_{ij}s are individual specific covariates and ϵ_{ij} are normally distributed errors that are possibly correlated between the two agents with a correlation parameter ρ. Finally β_js are the regression coefficients. The observed binary variable is then related to the latent variable according to:

$$Y_{ij} = \begin{cases} 1 & \text{if} \quad y_{ij}^* > 0, \\ 0 & \text{if} \quad y_{ij}^* \leq 0. \end{cases}$$

A key feature of a probit model is the assumption of a standard normal link function that associates the binary dependent variable to the covariates.

The probability of choosing an alternative in a probit model is given by

$$\begin{aligned} \Pr(Y_{ij} = 1) &= \Pr(y_{ij}^* > 0) \\ &= \Phi(x_{ij}\beta_j; \rho) \end{aligned}$$

where $\Phi(z)$ represents a standard normal cdf evaluated at z. In a bivariate probit model, however, we model the joint response as a bivariate standard normal cdf.

13.2.1 Levels of Observability

There are four possible outcomes in a bivariate probit model represented by individual cells in Table 13.1. Levels of observability refers to which of the outcomes are observed by the researcher. A general case is of a full observability where all the cells are observed. Meng and Schmidt (1985) consider an example of a durable good purchase decision and a credit decision. In this case an agent makes a decision on whether or not to purchase a durable good and also whether to use credit or not to make the purchase. An observer can see all outcomes in this scenario.

Table 13.1: Full observability.

	Agent 2 (Y_2)	
Agent 1 (Y_1)	$Y_1 = 1, Y_2 = 1$	$Y_1 = 1, Y_2 = 0$
	$Y_1 = 0, Y_2 = 1$	$Y_1 = 0, Y_2 = 0$

If the decisions are related then joint probability of various outcomes in a full observable case is given by

$$\begin{aligned} \Pr(Y_1 = 1, Y_2 = 1) &= \Phi_2(x_1\beta_1, x_2\beta_2; \rho), \\ \Pr(Y_1 = 1, Y_2 = 0) &= \Phi(x_1\beta_1) - \Phi_2(x_1\beta_1, x_2\beta_2; \rho), \\ \Pr(Y_1 = 0, Y_2 = 1) &= \Phi(x_2\beta_2) - \Phi_2(x_1\beta_1, x_2\beta_2; \rho), \\ \Pr(Y_1 = 0, Y_2 = 0) &= 1 - \Phi(x_1\beta_1) - \Phi(x_2\beta_2) + \Phi_2(x_1\beta_1, x_2\beta_2; \rho), \end{aligned}$$

where $\Phi_2(\cdot)$ represents a bivariate standard normal cdf and ρ is the correlation.

Other levels of observability include censored probit models where we observe Y_{i1}, $Z_i = Y_{i1}Y_{i2}$ and X_i. The decision of the second agent Y_{i2} is observed if and only if $Y_{i1} = 1$ since $Y_{i2} = Z_i$ and Z_i is observed. If $Y_{i1} = 0$ then we cannot observe Y_{i2}. So depending on the first outcome, the second outcome is censored.

13.2.2 Partial Observability

In this chapter we are interested in a partially observable model where only few cells are observed to the researcher as compared to the fully observable case. An example motivated in Poirier (1980) consists of a committee of two members voting for or against a motion where an outsider only observes whether the motion was passed or not. Passing of the motion means that both members voted in favor of the motion whereas failure to pass might represent any three of the remaining cases. Therefore an outsider only observes two outcomes as seen in Table 13.2. The shaded region represents a single outcome visible to an outsider.

Table 13.2: Partial observability.

	Agent 2 (Y_2)	
Agent 1 (Y_1)	$Y_1 = 1, Y_2 = 1$	$Y_1 = 1, Y_2 = 0$
	$Y_1 = 0, Y_2 = 1$	$Y_1 = 0, Y_2 = 0$

Let $Z_i = Y_{i1}Y_{i2}$ represent the binary outcome of a joint decision. Then $Z_i = 1$ is only observed when both agents vote in favor of the motion which is the unshaded region in Table 13.2. Whereas $Z_i = 0$ is observed for all remaining cases represented by the shaded region, i.e.,

$$Z_i = \begin{cases} 1 & \text{if } Y_{i1} = 1,\ Y_{i2} = 1, \\ 0 & \text{otherwise.} \end{cases}$$

The joint probability of various outcomes in a partial observability model is then given by:

$$\begin{aligned} \Pr(Z = 1) &= \Pr(Y_1 = 1, Y_2 = 1) \\ &= \Phi_2(x_1\beta_1, x_2\beta_2; \rho), \\ \Pr(Z = 0) &= \Pr(Y_1 = 0 \text{ or } Y_2 = 0) \\ &= 1 - \Phi_2(x_1\beta_1, x_2\beta_2; \rho). \end{aligned}$$

The log likelihood function is expressed as

$$L(\beta_1, \beta_2, \rho; Z) = \sum_{i=1}^{n} Z_i \ln[\Phi_2(x_{i1}\beta_1, x_{i2}\beta_2; \rho)] + (1 - Z_i) \ln[1 - \Phi_2(x_{i1}\beta_1, x_{i2}\beta_2; \rho)].$$

We next discuss identification issues arising in such models.

13.3 Identification in a Partially Observable Model

As mentioned in Poirier (1980), identification of this model depends on data configuration, with some perverse configurations leading to a likelihood with multiple modes. This is elaborated in Poirier (1980) with an example using the following model:

$$\begin{aligned} y_{i1}^* &= \beta_{01} + \beta_{11} x_{ic} + \epsilon_{i1}, \\ y_{i2}^* &= \beta_{02} + \beta_{12} x_{ic} + \epsilon_{i2}, \end{aligned}$$

where x_{ic} is common to both agents. Restrictions such as $\beta_{12} = 0$ and $\rho = 0$ are known to be valid, leaving three unknown parameters to be estimated. In such a scenario, if x_{ic} is a dummy variable that takes only two distinct data configuration then the parameters are not identified. Therefore, the exogeneous variable x_{ic} needs to take at least three distinct configurations, which is at least as many as there are unknown parameters.

A minimal identification then requires at least one unique exogenous variable for an agent that is not included in other agent's decision set. This also overcomes the labeling problem, which arises when both agents have the same set of elements. Then we can interchange the elements of one agent and the other to yield an observationally equivalent model. Under these conditions all parameters in the model are then formally identified.

We now discuss estimation of these models using numerical optimization methods. As mentioned earlier, we find applications of this model mostly confined to a scenario where independence is assumed among the pair of equations. Estimating the correlation parameter nonetheless is a difficult task in this case due to severely limited information predisposed to the researcher as compared to a fully observed bivariate probit model which dominates all other cases of observability in terms of information gained about individual decisions. We begin with a minimally identified model:

$$\begin{aligned} y_{i1}^* &= \beta_{01} + \beta_{11} x_{ic} + \epsilon_{i1}, \\ y_{i2}^* &= \beta_{02} + \beta_{12} x_{ic} + \beta_{22} x_i + \epsilon_{i2}, \end{aligned}$$

$$Z_i = \begin{cases} 1 & \text{if } y_{i1}^* > 0,\ y_{i2}^* > 0, \\ 0 & \text{otherwise.} \end{cases}$$

The regression coefficients for this model were fixed at $\beta_{01} = -0.5$, $\beta_{11} = 1.5$, $\beta_{02} = 0.25$, $\beta_{12} = -2$ and $\beta_{22} = 0.75$. For the Monte Carlo simulations we choose different values for the correlation parameter from the interval $[-0.9, 0.9]$ as discussed later. Finally, exogenous variables are drawn from a standard normal distribution.

Standard identification restrictions in the above model also impose a unit variance in the error terms. The errors ϵ_{ij} are drawn from a bivariate standard normal distribution with correlation varying across samples which we discuss later. The ML estimates of this model were obtained using BFGS which is a quasi-Newton method that approximates the Hessian matrix instead of using analytical second derivatives.

13.4 Monte Carlo Simulations

In this section we attempt to shed light on the difficulty of recovering the correlation parameter under a partially observable bivariate probit model. For this purpose, we perform Monte Carlo simulations from the minimally identified model discussed in the previous section. We consider 9 different values for the correlation parameter and for each value we generate 500 samples with a sample size of 1000 each. Thus for each value of the correlation parameter we obtain 500 Monte Carlo estimates.

Furthermore, using the same parameter values, we also generate draws from a fully observable bivariate probit model and graph the estimates along with the partially observable model. The empirical probability distribution plots of the correlation parameter obtained from Monte Carlo simulations in both the models can be seen in Figure 13.1.

Vertical dashed lines represent the true value of the correlation parameter that was used to generate simulated data. The solid line is the empirical distribution of the correlation parameter under a fully observable bivariate probit model while Monte Carlo estimates obtained under a partially observable model are represented by the dashed curve. We find that the empirical distribution plots for both the models look similar for cases when the correlation equals -0.9, -0.75 and -0.5 although the fully observable case does slightly better in following the true correlation parameter. The difference in these two cases is more obvious when the correlation parameter equals -0.25 and so on. The fully observable bivariate probit model estimates the true correlation parameter correctly and the empirical distribution obtained from the Monte Carlo samples is also skewed towards it. On the contrary, there is not much change in the distribution of estimated correlation under the partially observable case. In none of the samples were we able to estimate the correlation parameter correctly. We find this parameter to be very dependent on the starting values. In Figure 13.2, we plot the empirical distribution of the correlation parameter from the same model as before. However, in this case we choose our starting value to be 0.5 and get a different distribution than in the previous case, thus leading to imprecise estimates. In the following sections we develop Bayesian methods of partially observable bivariate probit model and estimate the model in a data set by Przeworski and Vreeland (2002) using ML as well as Bayesian techniques.

13.5 Bayesian Methodology

Bivariate probit models are estimated with Bayesian methods using data augmentation techniques introduced in Albert and Chib (1993). This allows us to model the joint posterior density of the augmented data along with model parameters through which conditional densities are obtained which can be conveniently sampled with MCMC samplers. We follow a technique as in Chib and Greenberg (1998) of sampling from the posterior density using a combination of Gibbs and Metropolis-Hastings algorithm. In this section we only list the conditional densities that are used in the sampling algorithm.

13.5.1 Conditional Posteriors

The conditional distribution for β and Y_j^* obtained from the joint posterior can be sampled using the Gibbs sampler. In the case of a fully observable model, the distribution of each conditional y_j^* would be distributed as a univariate truncated normal. In case of partial observability, when we observe the case $(y_1^* > 0, y_2^* > 0)$, a univariate truncated normal distribution can be used as a conditional density. However, since we are unable to distinguish the remaining three cases, conditional distributions are not easily available for sampling. We use a simple accept reject sampling in this case which will be described later. The conditional distribution of β is fairly straightforward and is formulated as follows:

$$P(\beta|Z, y_j^*, \rho) \propto \pi(\beta) \prod_{i=1}^{n} f(Z|x_{ij}\beta_j, \rho),$$

where $\pi(\beta)$ is the prior density of β assumed to be distributed as $N_k(\beta_0, B_0^{-1})$ where k is the total number of parameters including the intercept. Furthermore, we assume prior independence between β and ρ. Then the conditional distribution of β is given by

$$P(\beta|Z, y_j^*, \Sigma) \sim N_k(\hat{\beta}, B^{-1}),$$

where $\hat{\beta} = B^{-1}\left(B_0^{-1}\beta_0 + \sum_{i=1}^{n} X_i'\Sigma^{-1}y_i^*\right)$, $B = B_0^{-1} + \sum_{i=1}^{n} X_i'\Sigma^{-1}X_i$, and Σ is a 2×2 correlation matrix with 1 in the diagonals. X_i now is a $2 \times k$ block diagonal matrix where the upper block diagonal matrix contains the covariates of the first agent and the lower block diagonal matrix contains the covariates of the second agent. Similarly y_i^* is a 2×1 vector of individual latent data.

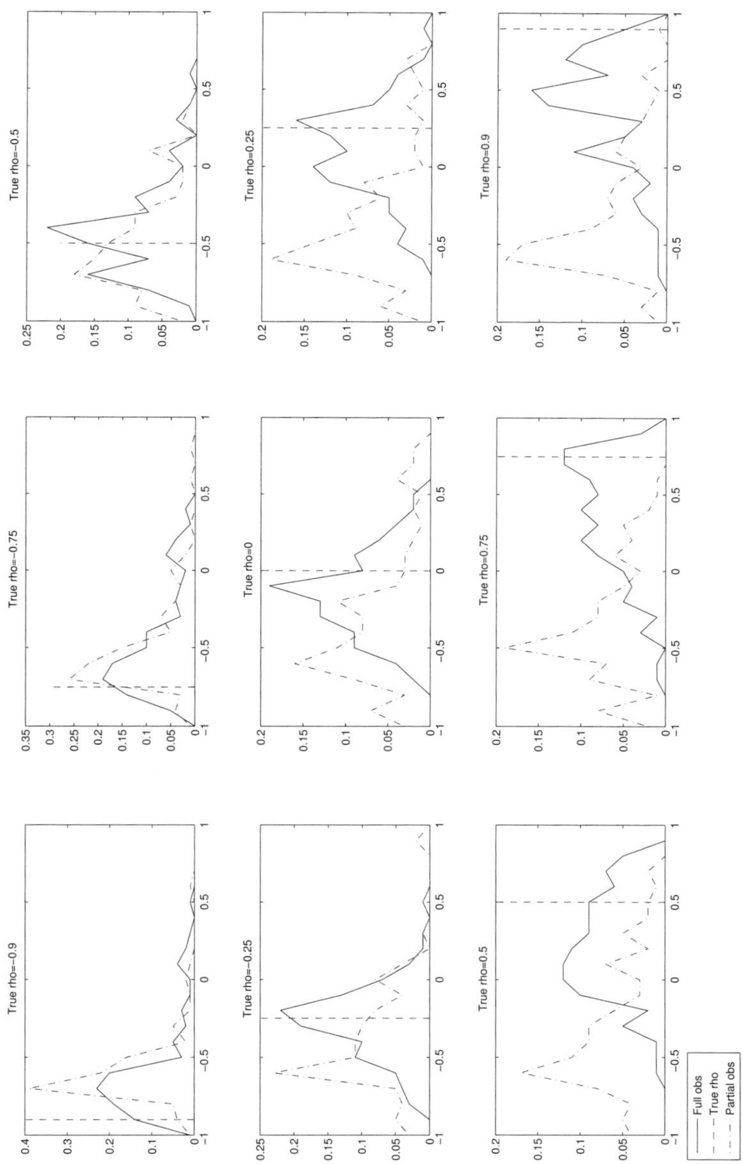

Figure 13.1: Empirical probability distribution of the correlation parameter.

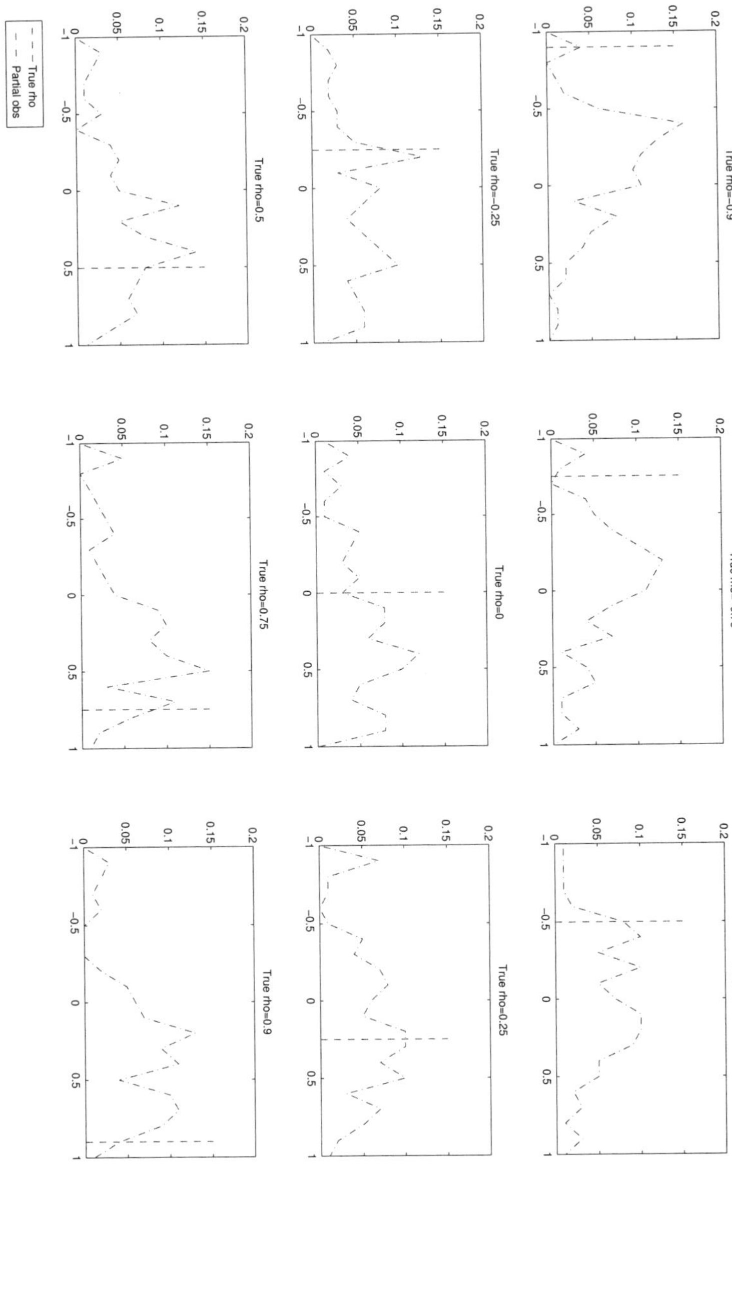

Figure 13.2: Empirical probability distribution of correlation parameter with different starting values.

13.5.1.1 Sampling Augmented Data An outsider observes a 'success' only when the augmented data y_1^* and y_2^* are both greater than 0. In such a case the conditional distribution is a univariate truncated normal distribution given by

$$y_{i1}^*|y_{i2}^* \sim TN_{[0,\infty]}(x_{i1}\beta_1 + \rho(y_{i2} - x_{i2}\beta_2), (1-\rho^2)),$$
$$y_{i2}^*|y_{i1}^* \sim TN_{[0,\infty]}(x_{i2}\beta_2 + \rho(y_{i1} - x_{i1}\beta_1), (1-\rho^2)).$$

However, when a 'failure' is observed one cannot distinguish which of the 3 possible cases ($[y_1^* > 0, y_2^* \leq 0]$,$[y_1^* \leq 0, y_2^* > 0]$,$[y_1^* \leq 0, y_2^* \leq 0]$) is responsible for the outcome. Hence we use a simple accept reject algorithm to sample the unobserved cases.

13.5.1.2 Sampling the Correlation Parameter The posterior conditional of the correlation parameter is given by

$$P(\rho|y_i^*, \beta, Z) \propto \pi(\rho) f(y_i^*|\beta, \Sigma).$$

The prior $\pi(\rho)$ is assumed to have a normal distribution expressed as $\pi(\rho) \propto I(\rho \in C)\exp\big[-\frac{(\rho-\rho_0)^2}{2v_0}\big]$. It is truncated to a region C (interval from -1 to 1) such that $I(\rho \in C) = 1$ if ρ lies within the region and 0 otherwise. ρ_0 and v_0 are prior mean and variance, respectively. Then the posterior conditional is given by

$$P(\rho|\beta, y_j^*, Z) \propto I(\rho \in C)\frac{1}{|\Sigma|^{\frac{n}{2}}} \exp\left[-\frac{1}{2}[(y_i^* - X_i\beta)'\Sigma^{-1}(y_i^* - X_i\beta)]\right]$$
$$\times \exp\left[-\frac{(\rho-\rho_0)^2}{2v_0}\right].$$

Simulating from the conditional posterior of ρ requires a Metropolis Hastings (MH) step since the it is not a standard distribution that we can simulate directly. We use tailored MH to sample the correlation parameter.

13.6 Application

In this section we implement our algorithm to a study by Przeworski and Vreeland (2002). The authors study a case of bilateral cooperation between the IMF and national governments during a year. Moreover, the binary dependent variable used is participation in IMF agreements and is coded 1 if a government is observed to participate in an IMF agreement. In this particular case, a government participates in an agreement with the IMF either for loan purposes or in order for some policy conditions to be imposed on them.

A country seeking loan faces a set of conditions imposed by the IMF which entails specific monetary and fiscal policy changes. Before agreeing on a loan agreement, the government has to accept all the necessary conditions. On the contrary, instead of loan purposes a country with poor financial or public infrastructure might also enter IMF agreements wanting those "conditions" imposed on them if they are beneficial for the country. A government can also reject IMF conditions and withdraw from the agreement depending on the country's economic condition.

The authors use a set of covariates that dictates a country's need for an IMF loan or its policy conditions. The covariates are as follows:

- Reserves - The average annual foreign reserves in terms of monthly imports to measure the need for a loan.
- Budget - Annual government budget surplus as a proportion of GDP.
- Debt Services - Annual debt services as a proportion of GDP.
- Investment - Real gross domestic investment as a proportion of GDP to capture the cost of deciding to reject the loan.
- Years Under - Number of past years a country has spend under IMF agreements.
- Number Under - Number of other countries around the world currently participating in an IMF agreement.
- Lagged Election - Dummy variable coded 1 if the previous year had a legislative election and 0 otherwise.

Furthermore, these variables also affect the IMF in the same way. As an example provided by the authors, the IMF might seek to assist countries that have a shortage of foreign reserves and also if some countries need fiscal or monetary discipline imposed upon them. However, identification of this bivariate model requires at least one variable that is assigned to only one of the agents. In their case, the authors use "BOP," the absolute size of the balance of payments as an extra variable in the IMF's equation. The reason being that IMF is concerned with world financial stability and for that purpose the IMF seeks out for countries with larger BOP deficits.

The bivariate system in latent representation then can be expressed as

$$\begin{aligned} E[Y_G^*|X'\beta] = \beta_{01} + \beta_{11}Reserves + \beta_{21}Budget + \beta_{31}DebtService \\ +\beta_{41}Investment + \beta_{51}YearsUnder + \beta_{61}NumberUnder \\ +\beta_{71}LaggedElection, \end{aligned}$$

$$\begin{aligned} E[Y_{IMF}^*|X'\beta] = \beta_{02} + \beta_{12}Reserves + \beta_{22}Budget + \beta_{32}DebtService \\ +\beta_{42}Investment + \beta_{52}YearsUnder + \beta_{62}NumberUnder \\ +\beta_{72}LaggedElection + \beta_{82}BOP, \end{aligned}$$

where Y_G and Y_{IMF} represents the latent indicators for government's and IMF's individual decision. The two equations are assumed to be correlated with coefficient ρ that is estimated in the model. The results are presented in Table 13.3.

Table 13.3: Posterior estimates from the bivariate model.

Variable	P&V (2002)	ML		Bayesian	
	Baseline	Unrestricted	$\rho = 0$	Unrestricted	$\rho = 0$
Government					
Constant	−1.68	−1.02	−1.03	0.64	−0.12
	(0.34)	(0.33)	(0.05)	(1.77)	(0.11)
Reserves	−0.41	−0.64	−0.33	−0.81	−0.27
	(0.34)	(0.19)	(0.04)	(0.95)	(0.38)
Budget	−0.82	−0.84	−0.71	−0.14	−0.63
	(0.16)	(0.11)	(0.15)	(0.64)	(1.36)
Debt Service	1.05	0.99	0.97	0.81	0.38
	(0.25)	(0.28)	(0.17)	(0.42)	(1.14)
Investment	−3.42	−4.17	−3.00	−2.67	−1.74
	(1.02)	(0.51)	(0.63)	(1.62)	(0.30)
Years Under	0.54	0.60	0.42	0.24	0.27
	(0.11)	(0.30)	(0.16)	(0.50)	(0.18)
Number Under	0.31	0.11	0.11	−0.05	0.25
	(0.10)	(0.08)	(0.02)	(0.26)	(0.68)
Lagged Election	0.31	0.31	0.22	0.05	0.02
	(0.17)	(0.17)	(0.19)	(0.36)	(0.50)
IMF					
Constant	5.55	3.41	4.94	0.42	0.13
	(1.41)	(0.39)	(0.21)	(0.55)	(0.52)
Reserves	−1.53	−1.59	−1.27	−0.41	0.47
	(0.53)	(0.32)	(0.05)	(0.44)	(0.35)
Budget	0.36	0.28	0.39	−0.57	−0.68
	(0.17)	(0.11)	(0.06)	(0.91)	(1.11)
Debt Service	0.35	0.47	0.20	0.91	0.46
	(0.20)	(0.24)	(0.21)	(0.64)	(0.80)
Investment	−3.41	0.45	−3.0	−3.19	−3.27
	(1.73)	(0.05)	(1.10)	(1.75)	(1.69)
Years Under	0.21	−0.02	0.05	0.43	0.49
	(0.16)	(0.28)	(0.02)	(0.57)	(0.75)
Number Under	−0.96	−0.59	−0.77	−0.07	−0.09
	(0.26)	(0.13)	(0.19)	(0.42)	(0.31)
Lagged Election	−0.19	−0.29	−0.12	0.10	0.10
	(0.22)	(0.23)	(0.02)	(0.46)	(0.54)
BOP	−0.38	−0.20	−0.30	−0.37	−0.37
	(0.13)	(0.32)	(0.07)	(0.23)	(0.24)
		$\hat{\rho} = 0.53$		$\hat{\rho} = -0.12$	
		(0.13)		(0.40)	
Log-likelihood	−545.92	−581.10	−579.20		

Przeworski and Vreeland (2002) estimate the model assuming independence of decisions between the country and IMF. The data set contains 1024 country-year observations. A country might have engaged in an agreement multiple times which are recorded as separate observations.

We estimate the model using maximum likelihood as well as Bayesian methods. For each method, we further estimate the model twice: once assuming independence and once with dependent decisions. The estimates of Przeworski and Vreeland (2002) are considered as the baseline model and are also reported in Table 13.3.

In the case of the maximum likelihood method, we estimate the correlation parameter to be about 0.53. The ML estimates obtained from the model with correlation are very similar to the baseline model except for the coefficients of "Investments" and "Years Under" for the IMF. We find these coefficients to be significantly different, as seen in Table 13.3. Furthermore, the third column reports results for a restricted model when $\rho = 0$ and we find that estimates in this case are similar to the baseline model.

Next we estimate the model using Bayesian MCMC methods. We use Gaussian priors for regression coefficients with mean 0 and variance of 10 and a truncated normal prior in the interval $[-0.9, 0.9]$ with mean 0 and a standard deviation of 0.25 for the correlation parameter. Furthermore $100,000$ draws were generated in the MCMC chain where 25% were discarded as burn in. We also use a thinning parameter where one in every 10 draws are stored in order to mitigate serial correlation in the draws.

The estimated correlation from the Bayesian method is -0.12 which is very different from that obtained from the maximum likelihood method. Nonetheless we find most of the estimates of the regression coefficients to be similar to the baseline model. The only differences with a change in sign are the coefficients on "Number Under" for the Government equation and "Budget" and "Lagged Election" variables in the IMF equation. Similarly under the assumption of independence, we find that the estimates from the Bayesian method are similar to the baseline model. The coefficients with an opposite sign in this case are variables "Reserves," "Budget," and "Lagged Election" in the IMF equation. Overall we find that estimation of correlation is nontrivial and can lead to varying estimates using different approaches.

13.7 Conclusion

In this chapter we examine a host of research in different fields that have an application of the partially observable bivariate probit model. Identifying restrictions for such a model have been provided in Poirier (1980). Subsequent work on the model is focused mostly on the application aspect. Based on the literature, a common assumption prevalent in most of applied work in this area is of independence among equations, i.e., the decisions of two participating agents are uncorrelated. While most results are based on this assumption, we also find other applications where the reported correlation estimates might be misleading.

In this chapter we investigate the correlation parameter by attempting to estimate it precisely through Monte Carlo simulation as well as Bayesian methods. We find that estimating the correlation parameter is nontrivial in a partially observable setting due to the severely limited information that is available to the researcher as compared to a fully observable case. Through Monte Carlo simulations we were able to recover the correlation parameter correctly in a fully observable case while estimating it in a partially observable case led to discouraging results. Moreover, we also find that in the latter case, the estimated correlation is dependent on the initial value. We also provide a Bayesian method of estimating a partially observable bivariate probit model using data augmentation. Finally, as an application, we estimate the model as in Przeworski and Vreeland (2002) where the authors study individual decisions made by the IMF and a country undergoing a bilateral loan agreement.

Appendix

Joint Posterior Let y_{i1}^* and y_{i2}^* be the latent variables with a joint distribution $N_2(x_i'\beta, \Sigma)$. Then Z_i are independent Bernoulli with probability given by $p_i = \Pr(Z_i = 1) = \Phi_2(x_{ij}'\beta_j, \Sigma)$. The joint augmented posterior density is expressed as

$$P(\beta, \Sigma, y_j^*|Z) \propto \pi(\beta, \Sigma)\ P(Z|y_j^*, \beta, \Sigma)\ P(y_j^*|\beta, \Sigma).$$

We also have

$$P(y_j^*|\beta, \Sigma) = \prod_{i=1}^{n} \phi_2(x_{ij}\beta_j, \Sigma)$$

and

$$P(Z|y_j^*, \beta, \Sigma) = \prod_{i=1}^{n} I(Z_i = 1)I(y_{i1} > 0 \text{ and } y_{i2} > 0) + I(Z_i = 0)I(y_{i1} \le 0 \text{ or } y_{i2} \le 0),$$

where $\phi_2(\cdot)$ is the bivariate normal pdf and $I(\cdot)$ is an indicator that equals 1 if the argument in the parentheses is true. Assuming independence in the priors, the joint posterior is given by

$$P(\beta, \Sigma, y_j^*|Z) \propto \pi(\beta)\ \pi(\Sigma) \prod_{i=1}^{n} \{I(y_{i1}^* > 0 \text{ and } y_{i2}^* > 0)I(Z_i = 1) + I(y_{i1}^* \le 0 \text{ or } y_{i2}^* \le 0)I(Y_i = 0)\}\phi_2(x_{ij}\beta_j, \Sigma).$$

Tailored Metropolis-Hastings

- At the mth iteration maximize

$$-\frac{n}{2}\log|\Sigma| - \frac{(\rho-\rho_0)^2}{2v_0}$$
$$-\frac{1}{2}\sum_{i=1}^{n}\left(y_i^{*(m-1)} - X_i\beta^{(m-1)}\right)'\Sigma^{-1}\left(y_i^{*(m-1)} - X_i\beta^{(m-1)}\right)$$

  with respect to ρ. Let the maximized value be $\hat{\rho}^m$ with an estimated variance of $\hat{S}^m$.

- Generate proposal values of ρ^* from $t(\nu, \hat{\rho}^m, \hat{S}^m)$ which is a standard MH algorithm. Chib and Greenberg (1998) add $P(\rho-\mu)$ to the proposed value and multiply the tuning parameter τ^2 to the estimated variance $\hat{S}$ to make it a tailored MH. P can be either 0, which makes it a standard MH chain, or -1, which makes it a tailored reflection chain that is useful for generating large moves. The value of τ^2 is fixed by experimentation.

- Then set

$$\alpha(\rho^{(m-1)}, \rho^*) = \frac{\pi(\rho^*)\ \prod_i f(y_i^*|\beta,\Sigma^*)I(\rho^* \in C)\ t(\rho^*|\nu,\hat{\rho}^m,\hat{S}^m)}{\pi(\rho)\ \prod_i f(y_i^*|\beta,\Sigma)I(\rho \in C)\ t(\rho^{(m-1)}|\nu,\hat{\rho}^m,\hat{S}^m)},$$

  where $t(t|\nu,\mu,\Sigma)$ denotes the density function (evaluated at t) of a univariate Student-t distribution with ν degrees of freedom, location vector μ and scale matrix Σ.

- Draw U from $U(0,1)$. If $U \leq \alpha(\rho^{(m-1)}, \rho^*)$, set $\rho^m = \rho^*$; else, set $\rho^m = \rho^{(m-1)}$.

Bibliography

Agostino, M. (2008). World bank conditional loans and private investments in recipient countries. *World Development 36*(10), 1692–1708.

Albert, J. H. and S. Chib (1993). Bayesian analysis of binary and polychotomous response data. *Journal of the American Statistical Association 88*(422), 669–679.

Boyes, W. J., D. L. Hofman, and A. L. Stuart (1989). An econometric analysis of the bank credit scoring problem. *Journal of Econometrics 40*(1), 3–14.

Chib, S. and E. Greenberg (1998). Analysis of multivariate probit models. *Biometrika 85*(2), 347–361.

Dwyer, G. P. and I. Hasan (2007). Suspension of payments, bank failures, and the nonbank public's losses. *Journal of Monetary Economics 54*(2), 565–580.

Firth, M., O. M. Rui, and W. Wu (2011). Cooking the books: Recipes and costs of falsified financial statements in China. *Journal of Corporate Finance 17*(2), 371–390.

Ghosh, A., M. Goretti, B. Joshi, A. Thomas, and J. Dalduendo (2008). Modeling aggregate use of IMF resources-analytical approaches and medium-term projections. *IMF Staff Papers 55*(1), 1–49.

Knight, M. and J. A. Santaella (1997). Economic determinants of IMF financial arrangements. *Journal of Development Economics 54*(2), 405–436.

Meng, C. and P. Schmidt (1985). On the cost of partial observability in the bivariate probit model. *International Economic Review 26*(1), 71–85.

Poirier, D. (1980). Partial observability in bivariate probit models. *Journal of Econometrics 12*(2), 209–217.

Przeworski, A. and J. Vreeland (2002). A statistical model of bilateral cooperation. *Political Analysis 10*(2), 101–112.

Stone, R. W. (2008). The scope of IMF conditionality. *International Organization 62*(4), 589–620.

CHAPTER 14

SCHOOL CHOICE EFFECTS IN TOKYO METROPOLITAN AREA: A BAYESIAN SPATIAL QUANTILE REGRESSION APPROACH

KAZUHIKO KAKAMU[1] AND HAJIME WAGO[2]

[1]Chiba University, Chiba, Japan.
[2]Kyoto Sangyo University, Kyoto, Japan.

This chapter examines the school choice effects on land prices from a Bayesian point of view. By utilizing spatial autoregressive quantile regression, we estimate a hedonic price model for the Tokyo 23-special wards in Japan in 2010. The empirical results capture the quantile dependent effects of the parameters, including the spatial interaction parameter, but cannot confirm the presence of school choice effects on land prices.

14.1 Introduction

In Japan, as in many other countries, the school choice system is one of the most controversial issues. Focusing on the Tokyo metropolitan area, the system is incorporated in a unique way. Tokyo is not only the capital of Japan but also a regional government center encompassing 23 special wards, 26 cities,

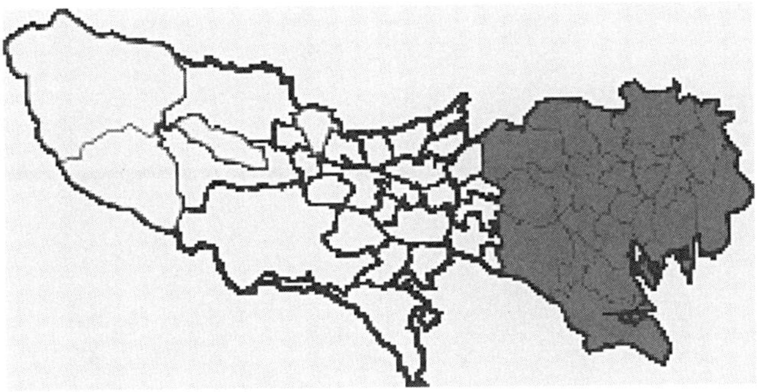

Figure 14.1: Tokyo 23-special wards.

5 towns and 8 villages. Reflecting the dense population, urban contiguity, and other realities of the 23-special ward area, a unique administrative system exists between the metropolitan government and the wards, which differs from the typical relationship between prefectures and municipalities. Therefore, the special wards are 23 municipalities that together make up the core and the most populous part of Tokyo. They are located on the east side of Tokyo (the highlighted area in Figure 14.1) and, for the purposes of this chapter, we define these 23 wards as the Tokyo metropolitan area. The individual systems are different in each ward and are classified into three types. One is the so called block system (BLOCK) in which the ward is divided into several blocks and the choice of schools is restricted to the block. Another one is the so-called adjoining system (ADJOINING), where the choice of schools is restricted to an adjoining parcel. Finally, the last one is the so-called whole-division system (WHOLE), where all schools in the ward can be chosen.

From a theoretical point of view, the school choice systems were analyzed by many researchers, e.g., using matching theory and other models. The focus has been on the fair allocation of students to the schools and on the construction of the efficient system. Research has concluded that the system led to the improvement of school quality. For example, Adbulkadiroğlu and Sönmez (2003), Erdil and Ergin (2008), Yasuda (2010) and others treated this problem. On the other hand, the empirical studies focused on good schools or good wards from a view point of the quality of school (e.g., Yoshida et al., 2009, and so on). However, in spite of the fact that the school quality is examined by many researchers, there is no empirical research that examines the school choice system. For example, Black (1999) and Downes and Zabel (2002) examined the school quality by hedonic approach proposed by Rosen (1974). Moreover, Brasington (1999) and Brasington and Haurin (2006) also considered the school quality using a hedonic approach, but they incorporate spatial effects into the model. Thus, we think it is important to examine, using a hedonic approach, whether introducing a school choice system is valued by residents or not.

In this chapter, we examine whether the school choice systems are valued by residents using a hedonic approach following previous studies which consider school quality. We employ a spatial model in the hedonic approach, however, our study is different in two respects. One is that we utilize the quantile regression approach. Quantile regression has been widely applied to the hedonic approach in recent studies (e.g., Coulson and McMillen, 2007; McMillen, 2008, and others). Therefore, we use a spatial quantile regression model, which is proposed by Kostov (2009) and widely applied to the hedonic approach. The other is that we use a Bayesian approach. The Bayesian quantile regression was first proposed by Yu and Moyeed (2001). After that, there are several Bayesian spatial quantile regression (e.g., Reich et al., 2011; Lum and Gelfand, 2012). However, the Bayesian spatial quantile regression approach following Kostov (2009) has not been proposed, even though this type of model is widely applied in empirical analysis. From the empirical results, we can confirm the effects of variables are quantile dependent including the spatial parameter but cannot confirm the school choice effects on land prices. Therefore, we find that spatial dependence plays an important role in land price determination, but that the school choice system is not appropriately introduced in Tokyo 23-special wards in Japan.

The rest of this chapter is organized as follows. In Section 14.2, we will introduce our econometric model to examine the school choice effects, which is based on a Bayesian quantile regression including spatial autoregressive process. Section 14.3 derives the full conditional distributions for our model. Section 14.4 presents the empirical results based on the land price data in Tokyo 23-special wards in Japan in 2010. Section 14.5 summarizes our results with remaining issues.

14.2 The Model

Let y_i and $\mathbf{x}_i$ be a dependent variable and a $1 \times k$ vector of covariates on the ith unit for $i = 1, \ldots, n$, respectively. Then, the quantile regression model (for an overview of quantile regression, see Koenker and Bassett, 1978; Koenker and Hallock, 2001; Koenker, 2005) given pth quantile is written as

$$y_i = \mathbf{x}_i \boldsymbol{\beta}_p + \epsilon_i,$$

where ϵ_i has the asymmetric Laplace distribution with density

$$f_p(\epsilon_i) = \frac{p(1-p)}{\sigma_p} \exp\left\{ -\frac{\epsilon_i}{\sigma_p}[p - I(\epsilon_i < 0)] \right\}, \tag{14.1}$$

and $I(\cdot)$ denotes the usual indicator function.

If let z_i and u_i have a standard exponential distribution and a standard normal distribution, ϵ_i in (14.1) can be represented as a location-scale mixture of normals given by

$$\epsilon_i = \theta_p v_i + \tau_p \sqrt{\sigma_p v_i} u_i, \tag{14.2}$$

where $v_i = \sigma_p z_i$, $\theta_p = \frac{1-2p}{p(1-p)}$, $\tau_p^2 = \frac{2}{p(1-p)}$. By using this representation, Kozumi and Kobayashi (2011) introduced a new Gibbs sampling approach.

In this chapter, we incorporate a spatial autoregressive process in the quantile regression model. Spatial interaction has been widely examined in several research areas. In the analyses of hedonic approach, it becomes the standard feature in land prices and there exist a lot of applications (e.g., Brasington, 1999; Brasington and Haurin, 2006, among others). One of the reasons to incorporate spatial interaction is an omitted variables problem (see LeSage and Pace, 2009). Therefore, we also incorporate the spatial autoregressive process. To introduce the spatial structure, let d_{ij} denote the distance between ith unit and jth unit. Then, the spatial weight of the jth unit in the ith unit is defined as $w_{ij} = \frac{f(d_{ij})}{\sum_{l=1}^{n} f(d_{il})}$ for $i \neq j$ and $w_{ij} = 0$ for $i = j$. Note that we have $\sum_{j=1}^{n} w_{ij} = 1$ for all i. Then, the spatial autoregressive linear regression model is given by

$$y_i = \sum_{j=1}^{n} \rho w_{ij} y_j + \mathbf{x}_i \boldsymbol{\beta} + \epsilon_i,$$

where ρ captures the intensity of spatial interaction.[1]

From the relationship given by (14.2) and by combining the spatial autoregressive model with the basic quantile regression model, the response y_i can be written as

$$y_i \quad = \quad \sum_{j=1}^{n} \rho_p w_{ij} y_j + \mathbf{x}_i \boldsymbol{\beta}_p + \theta_p v_i + \tau_p \sqrt{\sigma_p v_i}\, u_i, \qquad (14.3)$$

where $v_i \sim \mathcal{E}(\sigma_p)$ and $u_i \sim \mathcal{N}(0,1)$ are mutually independent, and $\mathcal{E}(a)$ denotes an exponential distribution with parameter a. Then, the likelihood function of the model is given by

$$\begin{aligned} & L(\mathbf{y}|\boldsymbol{\beta}_p, \sigma_p, \rho_p, \mathbf{v}, \mathbf{X}, \mathbf{W}) \\ & \quad \propto \quad |\mathbf{I} - \rho_p \mathbf{W}| \left\{ \prod_{i=1}^{n} v_i^{-\frac{1}{2}} \right\} \sigma_p^{-\frac{3n}{2}} \exp\left(- \sum_{i=1}^{n} \frac{e_i^2}{2\tau_p^2 \sigma_p v_i} \right), \qquad (14.4) \end{aligned}$$

where $\mathbf{y} = (y_1, \ldots, y_n)'$, $\mathbf{v} = (v_1, \ldots, v_n)'$, $\mathbf{X} = (\mathbf{x}_1', \ldots, \mathbf{x}_n')'$, $\mathbf{W} = \{w_{ij}\}$ and $e_i = y_i - \sum_{j=1}^{n} \rho_p w_{ij} y_j - \mathbf{x}_i \boldsymbol{\beta}_p - \theta_p v_i$.

[1]There are several kinds of spatial quantile models in a Bayesian framework (e.g., Reich et al., 2011; Lum and Gelfand, 2012). However, in the hedonic price analysis, the spatial autoregressive specification of Kostov (2009) is widely used, and we therefore consider this type of model in a Bayesian framework.

14.3 Posterior Analysis

14.3.1 The Joint Posterior Distribution

Given a prior density $\pi(\boldsymbol{\beta}_p, \sigma_p, \rho_p)$ and the likelihood function given in (14.4), the joint posterior distribution can be expressed as

$$\pi(\boldsymbol{\beta}_p, \sigma_p, \rho_p|\mathbf{y}, \mathbf{v}, \mathbf{X}, \mathbf{W}) \propto \pi(\boldsymbol{\beta}_p, \sigma_p, \rho_p)L(\mathbf{y}|\boldsymbol{\beta}_p, \sigma_p, \rho_p, \mathbf{v}, \mathbf{X}, \mathbf{W}). \quad (14.5)$$

To proceed with our Bayesian analysis, we assume the following prior distributions:

$$\boldsymbol{\beta}_p \sim \mathcal{N}(\boldsymbol{\beta}_{p0}, \boldsymbol{\Sigma}_{p0}), \;\; \sigma_p \sim \mathcal{IG}(\nu_{p0}/2, \lambda_{p0}/2), \;\; \rho_p \sim \mathcal{U}(-1, 1), \quad (14.6)$$

where $\mathcal{IG}(a, b)$ denotes an inverse gamma distribution with scale and shape parameters a and b. As is shown in Sun et al. (1999), it is well known that ρ_p must lie in the interval $(-1, 1)$, when the row-standardized weight matrix is utilized. Therefore, we restrict the prior space as $\rho_p \in (-1,\ 1)$. The details of the full conditional distributions of the parameters are discussed next.

14.3.2 Full Conditional Distributions for the Model

Sampling $\boldsymbol{\beta}_p$ and σ: Let $\pi(\boldsymbol{\beta}_p, \sigma_p, \rho_p)$ denote a prior distribution of the parameters given in (14.6). Given the likelihood function in (14.4) and the prior distributions in (14.6), the full conditional distributions of the parameters are proportional to

$$\pi(\boldsymbol{\beta}_p, \sigma_p, \rho_p)L(\mathbf{y}|\boldsymbol{\beta}_p, \sigma_p, \rho_p, \mathbf{v}, \mathbf{X}, \mathbf{W}).$$

The full conditional distributions of $\boldsymbol{\beta}_p$ and σ_p are

$$\boldsymbol{\beta}_p \sim \mathcal{N}(\hat{\boldsymbol{\beta}}_p, \hat{\boldsymbol{\Sigma}}_p), \quad \sigma_p \sim \mathcal{IG}(\hat{\nu}_p/2, \hat{\lambda}_p/2),$$

where

$$\begin{aligned}
\hat{\boldsymbol{\Sigma}}_p^{-1} &= \sum_{i=1}^{n} \frac{\mathbf{x}_i'\mathbf{x}_i}{\tau_p^2\sigma_p v_i} + \boldsymbol{\Sigma}_{p0}^{-1}, \\
\hat{\boldsymbol{\beta}}_p &= \hat{\boldsymbol{\Sigma}}_p \left\{ \sum_{i=1}^{n} \frac{\mathbf{x}_i'\left(y_i - \rho_p \sum_{j=1}^n w_{ij}y_j - \theta_p v_i\right)}{\tau_p^2\sigma_p v_i} + \boldsymbol{\Sigma}_{p0}^{-1}\boldsymbol{\beta}_{p0} \right\}, \\
\hat{\nu}_p &= 3n + \nu_{p0}, \\
\hat{\lambda}_p &= 2\sum_{i=1}^{n} v_i + \sum_{i=1}^{n} \frac{\left(y_i - \rho_p \sum_{j=1}^n w_{ij}y_j - \mathbf{x}_i\boldsymbol{\beta}_p - \theta_p v_i\right)^2}{\tau_p^2 v_i} + \lambda_{p0}.
\end{aligned}$$

These parameters are easily sampled by Gibbs sampler (see Gelfand and Smith, 1990).

Sampling v_i for $i = 1, \ldots, n$: The full conditional distribution of v_i for $i = 1, \ldots, n$ is given by

$$v_i \sim \mathcal{GIG}\left(\frac{1}{2}, \hat{\delta}_i, \hat{\gamma}_i\right),$$

where $\hat{\delta}_i^2 = \frac{\left(y_i - \rho_p \sum_{j=1}^n w_{ij} y_j - \mathbf{x}_i \boldsymbol{\beta}_p\right)^2}{\tau_p \sigma_p}$, and $\hat{\gamma}_i^2 = \frac{2}{\sigma_p} + \frac{\theta_p^2}{\tau_p^2 \sigma_p}$. Note that $\mathcal{GIG}(a, b, c)$ is a generalized inverse Gaussian distribution with parameters a, b and c (see Kozumi and Kobayashi, 2011, for details).
Sampling ρ_p: The full conditional distribution of ρ_p is given by

$$\pi(\rho_p) \propto |\mathbf{I} - \rho_p \mathbf{W}| \exp\left(-\sum_{i=1}^n \frac{\left(y_i - \sum_{j=1}^n \rho_p w_{ij} y_j - \mathbf{x}_i \boldsymbol{\beta}_p - \theta_p v_i\right)^2}{2\tau_p^2 \sigma_p v_i}\right),$$

which cannot be sampled by standard methods. Therefore, we adopt a random-walk Metropolis algorithm (see, e.g., Tierney, 1994).

Sample ρ_p^{new} from:

$$\rho_p^{new} = \rho_p^{old} + c_p \phi, \quad \phi \sim \mathcal{N}(0, 1).$$

The scalar c_p is called a tuning parameter, and ρ_p^{old} is the parameter of the previous sampling. Next, we evaluate the acceptance probability:

$$\alpha(\rho_p^{old}, \rho_p^{new}) = \min\left\{\frac{\pi(\rho_p^{new})}{\pi(\rho_p^{old})}, 1\right\}.$$

Finally, we set $\rho_p = \rho_p^{new}$ with probability $\alpha(\rho_p^{old}, \rho_p^{new})$, otherwise $\rho_p = \rho_p^{old}$. The scalar c_j is tuned to produce an acceptance rate between 40% and 60%, as suggested by Holloway et al. (2002).

The proposal density of ρ_p is not truncated to the interval $(-1, 1)$ given the constraint is part of the target density. Therefore, if the proposal value of ρ_p is not within the interval, the conditional posterior is zero and the proposal value is rejected with probability one (see Chib and Greenberg, 1998).

14.4 Empirical Analysis

Before examining empirics, we explain the data set which is used in this empirical analysis. We used data for the Tokyo 23-special wards in Japan in 2010 from the Land Value Publication prepared by the Ministry of Land, Infrastructure, Transport and Tourism (MLITT). There are 1299 observations and the variables are summarized in Table 14.1. These variables are standard characteristics in this data. As variables for school choice systems, we introduce three dummy variables (BLOCK, ADJOINING, and WHOLE). It

Table 14.1: Variable descriptions.

Variable	Description
	Structural Characteristics
SIZE	Area (100m^2) of land, which includes private roads.
SHAPES	Equals 1 if land is not square; o.w. equals 0.
DIRECTIONS	Equals 1 if the front road lies to the south of land; o.w. equals 0.
WIDTH	Width (m) of the front road.
PRIVATE	Equals 1 if the front road is private; o.w. equals 0.
GAS	Equals 1 if city gas is available; o.w. equals 0.
VOLUME	Legal ratio (10%) of *total* floor area to land area.
	School Choice System
BLOCK	Equals 1 if the block system is introduced; o.w. equals 0.
ADJOINING	Equals 1 if the adjoining system is introduced; o.w. equals 0.
WHOLE	Equals 1 if the whole-division system is introduced; o.w. equals 0.

Table 14.2: School choice system.

System	Ward
BLOCK	Shinagawa
ADJOINING	Katsushika, Minato, Meguro, Shinjuku, Suginami, Toshima
WHOLE	Adachi, Arakawa, Edogawa, Itabashi, Koto, Shibuya, Sumida
Unintroduced	Bunkyo, Chiyoda, Chuo, Nakano, Nerima, Ota, Setagaya, Taito

means that the points where the school choice systems are unintroduced is the baseline. The wards where the school choice system is introduced are summarized in Table 14.2. Thus, if the point belongs to Shinagawa, for example, the variable (BLOCK) takes 1 and the other variables (ADJOINING and WHOLE) takes 0. If the points belong to the unintroduced wards, all the variables (BLOCK, ADJOINING, and WHOLE) takes 0. Finally, we explain the weight matrix. As a weight matrix, we use a inverse square distance (d_{ij}^{-2}) between points

$$d_{ij} = radius \times \cos^{-1}[\cos|long_i - long_j| \cos lat_i \cos lat_j + \sin lat_i \sin lat_j],$$

where $radius$ is the Earth's radius, and lat and $long$ are respectively latitude and longitude for i and j (see Ertur and Koch, 2007, and so on).

We examined the proposed Bayesian spatial autoregressive quantile regression models for $p = 0.1$ to 0.9 by 0.1. For each quantile, we used the following same hyper-parameters to run the MCMC algorithm, using 11,000 iterations and discarding the first 1,000 iterations:

$$\boldsymbol{\beta}_{p0} = \mathbf{0}, \ \boldsymbol{\Sigma}_{p0} = 100 \times \mathbf{I}, \ \nu_{p0} = 10, \ \lambda_{p0} = 0.02.$$

All the results reported here are generated using Ox version 6.21 (see Doornik, 2006).

First, we examine the results for the structural characteristics. Figure 14.2 shows the posterior results of structural characteristics. Except for WIDTH, PRIVATE and GAS, quantile dependence of the parameters is captured. For these variables, the parameters are not quantile dependent and include zero in the 19% credible intervals in each quantile. Therefore, these variables are not important for the land price determination in Tokyo 23-special wards. On the other hand, the parameters of SIZE, SHAPES and DIRECTIONS are quantile dependent and do not include zero in the 95% credible intervals in each quantile. Moreover, the effects of these variables increase with the increase of the quantile. It means that the effects of these variables are important in high land price points, especially the increase of the effect of DIRECTIONS is drastic.

Next, we examine the effect of school choice. Figure 14.3 shows the posterior results of school choice effects. From the results, we can observe that the 95% credible intervals include zero in almost all the quantiles in BLOCK and ADJOINING. On the other hand, WHOLE shows negative effects under the 0.6th quantiles and includes zero in the 95% credible intervals over the 0.6th quantiles. As the benchmark are the locations where the school choice systems are not introduced, these results imply that school choice systems are not valued by the residents in Tokyo 23-special wards and sometimes make residents worse. From a theoretical point of view, it is well known that the school choice system plays an important role in school choice and performs well if the systems are appropriately introduced. Therefore, this result leads us to conclude that the present systems are not suitable to Tokyo 23-special

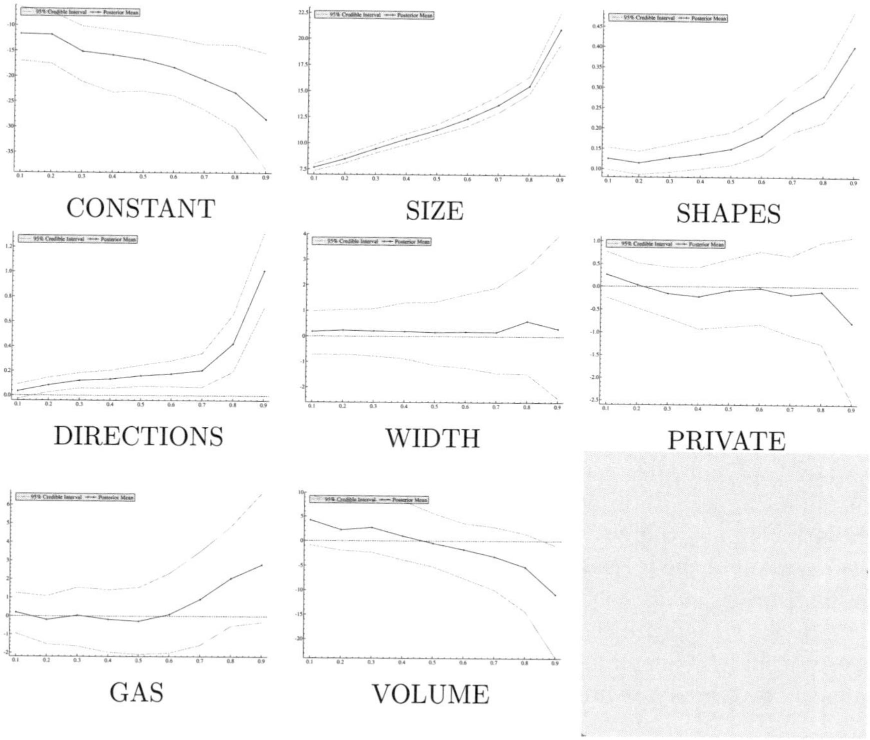

Figure 14.2: Posterior results for the structural characteristics.

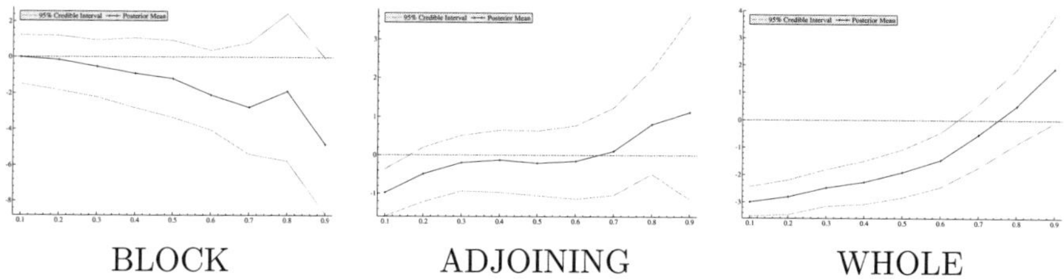

Figure 14.3: Posterior results for the school choice.

wards in Japan. For example, the cost to go to school might exceed the benefit from the system in case of WHOLE under the 0.6th quantiles. However, it might depend on the evaluation methods of the systems. Therefore, we think that other approaches, for example, contingent valuation methods (CVM), are required to evaluate the school choice systems in Tokyo 23-special wards in Japan.

Finally, we see the results for the spatial parameter ρ_p and scale parameter σ_p. Figure 14.4 shows the posterior results for these parameters. We find that

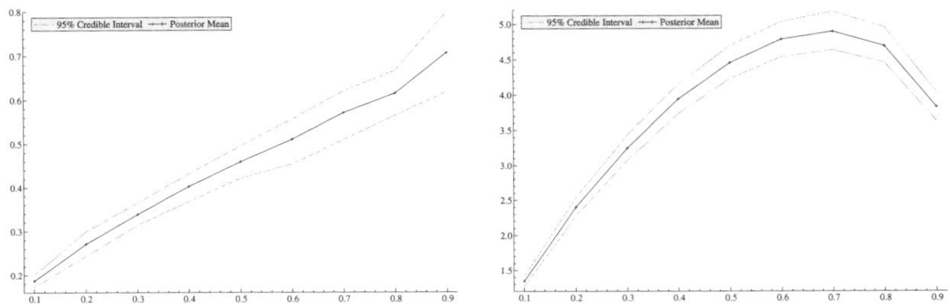

ρ_p σ_p

Figure 14.4: Posterior results for the other parameters.

the higher the quantile, the higher the spatial dependence. It means that the spillover on land price is powerful in high land price points. Therefore, spatial dependence plays an important role on the determinations of land prices. On the other hand, the scale parameter σ_p shows an inverse-U shape and the peak is around the 0.7th quantile. This suggests that whereas the uncertainty on land prices is low in the high and low land price areas, but that it might be difficult to capture the behavior of land prices around 0.7th quantile from these variables. Therefore, if we want to examine the effects around the 0.7th quantile accurately, it might be necessary to include more variables.

14.5 Conclusions

This chapter has proposed a Bayesian spatial autoregressive model and applied it to a hedonic price model to examine the school choice systems in Tokyo 23-special wards in Japan. Using a data set for 2010, we estimated the hedonic price functions. The empirical results show that the effects of the variables were quantile dependent in some cases. On the other hand, we could not confirm the importance of the school choice systems and WHOLE was estimated with negative values. Therefore, we are forced to conclude that the current systems did not work well in Tokyo 23-special wards in spite of the fact that it is well known that the school choice systems play an important role from a theoretical point of view. Thus, a detailed examination might be required to judge whether these results stem from the evaluation methods or whether the school choice system is not appropriate for Tokyo 23-special wards in Japan. However, we think that our proposed model and the results represent an interesting and important first step in this direction.

Bibliography

Adbulkadiroğlu, A. and T. Sönmez (2003). School choice: A mechanism design approach. *American Economic Review 93*, 729–747.

Black, S. E. (1999). Do better schools matter? Parental valuation of elementary education. *The Quarterly Journal of Economics 114*, 577–599.

Brasington, D. (1999). Which measures of school quality does the housing market value? *Journal of Real Estate Research 18*, 395–413.

Brasington, D. and D. R. Haurin (2006). Educational outcomes and house values: A test of the value added approach. *Journal of Regional Science 46*, 245–268.

Chib, S. and E. Greenberg (1998). Analysis of multivariate probit models. *Biometrika 88*, 347–361.

Coulson, N. E. and D. P. McMillen (2007). The dynamics of intraurban quantile house price indexes. *Urban Studies 44*, 1517–1537.

Doornik, J. A. (2006). *Ox: Object Oriented Matrix Programming Language.* London: Timberlake Consultants Press.

Downes, T. A. and J. E. Zabel (2002). The impact of school characteristics on house prices: Chicago 1987–1991. *Journal of Urban Economics 52*, 1–25.

Erdil, A. and H. Ergin (2008). What's the matter with tie-breaking? Improving efficiency in school choice. *American Economic Review 98*, 669–689.

Ertur, C. and W. Koch (2007). Growth, techonological interdependence and spatial externalities: Theory and evidence. *Journal of Applied Econometrics 22*, 1033–1062.

Gelfand, A. E. and A. F. M. Smith (1990). Sampling-based approaches to calculating marginal densities. *Journal of the American Statistical Association 85*, 398–409.

Holloway, G., B. Shankar, and S. Rahman (2002). Bayesian spatial probit estimation: A primer and an application to HYV rice adoption. *Agricultural Economics 27*, 383–402.

Koenker, R. (2005). *Quantile Regression.* New York: Cambridge University Press.

Koenker, R. and G. Bassett (1978). Regression quantiles. *Econometrica 46*, 33–50.

Koenker, R. and K. F. Hallock (2001). Quantile regression. *Journal of Economic Perspectives 15*, 143–156.

Kostov, P. (2009). A spatial quantile regression hedonic model of agricultural land prices. *Spatial Economic Analysis 4*, 53–72.

Kozumi, H. and G. Kobayashi (2011). Gibbs sampling methods for Bayesian quantile regression. *Journal of Statistical Computation and Simulation 81*, 1565–1578.

LeSage, J. P. and R. K. Pace (2009). *Introduction to Spatial Econometrics.* Chapman & Hall/CRC.

Lum, K. and A. E. Gelfand (2012). Spatial quantile multiple regression using the asymmetric Laplace process. *Bayesian Analysis 7*, 235–258.

McMillen, D. P. (2008). Changes in the distribution of house prices over time: Structural characteristics, neighborhood, or coefficients? *Journal of Urban Economics 64*, 573–589.

Reich, B. J., M. Fuentes, and D. B. Dunson (2011). Bayesian spatial quantile regression. *Journal of the American Statistical Association 106*, 6–20.

Rosen, S. (1974). Hedonic prices and implicit market: Product differentiation in pure competition. *Journal of Political Economy 82*, 34–55.

Sun, D., R. K. Tsutakawa, and P. L. Speckman (1999). Posterior distribution of hierarchical models using CAR(1) distributions. *Biometrika 86*, 341–350.

Tierney, L. (1994). Markov chains for exploring posterior distributions (with discussion). *Annals Statistics 22*, 1701–1762.

Yasuda, Y. (2010). *The Design of School Choice System: A Game Theoretic Approach.* NTT Publisher (in Japanese).

Yoshida, A., K. Kogure, and K. Ushijima (2009). School choice and student sorting: Evidence from Adachi ward in Japan. *Japanese Economic Review 60*, 446–472.

Yu, K. and R. A. Moyeed (2001). Bayesian quantile regression. *Statistics & Probability Letters 54*, 437–447.

Index

Bayesian Inference in the Social Sciences.
By Ivan Jeliazkov and Xin-She Yang